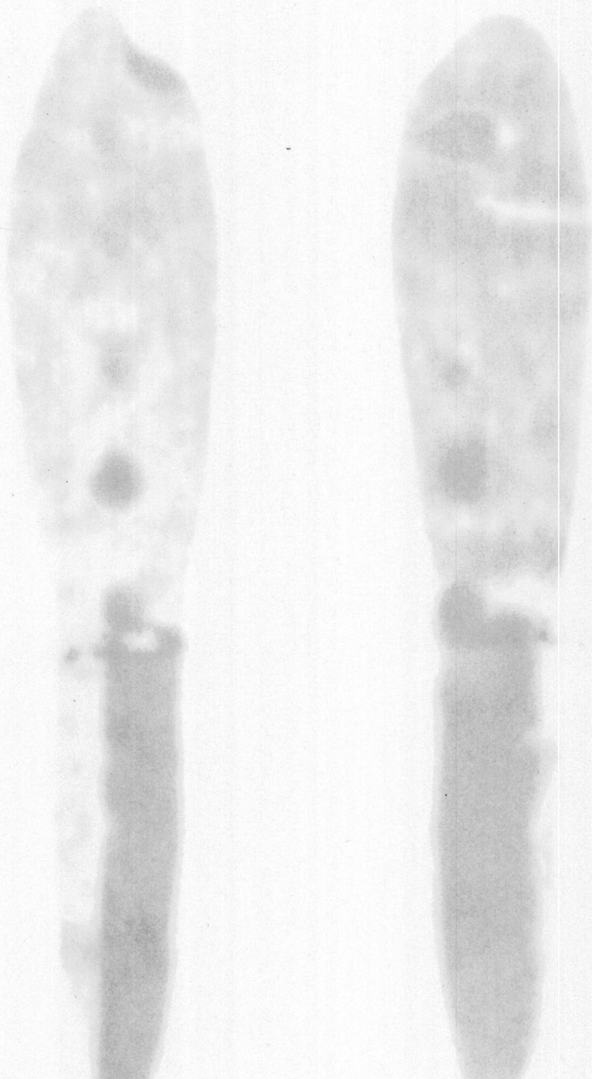

DISEASES OF TROPICAL AND SUBTROPICAL VEGETABLES AND OTHER PLANTS

DISEASES OF TROPICAL AND SUBTROPICAL VEGETABLES AND OTHER PLANTS

ALLYN AUSTIN COOK

HAFNER PRESS
A Division of Macmillan Publishing Co., Inc.
NEW YORK

Collier Macmillan Publishers
LONDON

Hafner Press
A Division of Macmillan Publishing Co., Inc.
866 Third Avenue, New York, N.Y. 10022

Collier Macmillan Canada, Ltd.

Library of Congress Catalog Card Number: 78-57055

Printed in the United States of America

printing number

1 2 3 4 5 6 7 8 9 10

Library of Congress Cataloging in Publication Data

Cook, Allyn Austin.
Diseases of tropical and subtropical vegetables and other plants.

Includes bibliographies and index.
1. Tropical crops--Diseases and pests.
2. Vegetables--Diseases and pests--Tropics. 3. Food crops--Diseases and pests--Tropics. I. Title.
SB608.T8C66 632 78-57055
ISBN 0-02-843080-8

CONTENTS

Preface

Plants considered in this volume can be divided into those grown for produce to be used in world trade and those grown primarily for subsistence or local home consumption. The beverage crops important throughout the world, such as cacao, coffee and tea, are subjected to the vagaries of disease incidence that derive from intensive monoculture. A large proportion of present-day production of these crops is conducted by sizeable farming interests, i.e. companies, corporations, etc., which often have sufficient physical and financial resources to approach disease control in an appropriate manner. Development of disease control programs that are efficient and economically feasible constitutes a major challenge to agriculturists under these circumstances.

Plants grown primarily in subsistence farming or for local use only present a vastly different challenge to agriculturists intent on improving production. The plants involved may not lend themselves readily to genetic improvement, and popular cultivars often have resulted from quite localized selection and culture. Further, the number of plant types to be considered and the limited economic value of each on a national scale may not permit sufficient individual attention to assure improvement in production. Climatic or other factors may preclude substitution of new crops or different cultivars for disease control, and personal tastes or local eating habits may further complicate improvement of crop or nutritional yield.

This volume is the second in a series about diseases of various crops grown in tropical and subtropical climates. It is a compilation of elemental information prepared to encourage the continuing search for better plant disease control imperative for agricultural improvement in tropical environments. Mention of any pesticidal material by trade name is not intended as an endorsement nor preferred recommendation over other material(s) of similar or identical chemical constitution.

Many colleagues have contributed generously in the preparation of this volume and I, hereby, recognize such assistance from G. M. Behncken, A. J. Cockbain, R. A. Conover, A. S. Costa, J. T. Dakwa, J. C. Delgado, M. de L. R. Duarte, I. Evans, O. P. Filho, R. Francki, E. R. French, H. Gallasch, R. Galton, G. E. Galvez, R. Gamez, C. O. Gourley, D. W. Hall, I. Hughes, W. V. Hutcheon, T. Inouye, A. M. Kabaara, W. J. Kaiser, P. J. Keane, A.

Kelman, J. W. Kimbrough, C. W. Kuhn, J. Lawrence, E. E. Leppik, J. C. Lozano, W. J. Martin, W. R. Mills, M. Moffett, S. A. de Moraes, D. M. Okioga, R. G. Orellana, G. K. Owusu, K. G. Pegg, D. L. Plucknett, G. S. Purss, C. J. Rodrigues, Jr., E. Schieber, D. E. Shaw, R. L. de Silva, G. Sowell, Jr., H. D. Thurston, E. Vargas, E. K. Wade, D. B. Ward, W. B. Weaver, R. J. Williams, A. G. Zagatto and F. W. Zettler.

Appreciation for assistance with preparation of the final draft is extended to Dr. Hugh L. Popenoe, Director of International Programs, and the Center for Tropical Agriculture, Institute of Food and Agricultural Sciences, University of Florida.

Grateful recognition is extended to Carolyn J. Cook for her help and patience during the preparation of the manuscript.

A

ALLSPICE (PIMENTO): *Pimenta officinalis* L.

Canker and Wilt (Fireblight, Dieback)

Diseases known as "fireblight" and "dieback" have been known in Jamaica since 1896 but were not attributed to a fungus until 1917. Losses apparently were much greater about the turn of the century, but the disease now is of minor import.

Infection occurs almost exclusively on mature trees through wounds and may progress in different ways according to the site of infection but typically proceeds from limb to limb until the tree eventually is killed. Progress of infection is much faster upward than downward, resulting in "fireblight" symptoms on the extremities of affected limbs followed by "dieback" or death of the extremities after the persistent leaves have fallen. Trees infected on the trunk below the first branching often die within a few months after infection. Infection sites typically appear as elliptical, sunken cankers with yellow-brown bark and discolored wood that extends as streaks 25 cm or more above the canker.

The causal fungus is *Ceratocystis fimbriata* Ell. and Halst. (see Cacao—Mal del machete).

Careful attention to avoid wounding limbs and application of fungicidal protectant to wounds and pruning injuries is recommended. Small limbs with infections and blighted twigs can be removed by pruning before the disease spreads into other limbs.

1. Hunt, J. 1956. Taxonomy of the genus *Ceratocystis*. *Lloydia 19:* 1–58.

2. Leather, R. I. 1966. A canker and wilt disease of pimento *(Pimenta officinalis)* caused by *Ceratocystis fimbriata* in Jamaica. *Trans. Brit. Mycol. Soc. 49:* 213–218.

3. Morgan-Jones, G. 1967. *Ceratocystis fimbriata. Commonwealth Mycol. Inst. Descriptions Pathogenic Fungi Bacteria 141*. 2 p.

Rust

In 1934, rust disease of allspice was reported from Jamaica, although the causal fungus had been described from Brazil in 1884. Within a short time, the full impact of the disease on allspice was apparent at altitudes above 600 m. Allspice cultivation at higher elevations in Jamaica, where spice, oil and liqueur were produced, was essentially lost within 2 years after the disease was found, and groves in those areas have been abandoned completely. The disease remains confined to Jamaica, where only minor losses in yield are sustained at altitudes below 450 m.

Typical symptoms include bright yellow clumps of urediospores mostly on the lower surfaces of young leaves, as well as on inflorescences and young twigs. Urediospores appear powdery *en masse,* and leaf lesions may remain isolated with little or no distortion of leaves that persist for the usual period, but lesions that coalesce to cover the entire surface cause death of the leaf and early defoliation. Successive leaves are killed in like manner.

Inflorescences are quite susceptible to infection, which causes them to become discolored and die. Infected flowers usually fall before fruit has set, although young fruits may persist for a short time after infection before they, too, become discolored and fall. Blossoms also may be killed from pedicel lesions.

Young twigs are infected systemically and sporulation may occur over the entire surface of the new twig growth. Dieback quickly follows. Repeated infection of twigs and defoliation result in progressive weakening of affected trees, which eventually are killed.

The causal fungus is *Puccinia psidii* Wint., that produces ellipsoid, finely echinulate urediospores 16–19 × 21–26 μm and occasional ellipsoid to cylindrical teliospores 19–22 × 30–48 μm. At least 3 pathotypes of the fungus, based on host range, are recognized, although the type originally described as a pathogen of guava (*Psidium guajava* L.) does not occur in Jamaica. The fungus type that is infectious to allspice also attacks bay rum (*Pimenta acris* Kostel), while a second type present in Jamaica infects species of *Eugenia* including rose-apple (*E. jambos* L.) and otaheite apple (*E. malaccensis* L.). The pathotypes in Jamaica do not infect guava, nor are they pathogenic on species other than those included in the *Pimenta* and *Eugenia* genera, respectively.

Dissemination of the fungus depends largely on spattering rain, possibly by mechanical agitation of the leaf surfaces to dislodge spores into the air rather than actual dispersal in drops of rain. Occurrence of the disease is closely correlated with ambient temperature, which also affects germination of urediospores. Although germination may occur from 9–28 C, maximum germination occurs at 15 C and progressively less up to 20 C with insignificant germination above the latter temperature. Night temperature of approximately 16–17 C regularly occurs at elevations of 450 m (and above) during the cooler months of the year in Jamaica, while corresponding temperatures at sea level (up to 180 m) are usually above 20 C. Spore germination also is highest on leaves 10 days old and becomes progressively less as leaves age until there is practically no germination (or infection) of leaves more than 45 days old.

Influence of environmental conditions, particularly temperature, has served to eliminate production of allspice in areas where rust is a problem. Sanitation, including removal of debris from infected plantings, may help to reduce disease incidence, although low volume, foliar application of Dithane Z-78 has been reported to control rust.

1. Hunt, P. 1965. Some aspects of the biology of the rust of pimento. *Proc. Am. Soc. Hort. Sci., Caribbean Reg. 9:* 46–59.

2. Laundon, G. F. and J. M. Waterston. 1965. *Puccinia psidii. Commonwealth Mycol. Inst. Descriptions Pathogenic Fungi Bacteria 56.* 1 p.

3. MacLachlan, J. D. 1936. The pimento rust disease. *J. Jamaica Agr. Soc. 40:* 277–281.

4. MacLachlan, J. D. 1938. A rust of the pimento tree in Jamaica, B. W. I. *Phytopathology 28:* 157–170.

5. Smith, F. E. V. 1935. Rust disease of pimento. *J. Jamaica Agr. Soc. 39:* 401–411.

(EDIBLE) AROIDS: SURAN, PONGAPONG, ELEPHANT FOOT YAM, *Amorphophallus campanulatus* Blume; TA'AMU, APE, KAPE, MAN KACHU, *Alocasia macrorrhiza* Schott; BABAI, GALLAN, *Cyrtosperma chamissonis* (Schott) Merrill; ARVI, KELADI, GABI, KALO, TARO, TALO, DALO, DASHEEN, EDDOE, COCOYAM, CULCAS, *Colocasia esculenta* (L.) Schott; YAUTIA, TANNIA, TANIER, COCOYAM, MACABO, MAFAFFA, MALANGA, *Xanthosoma sagittifolium* (L.) Schott

Bacterial leaf spot

Dark green, mostly round spots to a cm in diameter which later coalesced into large, light yellow to brown lesions were observed in India on taro and *Alocasia indica* Schott. Irregular, brown, oily spots surrounded by yellow halos and often delimited by veins were described from malanga leaves in

Guadeloupe in 1974. The causal bacterium was gram negative, aerobic with polar flagella and identified as *Xanthomonas diffenbachiae* (McCulloch & Pirone) Dowson. (The name *X. campestris* var. *aracearum* Berniac has been suggested for all xanthomonad pathogens of the *Araceae*.)

1. Asthana, R. P. 1946. Bacterial leaf-spot on *Arum*. *Current Sci. 15:* 356.

2. Berniac, M. 1974. Une maladie bacterienne de *Xanthosoma sagittifolium* (L.) Schott. *Ann. Phytopathol. 6:* 197–202.

3. Lelliott, R. A. 1971. The genus *Xanthomonas*. *Proc. Third Intern. Conf. Plant Pathogenic Bacteria (Wageningen, 1971):* 269–272.

Pythium rot (Currutaca)

One of the most widely distributed diseases of taro is *Pythium* rot which has been known to occur in Hawaii since 1919 and has been found more recently in many other areas where this crop is grown. Harvest losses from 10 percent upward have been reported, but the disease also can cause damage in young plantings, necessitating additional expense for replanting and delay in harvest.

Affected plants become stunted with shortened petioles, chlorotic laminae and complete deterioration of the root system, which causes plants to be easily pulled from the soil. In fact, root damage may be so extensive that plants will fall by themselves. However, the phase of the disease from which the common name derives is a soft, wet rot that affects varying portions of the corm and is characterized by a foul, acrid smell. The rot, which may assume coloration from yellowish to purple, often begins at the base of the corm, progresses upward and remains sharply demarcated from healthy corm tissues.

Four species of *Pythium* [*P. aphanidermatum* (Edson) Fitzpatrick, *P. graminicola* Subramanian, *P. splendens* Br. and *P. irregulare* Buisman] have been identified as causal agents. Inoculum may be carried with planting materials or may persist in soil between crops. Environmental conditions, including abundant moisture and warm temperatures, contribute to the severity of *Pythium* rot, which often is most serious in areas with slow water movement. *Pythium* rot is not a problem in upland taro.

Removal of all diseased plant material from the field after harvest is completed will serve to reduce the available inoculum. The causal fungi can be effectively eliminated by plowing and drying the soil and by cultivation of taro in rotation with other crops to reduce (and maintain low) levels of inoculum. Extreme care should be taken to select taro planting materials free of *Pythium* infection. Varietal differences in susceptibility to *Pythium* rot have been reported but remain to be thoroughly investigated. Addition of Captan 50 W (90 kg/ha) to the soil has been found to provide control in non-alkaline soils.

1. Alvarez-Garcia, L. A. and A. Cortés-Monllor. 1971. Currutaca: a *Pythium* soft rot of *Xanthosoma* and *Colocasia* spp. in Puerto Rico. *J. Agr. Univ. Puerto Rico 55:* 78–84.

2. Jackson, G. V. H. and D. E. Gollifer. 1975. Storage rots of taro *(Colocasia esculenta)* in the British Solomon Islands. *Ann. Appl. Biol. 80:* 217–230.

3. Parris, G. K. 1941. Diseases of taro in Hawaii and their control. *Hawaii agr. Expt. Sta. Circ. 13.* 29 p.

4. Plucknett, D. L., R. S. de la Pena and F. Obrero. 1970. Taro *(Colocasia esculenta). Field Crop Abstracts 23:* 413–426.

5. Trujillo, E. E. 1969. Diseases of the genus *Colocasia* in the Pacific area and their control. *Proc. Intern. Symp. Trop. Root Crops (Trinidad, 1967) 2:* 13–18.

6. Waterhouse, G. M. and J. M. Waterston. 1964. *Pythium aphanidermatum. Commonwealth Mycol. Inst. Descriptions Pathogenic Fungi Bacteria 36.* 2 p.

7. Waterhouse, G. M. and J. M. Waterston. 1964. *Pythium graminicola. Commonwealth Mycol. Inst. Descriptions Pathogenic Fungi Bacteria 38.* 2 p.

8. Waterhouse, G. M. and J. M. Waterston. 1966. *Pythium splendens. Commonwealth Mycol. Inst. Descriptions Pathogenic Fungi Bacteria 120.* 2 p.

Leaf blight (*Phytophthora* spot)

Blight of taro leaves was described first from Java in 1900 but soon thereafter was found throughout much of southeastern Asia and many of the islands in the southwestern Pacific area. The disease commonly occurs in wetland taro, and frequently in upland taro as well, and has been considered a limiting factor in some locations accounting for losses estimated at 25–50 percent.

Leaf lesions develop as small, dark, mostly round spots that enlarge rapidly to a diameter of 1–2 cm. As the spots enlarge, a yellowish liquid exudes from the centers of infected areas and the spots assume various

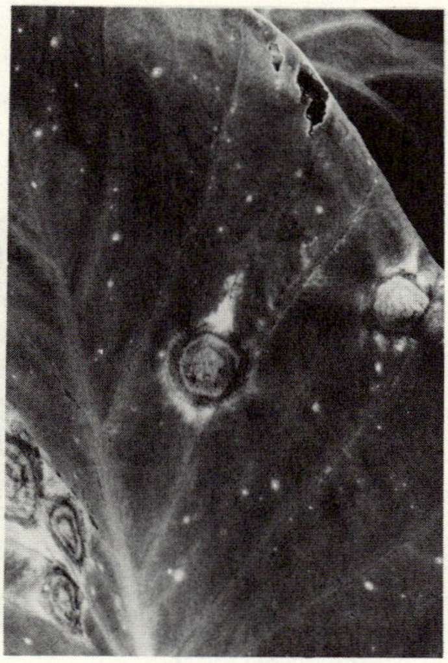

Fig. 1. Leaf blight (*Phytophthora* spot) of taro. (Courtesy of D. L. Plucknett, University of Hawaii, Honolulu.)

shapes, often are zonate with yellow margins and frequently coalesce (Fig. 1). Infections may occur on any part of the laminae but develop more often along the margins and later tear, imparting a ragged appearance to affected leaves.

The blight disease is caused by *Phytophthora colocasiae* Rac., which produces short, slender sporangiophores (2–4 μm) on which are borne elliptical, elongate sporangia (23 × 45–60 μm) that typically retain a short pedicel when they fall. The reniform swarmspores, each with 2 cilia, form inside the sporangium and may germinate to form 1–2 new swarmspores. Oogonia range to 35 μm in diameter with antheridia 11 × 11–13 μm and chlamydospores that usually measure 26–30 μm. The fungus sporulates (in lesions on detached leaves) most readily at 21 C and 100 percent humidity. Sporulation does not occur under conditions of less than 90 percent relative humidity. Sporangia are disseminated by wind or splashing rain, and some germinate (indirectly) within 2 hours (at 21 C), and some zoospores germinate within 30 minutes after release.

Interplant spacing of 75 cm served to reduce the severity of leaf blight and increase yield because larger corms were produced, but 50 percent improvement in yields was obtained in Hawaii through application of copper fungicides. In recent trials, Dithane M-45 (approx. 0.9 kg/ha) applied at 5–7 day intervals provided best disease control. Resistance to leaf blight in cv. Ahina offers some promise of developing additional resistant types of *Colocasia esculenta*, and *Xanthosoma* sp., which are comparable in nutritional value, also are resistant.

1. Bergquist, R. R. 1974. Effect of fungicide rate, spray interval, timing of spray application, and precipitation in relation to control of *Phytophthora* leaf blight of taro. *Ann. Bot. 38:* 213–221.

2. Deshmukh, M. J. and K. N. Chhibber. 1960. Field resistance to blight (*Phytophthora colocasia* Rac.) in *Colocasia antiquorum* Schott. *Current Sci. 29:* 320–321.

3. Hicks, P. G. 1967. Resistance of *Colocasia esculenta* to leaf blight caused by *Phytophthora colocasiae. Papua New Guinea Agr. J. 19:* 1–4.

4. Parris, G. K. 1941. Diseases of taro in Hawaii and their control. *Hawaii Agr. Expt. Sta. Circ. 18.* 29 p.

5. Plucknett, D. L., R. S. de la Pena and F. Obrero. 1970. Taro (*Colocasia esculenta*). *Field Crop Abstracts 23:* 413–426.

6. Trujillo, E. E. 1965. The effects of humidity and temperature on *Phytophthora* blight of taro. *Phytopathology 55:* 183–188.

7. Trujillo, E. E. 1967. Diseases of the genus *Colocasia* in the Pacific area and their control. *Proc. Intern. Symp. Trop. Root Crops (Trinidad, 1967) 2:* 13–19.

8. Trujillo, E. E. and M. Aragaki. 1964. Taro blight and its control. *Hawaii Farm Sci. 13:* 11–13.

9. Waterhouse, G. M. 1963. Key to the species of *Phytophthora* de Bary. *Commonwealth Mycol. Inst. Mycol. Paper 92.* 22 p.

Sclerotium rot

A minor disease of upland taro, which occurs in Fiji, Hawaii and the Philippines, is characterized by stunted plants on which the lower portion of the corm is often rotted. The presence of numerous light brown, spherical sclerotia, many of which remain on pulled plants along with some soil in a fragile network of white mycelium, is typical and makes the disease easily recognizable.

The disease is caused by *Sclerotium rolfsii* Sacc. [*Botryobasidium rolfsii* (Curzi) Venkat.], which is pathogenic on many plants and can persist in the soil as mycelium but more commonly as sclerotia, which germinate under warm, wet conditions.

Soil drenches of either Terraclor (pentachloronitrobenzene) or Botran (2,6-dichloro-4-nitroaniline) (11–22 kg/ha) have been reported to provide effective control.

1. Mordue, J. E. M. 1974. *Corticium rolfsii. Commonwealth Mycol. Inst. Descriptions Pathogenic Fungi Bacteria 410.* 2 p.

2. Parris, G. K. 1941. Diseases of taro in Hawaii and their control. *Hawaii Agr. Expt. Sta. Circ. 18.* 29 p.

3. Plucknett, D. L., R. S. de la Pena and F. Obrero. 1970. Taro (*Colocasia esculenta*). *Field Crop Abstracts 23:* 413–426.

Cladosporium leaf spot (Brown spot)

Spots mostly round and 5–10 mm in diameter colored dark brown, but often with lighter centers (Fig. 2), occur on taro throughout the Pacific basin. Lesions are found mostly on older leaves of both wetland and upland taro

Fig. 2. *Cladosporium* leaf spot (brown spot) of taro. (Courtesy of D. L. Plucknett, University of Hawaii, Honolulu.)

and are caused by *Cladosporium colocasiae* Sawada. The disease apparently causes no appreciable economic losses and does not warrant control, although foliage sprays for taro blight have been reported effective.

1. Bugnicourt, R. 1958. Contribution a l'étude de *Cladosporium colocasiae* Sawada. *Rev. Mycol. 23:* 233–236.

2. Graham, K. 1965. Dalo diseases. *Fiji Farmer 1:* 54–56.

3. Parris, G. K. 1941. Diseases of taro in Hawaii and their control. *Hawaii Agr. Expt. Sta. Circ. 18.* 29 p.

4. Plucknett, D. L., R. S. de la Pena and F. Obrero. 1970. Taro (*Colocasia esculenta*). *Field Crop Abstracts 23:* 413–426.

5. Trujillo, E. E. 1967. Diseases of the genus *Colocasia* in the Pacific area and their control. *Proc. Intern. Symp. Trop. Root Crops (Trinidad, 1967) 2:* 13–19.

Phyllosticta leaf spot (Shot hole)

Brown, necrotic, round to irregular leaf lesions 5–25 mm or more in diameter have been noted on upland taro in Fiji and Hawaii and are caused by *Phyllosticta colocasiophila* Weedon. The centers of lesions, which may coalesce, often rot and fall, imparting a ragged or "shot hole" appearance to affected leaves. Pycnidiospores in white cirri are extruded from small, dark pycnidia in wet weather and are spread to other leaves by spattering rain. Disease development is favored by extended periods of wet weather but usually is not serious enough to justify applications of fungicidal spray for control.

1. Graham, K. 1965. Dalo diseases. *Fiji Farmer 1:* 54–56.

2. Parris, G. K. 1941. Diseases of taro in Hawaii and their control. *Hawaii Agr. Expt. Sta. Circ. 18.* 29 p.

3. Plucknett, D. L., R. S. de la Pena and F. Obrero. 1970. Taro (*Colocasia esculenta*). *Field Crop Abstracts 23:* 413–426.

Alomae-Bobone

Two diseases of taro were reported and described from the Solomon Islands in 1972, although similar symptoms had been noted earlier (1960). These diseases are not known to occur elsewhere except in Papua New Guinea (D. E. Shaw, pers. comm.). Incidence of plants with "alomae" symptoms ranged from 39 percent to 83 percent in cultivar trials in 1970–71 and caused severe yield reductions. Plants affected by "bobone" ranged from 0.4 percent in 1970 to 15 percent in 1972, but about half of the plants in the latter trial recovered with no appreciable yield loss.

Symptoms of alomae include an early, conspicuous, but sometimes indistinct, feathery mottle on leaves in which veins thicken and hypertrophy.

Young leaves often fail to open, and apical necrosis progresses through the petioles to kill the entire plant, leaving only curved, shortened, rotted petioles.

Plants affected by bobone disease display similar symptoms except that the foliage remains green and does not become necrotic, and plants are more noticeably reduced in size. Affected plants typically recover.

Both diseases are considered to be virus-induced. Bacilliform particles of 2 sizes, 30 × 140 nm and 55 × 280 nm, are present in alomae plants, but only the latter have been found in bobone plants. The delphacid *Tarophagus proserpina* (Kirk) is suspected to be the vector of the larger particle but the vector of the smaller particle has not been determined yet.

Plants that produce larger corms and leaves but fewer cormels are referred to as "male" taro by local growers who believe such plants to be susceptible to alomae but resistant to bobone. Other plants, referred to as "female" taro, are resistant to alomae but susceptible to bobone.

Relation between the diseases has not been clearly established. No control, other than selection of male vs. female taro cultivars, has been devised, but planting materials should be taken only from healthy plants.

1. Gollifer, D. E. and J. F. Brown. 1972. Virus diseases of *Colocasia esculenta* in the British Solomon Islands. *Plant Dis. Reptr. 56:* 597–599.

2. Jackson, G. V. H. and D. E. Gollifer. 1975. Disease and pest problems of taro [*Colocasia esculenta* (L.) Schott] in the British Solomon Islands. *PANS 21:* 45–53.

3. Kenten, R. H. and R. D. Woods. 1973. Viruses of *Colocasia esculenta* and *Xanthosoma sagittifolium*. *PANS 19:* 38–41.

Dasheen mosaic

A number of araceous plants collected in Florida (United States) and calla lily [*Zantedeschia elliottiana* (Knight ex Watson) Engl.] from the Netherlands were reported to be infected by virus in 1970. The same disease was reported later from Puerto Rico and Venezuela and probably is widespread wherever such plants are grown throughout the world. Effects on yield have not been assessed.

Foliar symptoms vary from complete absence of visible symptoms through differing degrees of feathery mottle, particularly along parts of main veins (Fig. 3), to extreme distortion of laminae with little or no mottle. Individual leaves on the same plant also may exhibit similar extreme variations in symptom severity, but absence of symptoms in one or more leaves may not be an accurate indication of permanent recovery from infection.

Dasheen mosaic is caused by a virus characterized by flexuous particles 750 nm in length. Transmission of the disease has been accomplished by mechanical inoculation as well as *Myzus persicae* (Sulz.) and *Aphis craccivora*

Fig. 3. Dasheen mosaic on *Xanthosoma* sp. (Courtesy of F. W. Zettler, University of Florida, Gainesville.)

Koch, but only to *Philodendron selloum* C. Koch and *Z. elliottiana* and not to any non-araceous species.

No satisfactory explanation has been offered for the erratic nature of leaf symptoms and the epidemiological significance of this phenomenon. Geographic distribution of the disease may have resulted incidentally from commerce involving vegetatively propagated, araceous ornamental plants such as caladium (*Caladium hortulanum* Birdsey), which exhibits only indistinct symptoms, as well as subsistence plants including taro.

Propagating material should be taken only from healthy plants. Virus-free plants of taro and cocoyam (*Xanthosoma sagittifolium*) have been propagated successfully from excised shoot-tips.

1. Alconero, R. 1972. Hot water treatments of corms of *Xanthosoma* spp. infected with dasheen mosaic virus. *Plant Dis. Reptr. 56:* 320–321.

2. Alconero, R. and F. W. Zettler. 1971. Virus infections of *Colocasia* and *Xanthosoma* in Puerto Rico. *Plant Dis. Reptr. 55:* 506–508.

3. Debrot, E. A. and A. Ordosgoitti. 1974. Dasheen mosaic virus infection of *Colocasia* and *Xanthosoma* in Venezuela. *Plant Dis. Reptr. 58:* 1032–1034.

4. Hartman, R. D. 1974. Dasheen mosaic virus and other phytopathogens eliminated from caladium, taro, and cocoyam by culture of shoot tips. *Phytopathology 64:* 237–240.

5. Tooyama, A. 1975. Indicator plants for dasheen mosaic virus. *Ann. Phytopathol. Soc. Japan 41:* 504–505.

6. Zettler, F. W., M. J. Foxe, R. D. Hartman, J. R. Edwardson, and R. G. Christie. 1970. Filamentous viruses infecting dasheen and other araceous plants. *Phytopathology 60:* 983–987.

Hard rot (Dry rot)

Corms affected by hard rot, which may cause 5–100 percent losses, are distinguishable by a furrowed, coarse, thickened skin. Internal tissues of diseased corms are woody and discolored yellow to brown with scattered, easily extractable lesions ("guava seeds"), but, in advanced stages, the interior of the corm is transformed into a blackened mass.

Neither the cause nor control for hard rot, which has been reported only from Hawaii, has been determined, but the disease does not occur in upland taro.

1. Parris, G. K. 1941. Diseases of taro in Hawaii and their control. *Hawaii Agr. Expt. Sta. Circ. 18.* 29 p.

2. Trujillo, E. E. 1967. Diseases of the genus *Colocasia* in the Pacific area and their control. *Proc Intern. Symp. Trop. Root Crops (Trinidad, 1967) 2:* 13–19.

Mineral deficiency symptoms

New leaves on plants deficient in *nitrogen* are reduced in size, yellow-green and develop a rapid, progressive light brown necrosis that first appears at leaf apices. Older leaves also become chlorotic and later necrotic as the deficiency becomes more acute.

Potassium deficiency becomes evident either as interveinal chlorosis that quickly turns into necrosis (with necrotic margins defined by an orange band), or a bright yellow, diffuse chlorosis with retention of some green coloration along lateral veins to produce a "feathering" effect. In both cases, symptoms develop along entire leaf margins, and dark colored necrotic spots may appear between lateral veins near leaf margins.

Symptoms of *magnesium* deficiency are first apparent on the oldest leaves which develop interveinal chlorosis with bands of tissue along lateral veins that remain green. Entire leaves gradually become chlorotic with a transitional feathery appearance, and interveinal tissues become necrotic but do not dry as quickly as with leaves deficient in nitrogen.

Early symptoms of *calcium* deficiency resemble those of nitrogen deficiency in that leaf size is reduced and the apical tips of leaves change color (orange-yellow). Progressive chlorosis is followed by necrosis that proceeds in the direction of the leaf bases, but healthy tissues of the lamina always are separated from necrotic by a distinct yellow or orange transition zone.

A chlorotic mottle on young leaves reduced in size is the first characteristic of *sulfur* deficiency. Such patterns become more pronounced as the chlorosis increases in severity, finally resulting in a light colored necrosis of the lamina that first appears at the leaf apex.

1. Bull, R. A. 1960. Macronutrient deficiency symptoms in the cocoyam (*Xanthosoma* sp.). *J. West Afr. Inst. Oil Palm Res. 3:* 181–186.

OTHER DISEASES

Phytophthora corm and root rot

Two unidentified species of *Phytophthora* in Fiji caused wilting of leaves after roots developed a brown or purple, firm, watery rot. In later stages of development, affected tissues emitted an offensive odor and crumbled easily. Planting on well-drained sites in rotation with other crops was recommended to reduce losses.

1. Graham, K. 1965. Dalo diseases. *Fiji Farmer 1:* 54–56.

Endoconidiophora (*Ceratocystis*) black rot of corms

1. Trujillo, E. E. 1967. Diseases of the genus *Colocasia* in the Pacific area and their control. *Proc. Intern. Symp. Trop. Root Crops (Trinidad, 1967) 2:* 13–19.

Rhizoctonia leaf rot [*Rhizoctonia bataticola* (Taub.) Butl.=*Macrophomina phaseolina* (Tassi) Goidanich] (See Bean—Charcoal rot)

Irregular, small brown lesions that became necrotic and enlarged to involve entire leaves and extended into the petioles on younger leaves and caused leaf rot and fall were reported in 1968 from India.

1. Grover, R. K. and R. D. Prasher. 1968. *Rhizoctonia* leaf rot of colocasia in India. *FAO Plant Protect. Bull. 16:* 6–7.

2. Holliday, P. and E. Punithalingam. 1970. *Macrophomina phaseolina. Commonwealth Mycol. Inst. Descriptions Pathogenic Fungi Bacteria 275.* 2 p.

Choanephora cucurbitarum Thaxter and *C. trispora* Thaxter

1. Sinha, S. 1940. A wet rot of leaves of *Colocasia antiquorum* due to secondary infection by *Choanephora cucurbitarum* Thaxter and *Choanephora trispora* Thaxter sp. n. (*Blakeslea trispora* Thaxter). *Proc. Indian Acad. Sci., Sect. B., 11:* 167–176.

Johnstonia colocasiae M. B. Ellis

1. Ellis, M. B. 1971. Dematiaceous Hyphomycetes. X. p. 14–15. *Commonwealth Mycol. Inst. Mycol. Paper 125.*

Loliloli

1. Parris, G. K. 1941. Diseases of taro in Hawaii and their control. *Hawaii Agr. Expt. Sta. Circ. 18.* 29 p.

Pleosphaerulina colocasiae Rangel

Leptosphaeria colocasiae Unamuno

1. Plucknett, D. L., R. S. de la Pena and F. Obrero. 1970. Taro (*Colocasia esculenta*). *Field Crop Abstracts 23:* 413–426.

Veinbanding

1. Graham, K. 1965. Dalo diseases. *Fiji Farmer 1:* 54–56.

ARRACACHA: *Arracacia xanthorrhiza* Bancr.

Gloeosporium sp.

1. Muñoz, F. H. 1968. El cultivo de la arracacha en la sabana de Bogotá. *Agr. Trop. (Colombia) 24:* 139–146.

ARROWROOT: *Maranta arundinacea* L.

Bacterial wilt

A vascular wilt disease of arrowroot (Fig. 4) was reported from Brazil in 1962, and the bacterial pathogen was identified as *Xanthomonas marantae* Zagatto and Pereira. Cells of the monotrichous, gram negative bacterium were 0.6 × 2.8 μm, formed round, raised yellow colonies, hydrolyzed starch, liquefied gelatin, reduced litmus and optimal temperature for growth was between 25–30 C.

Fig. 4. Bacterial wilt of arrowroot. (Courtesy of A. G. Zagatto, Biological Institute, São Paulo.)

Use of healthy propagating materials was recommended for control. New plantings should be isolated from areas where the disease has occurred. Infected plants should be rogued, including the roots, and burned.

1. Issa, E. 1962. "Murcha bacteriana" nova doença da araruta. *Biologico 28:* 168–169.

2. Zagatto, A. G. and A. L. G. Pereira. 1963. Estudo do organismo causador da "murcha bacteriana" da araruta (*Maranta arundinacea* L.). *Arq. Inst. Biol. 30:* 33–36.

Banded leaf blight

Banded leaf blight of arrowroot was observed in North Malabar (India) in 1946 and described 2 years later after growers had experienced considerable losses. The disease can be of some consequence under favorable conditions but has been reported only from India.

Evidence of infection is small, water-soaked spots that may enlarge rapidly by the formation of successive but sometimes discontinuous, concentric zones. Aerial mycelium quickly covers both leaf surfaces, and basidia and basidiospores are produced on the scant, whitish, superficial growth. Necrotic green to yellow or brown areas, separated by bands of healthy green areas, develop to impart an appearance of irregular bands that extend transversely across the entire leaf blade or only one side. The aerial surface mycelium produces sclerotia as it grows older and may spread to neighboring leaves, but the infection does not often extend to the leaf sheath or to the shoot. Eventually, affected leaves become completely necrotic and are rotted by secondary organisms.

Thanatephorus cucumeris (Frank) Donk was identified as the causal fungus and found to produce branched hyphae 7–10 μm in width (clamp connections were not observed) and basidia, from clumps of mycelium, on which were formed oblong, single-celled basidiospores 4–7 × 7–11 μm. The fungus was infectious to leaves on a number of plants, including rice, corn, sugarcane, pigeon pea and citrus.

The disease is only troublesome during periods of high rainfall and can be satisfactorily controlled by foliar spray (both leaf surfaces), just prior to the onset of high rainfall seasons, with Bordeaux mixture (1 percent), but other fungicidal materials, viz. zineb, maneb or copper, should be effective.

1. Ramakrishnan, K. and T. S. Ramakrishnan. 1948. Banded leaf blight of arrowroot, *Maranta arundinacea. Indian Phytopathol. 1:* 129–136.

Mosaic

Reported to occur in the Philippines.

1. Espeleta, V. P. and F. L. Nuque. 1959. Synergism of *Maranta arundinacea* mosaic virus and Coix lachrymajobi mosaic virus in the abaca plant. *Araneta J. Agr. 6:* 45–50.

"Burning" disease

"Cigar" or "Long root"

1. Briant, A. K. 1933. Maladies affecting arrowroot in St. Vincent. *Trop. Agr. (Trinidad) 10:* 183–188.

B

BAMBARRA GROUNDNUT: *Voandzeia subterranea* (L.) Thou.

Fusarium wilt

Wilt disease, caused by *Fusarium oxysporum* f. sp. *voandzeiae* Armstrong, Armstrong & Billington, was reported from Kenya in 1961 and Tanzania in 1972. Symptoms first appear about 50 days after sowing as stunted plants with vascular systems discolored red-brown to purple. Leaflets soon develop interveinal chlorosis followed by general yellowing and progressive necrosis of the foliage, wilt and often death of affected plants. The causal fungus also is pathogenic on bean and cowpea.

1. Armstrong, G. M., J. K. Armstrong and R. V. Billington. 1975. *Fusarium oxysporum* forma specialis *voandzeiae*, a new form species causing wilt of Bambarra groundnut. *Mycologia 62:* 709–714.

2. Ebbels, D. L. and R. V. Billington. 1972. *Fusarium* wilt of *Voandzeia subterranea* in Tanzania. *Trans. Brit. Mycol. Soc. 58:* 336–338.

3. Nattrass, E. M. 1961. Host lists of Kenya fungi and bacteria. *Commonwealth Mycol. Inst. Mycol. Paper 81.* 46 p.

BEAN: *Phaseolus vulgaris* L.

Bacterial blight diseases

The first report of "common blight" of bean was from the United States (New York), and description of the disease appeared the same year (1892) but from another state, New Jersey. Worldwide occurrence makes blight one of the most important diseases of bean, although sizeable differences in seasonal losses have been observed. Reduction in yield has been estimated at more than 50 percent of the crop and millions of pounds of beans in some locations in some years.

Lesions on leaves appear first as small, angular, water-soaked spots that quickly coalesce and enlarge into irregular, dark areas 10–50 mm in diameter. Tissue in the affected areas becomes flaccid, dries and turns dark in color, and a noticeable, but relatively narrow, yellow margin of variable width develops around each area (Fig. 5). The unaffected portion of a leaf also may become chlorotic and defoliation results if infection of leaves becomes extensive. Young leaves are more susceptible to infection than old leaves.

Fig. 5. Common blight of bean. (Courtesy of K. G. Pegg, Department of Primary Industries, Indooroopilly.)

Elongated, longitudinal, water-soaked streaks that later become sunken, and dry, develop on stems and petioles. Yellow exudate may form after the diseased tissues have been exposed by cracking of the necrotic tissue in stems or petioles.

Spots on pods are first small and circular, dark, and appear either wet or greasy. They soon enlarge to 5 mm or more in diameter, become dark in color, and drops of exudate may develop before the lesions dry. Seeds produced in diseased pods exhibit symptoms ranging from no apparent effect through various degrees of wrinkling and discoloration and reduction in size of beans to complete rot and shrivel in pods affected when young. Discoloration is most readily noticed on cultivars with light colored seedcoats.

The bacterium that causes common blight is *Xanthomonas phaseoli* (E. F. Smith) Dowson, characterized by aerobic, gram negative rods 0.4–0.9 × 0.6–2.6 μm that are motile by single, polar flagella. Cells produce an extracellular slime layer of heteropolysaccharide and form yellow, wet, shiny colonies *in vitro* but no diffusible brown pigment on agar that contains tyrosine. A form of the organism, identified as *X. phaseoli* var. *fuscans* (Burk.) Starr & Burk., produces dark pigment *in vitro* on agar with tyrosine and causes symptoms referred to as "fuscous blight," which are indistinguishable from common blight.

A distinct bacterium, *Pseudomonas phaseolicola* (Burk.) Dowson, characterized by aerobic, gram negative rods, single or in pairs with 1–6 polar or bipolar flagella, and which produces a green, diffusible pigment on medium B of King, causes "halo blight" of bean. Foliar symptoms are similar to those of common blight except that leaf lesions are surrounded by much larger yellow "halos" (to 2.5 cm in diameter), and exudate from pod lesions is light colored to nearly white instead of yellow as with common blight (Fig. 5). Halo symptoms are most evident at 20 C or below and do not appear at 28 C or above, which makes distinction between common and halo blight symptoms in the field practically impossible at the higher temperature. Development of halo blight, strictly a foliar disease in the field, is enhanced by cool temperatures, whereas common blight is favored by warm temperatures.

Infection by the common blight bacterium typically occurs through leaf stomata, cotyledons or wounds, and the bacteria remain localized for a while in substomatal cavities and intercellular spaces before eventually invading the vascular system and becoming systemic. After infection has progressed to the stem, some wilting of plants may result, but penetration through the pod to the seed via the funiculus assures internal seed contamination. Seedlings from infected seed are invaded through rifts that develop in the cotyledonary surfaces during germination.

Bacterial cells in exudate are readily disseminated within fields by spattering rain and can be carried greater distances by winds that accom-

Fig. 6. Halo blight of bean. (Courtesy of K. G. Pegg, Department of Primary Industries, Indooroopilly.)

pany rainstorms. Wounds created by blown particles of sand or soil or hail serve as infection sites, and some insects (grasshoppers and the Mexican bean beetle, *Epilachna varivestis* Muls.) have been found to transmit the common blight bacterium. The most important means by which the disease is disseminated over long distances or from one season to the next, however, is through infected seed where the bacterium may remain viable for 10 years. There is some evidence that the bacterium can survive between seasons on alternative hosts or in the soil, but survival in debris has been demonstrated.

Only crop rotation to avoid planting in fields where beans have been grown, or common blight has been a problem recently, and strict attention to use of clean seed can be recommended for control of common blight. Differences in susceptibility of cultivars has been noted with the cvs. Jules and Tara reported to be tolerant. Several varieties are resistant to halo blight.

1. Basu, P. K. and V. R. Wallen. 1967. Factors affecting virulence and pigment production of *Xanthomonas phaseoli* var. *fuscans*. *Can. J. Bot. 45:* 2367–2374.

2. Beach, S. A. 1892. Some bean diseases. *N. Y. State Agr. Expt. Sta. Bull. 48:* 305–333.

3. Boelema, B. H. 1967. Fuscous blight of beans in South Africa. *S. Afr. J. Agr. Sci. 10:* 1059–1063.

4. Claflin, L. E., D. L. Stuteville and D. V. Armbrust. 1973. Wind-blown soil in the epidemiology of bacterial leaf spot of alfalfa and common blight of bean. *Phytopathology 63:* 1417–1419.

5. Coyne, D. P. and M. L. Schuster. 1970. "Jules," a Great Northern dry bean variety tolerant to common blight bacterium (*Xanthomonas phaseoli*). *Plant Dis. Reptr. 54:* 557–559.

6. Coyne, D. P., M. L. Schuster and S. Al-Yasiri. 1963. Reaction studies of bean species and varieties to common blight and bacterial wilt. *Plant Dis. Reptr. 47:* 534–537.

7. Coyne, D. P., M. L. Schuster and L. Harris. 1965. Inheritance, heritability, and response to selection for common blight (*Xanthomonas phaseoli*) tolerance in *Phaseolus vulgaris* field bean crosses. *Proc. Am. Soc. Hort. Sci. 86:* 373–379.

8. Coyne, D. P., M. L. Schuster and L. Shaughnessy. 1966. Inheritance of reaction to halo blight and common blight bacteria in a *Phaseolus vulgaris* variety cross. *Plant Dis. Reptr. 50:* 29–32.

9. Epton, H. A. S. and B. J. Deverall. 1965. Physiological races of *Pseudomonas phaseolicola* causing halo blight of bean. *Plant Pathol. 14:* 53–54.

10. Goss, R. W. 1940. The relation of temperature to common and halo blight of beans. *Phytopathology 30:* 258–264.

11. Halsted, B. D. 1892. A bacterium of *Phaseolus. N. J. Agr. Expt. Sta. Ann. Rep. 13:* 283–285.

12. Hayward, A. C. and J. M. Waterson. 1965. *Pseudomonas phaseolicola* (Burkholder) Dowson. *Commonwealth Mycol. Inst. Descriptions Pathogenic Fungi Bacteria 45.* 2 p.

13. Hayward, A. C. and J. M. Waterston. 1965. *Xanthomonas phaseoli* (E. F. Smith) Dowson. *Commonwealth Mycol. Inst. Descriptions Pathogenic Fungi Bacteria 48.* 2 p.

14. Hayward, A. C. and J. M. Waterston. 1965. *Xanthomonas phaseoli* var. *fuscans* (Burkholder) Starr & Burkholder. *Commonwealth Mycol. Inst. Descriptions Pathogenic Fungi Bacteria 49.* 1 p.

15. Omer, M. E. H. and R. K. S. Wood. 1969. Growth of *Pseudomonas phaseolicola* in susceptible and in resistant bean plants. *Ann. Appl. Biol. 63:* 103–116.

16. Patel, P. N. and J. C. Walker. 1963. Relation of air temperature and age and nutrition of the host to the development of halo and common bacterial blights of bean. *Phytopathology 53:* 407–411.

17. Schuster, M. L. 1955. A method for testing resistance of beans to bacterial blights. *Phytopathology 45:* 519–520.

18. Sutton, M. D. and V. R. Wallen. 1970. Epidemiological and ecological relations of *Xanthomonas phaseoli* and *X. phaseoli* var. *fuscans* on beans in southwestern Ontario, 1961–1968. *Can. J. Bot. 48:* 1329–1334.

19. Wallen, V. R. and M. D. Sutton. 1965. *Xanthomonas phaseoli* var. *fuscans* (Burkholder) Starr & Burkholder on field bean in Ontario. *Can. J. Bot. 43:* 437–446.

20. Watson, D. R. W. 1970. Bean common blight and fuscous blight in New Zealand. *Plant Dis. Reptr. 54:* 1068–1072.

21. Zaumeyer, W. J. and H. R. Thomas. 1957. A monographic study of bean diseases and methods for their control. *U. S. Dept. Agr. Tech. Bull. 868.* 255 p.

Pythium Blight (Wilt, Stem rot)

A number of similar disease syndromes observed in the United States have been ascribed to related fungi since the description of "bean wilt" in 1931. Disease incidence has been associated, generally, with particular environmental conditions, but worldwide occurrence and the potential for sizeable losses combine to make this complex of diseases a continuing threat to successful production.

Wilt symptoms include water-soaking of the stem near the soil line that progresses upward to include branches and leaves with little discoloration of the softened cortical tissues. Foliar wilt first appears only during the warmer period of the day with return to full turgidity later in the day and at night. Affected plants subsequently develop continuous wilt symptoms and are killed.

Symptoms attributed to *Pythium* blight are quite similar, although infection occurs at a lower stem node, from where disease development progresses both upward and downward. Leaves and growing points are subsequently affected by a water-soaked, soft, wet deterioration or "blight" that often spreads to flowers and pods. Infection may originate in leaves, from where the disease progresses through the petioles into the stem. White mycelial growth may develop over parts of plants affected by either wilt or blight symptoms under favorable environmental conditions of temperature and humidity.

Several species of *Pythium* have been associated with blight or wilt symptoms in bean plants, but the species identified more frequently are *P. aphanidermatum, P. debaryanum* Hesse, *P. ultimum* Trow. and *P. myriotylum* Drechs. Each of these fungus species also can cause seedling damping-off and root rot symptoms (see *Rhizoctonia* root rot). All species of the genus thrive under conditions of high moisture, which can result from either excessive rainfall or intermittent but frequent periods of high humidity. *P. debaryanum* and *P. ultimum* are found more commonly under cool weather conditions, while *P. aphanidermatum* occurs at warmer temperatures and *P. myriotylum* at still higher temperatures.

Spherical sporangia of *P. debaryanum* are terminal, but conidia are terminal or intercalary and germinate directly by formation of a germ tube. Oogonia 23 μm in diameter are smooth, spherical and terminal.

Young hyphae of *P. myriotylum* are 2.5–4 μm in width while older hyphae measure 7 μm. Lobulate sporangia are terminal or intercalary and germinate to produce 3–40 reniform, biciliate zoospores 9–16 μm in diameter that germinate by forming single germ tubes. Smooth oogonia 15–44 μm in diameter are terminal or intercalary with 1–10 clavate antheridia 4–8 × 8–30 μm and colorless to yellow oospores measure 12–37 μm in diameter.

P. ultimum produces conidia 12–28 μm in diameter that are usually terminal and spherical while oogonia 19.6–22.9 μm in diameter are terminal and spherical also. There is usually one curved antheridium for each oogonium, and the spherical oospores, 14.7–18.3 μm in diameter, are smooth with 2-layered walls and germinate by 1 or more germ tubes. Zoospores are never produced.

Blight and related diseases are not usually severe enough to justify control measures, but planting in well-drained locations and thin seeding to provide maximum aeration and reduce the probability of infections spreading to neighboring plants are recommended. Resistance to seedling damping-off (caused by *P. ultimum*) and to blight (caused by *P. debaryanum, P. aphanidermatum* and *P. ultimum*) has been reported.

1. Adegbola, M. O. K. and D. J. Hagedorn. 1969. Symptomatology and epidemiology of *Pythium* bean blight. *Phytopathology 59:* 1113–1118.

2. Adegbola, M. O. K. and D. J. Hagedorn. 1969. Host–parasite relations in *Pythium* bean blight. *Phytopathology 59:* 1484–1487.

3. Adegbola, M. O. K. and D. J. Hagedorn. 1970. Host resistance and pathogen virulence in *Pythium* blight of bean. *Phytopathology 60:* 1477–1479.

4. Dickson, M. H. and G. S. Abawi. 1974. Resistance to *Pythium* in white-seeded beans (*Phaseolus vulgaris*). *Plant Dis. Reptr. 58:* 774–776.

5. Drechsler, S. C. 1952. Bean stem rot in Maryland and Delaware caused by several *Pythium* species. *Plant Dis. Reptr. 36:* 13.

6. Gay, J. D. 1969. Effects of temperature and moisture on snap bean damping-off caused by three isolates of *Pythium myriotylum. Plant Dis. Reptr. 53:* 707–709.

7. Gay, J. D. and S. M. McCarter. 1968. Stem rot of snap beans in southern Georgia caused by *Pythium myriotylum. Plant Dis. Reptr. 52:* 416.

8. Harter, L. L. and W. J. Zaumeyer. 1931. *Pythium butleri*—the cause of a bean wilt. *Phytopathology 21:* 991–994.

9. Kim, S. H., J. G. Kantzes and L. O. Weaver. 1974. Infection of aboveground parts of bean by *Pythium aphanidermatum. Phytopathology 64:* 373–380.

10. Kraft, J. M. and D. W. Burke. 1971. *Pythium ultimum* as root pathogen of beans and peas in Washington. *Plant Dis. Reptr. 55:* 1056–1060.

11. Stanghellini, M. E. and J. G. Hancock. 1971. Radial extent of the bean spermosphere and its relation to the behavior of *Pythium ultimum. Phytopathology 61:* 165–168.

12. Waterhouse, G. M. and J. M. Waterston. 1964. *Pythium aphanidermatum. Commonwealth Mycol. Inst. Descriptions Pathogenic Fungi Bacteria 36.* 2 p.

13. Waterhouse, G. M. and J. M. Waterston. 1966. *Pythium myriotylum. Commonwealth Mycol. Inst. Descriptions Pathogenic Fungi Bacteria 118.* 2 p.

Powdery mildew

The fungus that causes powdery mildew of bean was originally described in 1805 from France but has been found since that time to occur throughout the world. Serious losses have been reported but vary considerably according to season, locality and year.

Infection is first apparent on leaves as small, slightly darkened areas which later become white, powdery spots that enlarge and may coalesce to nearly cover the entire laminae. Affected areas retain a white, powdery appearance, and infection may spread to involve all aerial parts of the plants, including stems and pods. Severly diseased leaves may become chlorotic and distorted before falling. Pods, also, may be reduced in size and misshapen and plant growth reduced by defoliation.

Mildew is caused by *Erysiphe polygoni* DC em. Salm. that produces external mycelium on which are borne oval, hyaline, unicellular conidia 14–18 × 25–30 μm on simple, erect, hyaline, septate hyphae. The sexual stage, infrequently found on bean, consists of nearly spherical perithecia in which are borne 2–8 asci 30–45 × 45–70 μm, each of which contain 3–8 ascospores 9–14 × 20–25 μm. Numerous morphologically indistinguishable pathotypes of the fungus have been identified from various hosts.

Conidia are easily detached and readily disseminated by wind and water but probably do not remain viable for long under adverse conditions. The fungus is widely distributed throughout the world on many crop plants, which serve to provide inoculum, even though infectivity of pathotypes may be limited to related plant species. Although conidia will germinate only under conditions of high humidity, disease development often is most severe under relatively dry conditions with cool temperatures, but late-season infection usually does not seriously reduce yield.

Different cultivars, including Great Northern, Stringless Green Refugee, Pinto, Idaho Refugee and Topcrop, have been found less severely affected at vrious locations. Dusting with 1.5–5.0 percent dinitrocaprylphenyl crotonate (Karathane) has been found to provide excellent control, and spraying with actidione has been used successfully.

1. Cook, H. T. 1931. Nature of powdery-mildew injury to snap beans in Virginia in 1930. *Phytopathology 21:* 116.

2. Cook, H. T. 1931. Powdery mildew disease of snap beans. *Virginia Truck Expt. Sta. Bull. 74:* 931–940.

3. Felber, I. M. and C. L. Hamner. 1948. Control of mildew on bean plants by means of an antibiotic. *Bot. Gaz. 110:* 324–325.

4. Keil, H. L., H. P. Froehlich and F. B. Maughan. 1953. Efficacy of certain organic compounds in control of bean powdery mildew under laboratory conditions. *Phytopathology 43:* 477.

5. Moore, W. D. 1936. Powdery mildew (*Erysiphe polygoni*) on garden snap beans. *Phytopathology 26:* 1135–1144.

6. Yarwood, C. E. 1936. The tolerance of *Erysiphe polygoni* and certain other powdery mildews to low humidity. *Phytopathology 26:* 845–859.

7. Yarwood, C. E. 1949. Effect of soil moisture and nutrient concentration on the development of bean powdery mildew. *Phytopathology 39:* 780–788.

8. Zaumeyer, W. J. and H. R. Thomas. 1957. A monographic study of bean diseases and methods for their control. *U. S. Dept. Agr. Tech. Bull. 868.* 255 p.

Rust

Rust disease of bean was first described from Germany in 1795 but now occurs throughout the world in nearly every location where beans are grown. In many early accounts of the disease, it was not considered a serious problem but later was found to cause extensive losses in some localities. Despite concerted efforts to control rust, it remains a major threat to successful bean production.

Symptoms of the disease are confined largely to leaves, but young stems and branches are infrequently attacked. Lesions usually are noted first as small, slightly raised spots that are light in color and more often on lower leaf surfaces. These sori gradually enlarge to 1–2 mm in size and typically rupture the epidermis 8–10 days after infection to expose red-brown pustules of

Fig. 7. Rust on bean leaves. (Courtesy of G. Galvez, Centro Internacional de Agricultura Tropical, Cali.)

urediospores (Fig. 7). A ring of secondary sori may develop around the original infection locus on susceptible varieties and late-season production of teliospores, rarely produced in tropical climates, imparts a nearly black color to sori on plants grown in temperate climates.

Uromyces phaseoli var. *typica* Arth. is an autoecious rust fungus for which the urediospore (8–24 × 20–37 μm, brown, unicellular, frequently elliptical and echinulate with thin walls) is the predominant spore form, but dark brown, thick-walled, unicellular, mostly smooth, elliptical teliospores 20–28 × 25–35 μm are produced in temperate climates. Aecia with elliptical aeciospores 16–20 × 20–26 μm are rarely found.

Germination of urediospores can occur immediately upon maturation, and infection via indirect penetration through stomata can become apparent after 5 days. Under conditions favorable for disease development, the entire infection cycle can be completed within 15 days, providing opportunity for several such cycles and tremendous increase of inoculum during a single crop season. There is no evidence that inoculum is seed-borne. Urediospores are viable only for a short time but are readily wind-borne and are largely responsible for dissemination of the fungus. They will germinate over a range of temperatures from 10–25 C, but maximum germ tube growth was found to occur at 20 C. A resting period is required to induce germination of teliospores, which are much more tolerant of inclement weather conditions but are of no etiological importance in tropical environments. Several pathotypes, i.e. physiologic races, have been identified on the basis of pustule size (degree of susceptibility) induced on specific bean cultivars, but a number of other *Phaseolus* species, including *P. coccineus* L., *P. acutifolius* var. *latifolius* Freeman and lima bean (*P. lunatus* L.) have been found susceptible to infection.

The existence of pathotypes of the rust fungus wherever beans are grown has complicated attempts to control the disease by developing resistant varieties. However, varieties specifically resistant to prevalent races of the fungus have been developed in Australia, Mexico and the United States. Usefulness of these varieties has been limited by geographic distribution of the particular races of the fungus to which the varieties are resistant. The appearance of new races of the fungus may necessitate use of cultivars with different resistances. Recent attempts to find cultivars with nonspecific resistance are still in progress. Beans should not be planted on land where rust was serious the preceding season or near fields or straw stacks with rust-infected debris. Repeated applications of sulfur dust beginning when plants are small or before the disease has appeared have provided effective control, but other materials (maneb at 3 kg/ha, fentin acetate at 800 gr/ha and oxycarboxin at 1.5 kg/ha) are more widely used now.

1. Andrus, C. F. 1931. The mechanism of sex in *Uromyces appendiculatus* and *U. vignae*. *J. Agr. Res. 42:* 559–587.

2. Augustin, E., D. P. Coyne and M. L. Schuster. 1972. Inheritance of resistance in *Phaseolus vulgaris* to *Uromyces phaseoli typica* brazilian rust race B and of plant habit. *J. Am. Soc. Hort. Sci. 97:* 526–529.

3. Cass Smith, W. P., H. L. Harvey and O. M. Goss. 1951. "Westralia"—a new rust resistant pole bean. *Western Australia Dept. Agr. J. 28:* 317–324.

4. Christen, R. and E. Echandi. 1967. Razas fisiológicas más comunes de la roya *Uromyces phaseoli* var. *phaseoli* en Costa Rica y evaluación de la resistencia de algunos cultivares de fríjol a la roya. *Turrialba 17:* 7–10.

5. Crispin, A. and S. Dongo. 1962. New Physiologic races of bean rust, *Uromyces phaseoli typica*, from Mexico. *Plant Dis. Reptr. 46:* 411–413.

6. Fisher, H. H. 1952. New physiologic races of bean rust (*Uromyces phaseoli typica*). *Plant Dis. Reptr. 36:* 103–105.

7. Harter, L. L. and W. J. Zaumeyer. 1941. Differentiation of physiologic races of *Uromyces phaseoli typica* on bean. *J. Agr. Res. 62:* 717–732.

8. Waterhouse, W. L. 1953. Australian rust studies. XIII. Specialization of *Uromyces phaseoli* (Pers.) Wint. in Australia. *Proc. Linn. Soc. New S. Wales 78:* 226–232.

9. Williams, P. H. 1975. Genetics of resistance in plants. *Genetics 79:* 409–419.

10. Wingard, S. A. 1935. Host-parasite relationship in bean rust. *Phytopathology 25:* 39.

11. Zaumeyer, W. J. 1946. Field control of bean rust with sulfur. *Phytopathology 36:* 689.

12. Zaumeyer, W. J. and L. L. Harter. 1941. Inheritance of resistance to six physiologic races of bean rust. *J. Agr. Res. 63:* 599–622.

13. Zaumeyer, W. J. and H. R. Thomas. 1957. A monographic study of bean diseases and methods for their control. *U. S. Dept. Agr. Tech. Bull. 868.* 255 p.

Web blight

The disease of bean known as "web blight" is caused by a fungus that was first described in 1917 as a pathogen of fig but soon thereafter was reported to affect bean. Occurrence of the disease has been reported from Brazil and southeast Asia (Burma, Japan, Phillipines, Sri Lanka) in addition to Puerto Rico, several locations in southern United States and Costa Rica. Serious losses have been sustained when conditions favor development of the disease.

Lesions appear on leaves as small, water-soaked spots that may enlarge to 3 cm in diameter often with superficial, light brown hyphae in lesioned areas on both leaf surfaces. The affected areas are first lighter in color than healthy tissues and resemble having been scalded. Later diseased tissues become light brown with dark borders, while the superficial hyphae grow rapidly over the entire plant surface, tying together leaves, stems, flowers and pods with a web of hyphal strands. Pods may be attacked in any stage with infections appearing light brown and irregular at first and later becoming more circular and distinctly sunken with dark borders. Lesions may extend across the entire width of the pod and be 2 cm or more in length. Seeds under lesions on the pod wall are always discolored.

The causal fungus *Thanatephorus microsclerotia* (Weber) Donk (*Rhizoctonia microslerotia* Matz) produces sub-glubose, superficial sclerotia 0.2–0.5 mm in diameter that are light colored when young and dark brown when mature. Oval, hyaline basidiospores 4–6 × 7–10 μm are borne on clavate basidia.

The fungus may persist in the soil and is pathogenic on a number of hosts other than bean. Disease development is most severe during periods of high temperature and high humidity with sclerotia usually serving as primary inoculum, but basidiospores also have been found to cause infection under controlled conditions.

No completely satisfactory control has been developed other than rotation with crops not susceptible to infection, i.e. grasses, maize (*Zea mays* L.) and tobacco (*Nicotiana tabacum* L.), and planting in rows to provide maximum aeration. Planting should be timed to avoid high temperatures and high rainfall.

1. Atkins, J. G., Jr., and W. D. Lewis. 1952. *Rhizoctonia* aerial blight of soybeans in Louisiana. *Phytopathology 42:* 1.

2. Echandi, E. 1965. Basidiospore infection by *Pellicularia filamentosa* (=*Corticium microsclerotia*), the incitant of web blight of bean. *Phytopathology 55:* 698–699.

3. Matz, J. 1917. A *Rhizoctonia* of the fig. *Phytopathology 7:* 110–118.

4. Weber, G. F. 1939. Web-blight, a disease of beans caused by *Corticium microsclerotia*. *Phytopathology 29:* 559–575.

5. Zaumeyer, W. J. and H. R. Thomas. 1957. A monographic study of bean diseases and methods for their control. *U. S. Dept. Agr. Tech. Bull. 868.* 255 p.

Root rots

Fusarium root rot, also referred to as dry root rot, originally was described in 1916 from New York (United States). Since that time the disease has been reported from Europe, South America and Australia and has caused extensive losses on occasion.

Early symptoms of infection develop soon after plants appear aboveground and consist of a reddening of the taproot that increases in intensity and extent to involve most of the taproot. Discoloration becomes brownish later, longitudinal fissures may develop and lateral roots usually are killed, but fibrous roots that develop above the diseased portion of the taproot and just below the soil line often serve to maintain the plant. Plants affected early may be retarded in development and later drop yellowed leaves if the weather is dry, but there usually is no evidence of wilting. Under favorable conditions, diseased plants may recover sufficiently to produce nearly a normal harvest.

Macroconidia of *Fusarium solani* f. *phaseoli* (Burk.) Snyder & Hansen are mostly 3–4 septate, hyaline, and measure 5 × 45–50 μm. Terminal or intercalary chlamydospores borne singly or in short chains are about 11 μm in diameter. Microconidia are rare. The fungus can survive saprophytically and in soil for an extended period as well as in plant debris and compost, but it is not borne within bean seed, although it is possible that spores could be carried on the surface of seeds.

No completely satisfactory control method has been devised but nabam (9.4 ℓ of 25 percent/ha), formaldehyde (14 ℓ/ha) and chlorobromopropene (1.9–18.7 ℓ/ha) applied in the row at planting have reduced infection. Beans should be planted only on fields not recently used for beans and an extended crop rotation schedule (more than 5 years) should prove beneficial. Differences in susceptibility of varieties have been noted, but no satisfactory resistance has been reported.

Rhizoctonia root rot occurs wherever beans are grown and has been referred to as damping-off, hollow stem, stem canker and stem rot. Losses vary according to season and location but usually do not exceed 10 percent.

Young bean plants affected by damping-off symptoms are usually killed quickly while stem or root infection of older plants may result in a longer disease cycle. Infection may occur above or below soil level but typically results in a brownish, longitudinal, often sunken canker. Young plants usually are killed while older plants may persist, although yield will be noticeably reduced. The pith of older plants may be invaded, inducing a dark red discoloration and sclerotia may be produced in the stems.

The disease is caused by *Thanatephorus cucumeris* (*Rhizoctonia solani* Kühn), a soil-inhabiting pathogen which is infectious to bean from 9–30 C, but more severe disease development occurs in warm weather. Inoculum may be seed-borne and disease incidence exacerbated by planting soon after turning under green vegetation.

No control procedure is completely satisfactory, but only disease-free seed should be planted in locations not recently planted to beans. Soil application of pentachloronitrobenzene has been shown to reduce disease severity, and incidence of disease can be reduced by planting during cool seasons. Some varietal differences in susceptibility have been reported.

1. Adams, P. B., J. A. Lewis and G. C. Papavizas. 1968. Survival of root-infecting fungi in soil. IX. Mechanism of control of *Fusarium* root rot with spent coffee grounds. *Phytopathology 58:* 1603–1608.

2. Bristow, P. R., J. Katan and J. L. Lockwood. 1973. Control of *Rhizoctonia solani* by pentachloronitrobenzene accumulated from soil by bean plants. *Phytopathology 63:* 808–813.

3. Burke, D. W. 1965. Plant spacing and *Fusarium* root rot of beans. *Phytopathology 55:* 757–759

4. Burke, D. W. 1965. The near immobility of *Fusarium solani* f. *phaseoli* in natural soils. *Phytopathology 55:* 1188–1190.

5. Burke, D. W. and A. W. Barker. 1966. Importance of lateral roots in *Fusarium* root rot of beans. *Phytopathology 56:* 292–294.

6. Cook, R. J. and W. C. Snyder. 1965. Influence of host exudates on growth and survival of germlings of *Fusarium solani* f. *phaseoli* in soil. *Phytopathology 55:* 1021–1025.

7. Dodman, R. L., K. R. Barker and J. C. Walker. 1968. A detailed study of the different modes of penetration by *Rhizoctonia solani. Phytopathology 58:* 1271–1276.

8. Leach, J. G. and W. C. Snyder. 1947. Localized chemical applications to the soil and their effects upon root rots of beans and peas. *Phytopathology 37:* 363.

9. Luttrell, E. S. and K. H. Garren. 1952. Blights of snap bean in Georgia. *Phytopathology 42:* 607–613.

10. Maloy, O. C. and W. H. Burkholder. 1959. Some effects of crop rotation on the *Fusarium* root rot of bean. *Phytopathology 49:* 583–587.

11. McLean, D. M., J. C. Hoffman and G. B. Brown. 1968. Greenhouse studies on resistance of snap beans to *Rhizoctonia solani. Plant Dis. Reptr. 52:* 486–488.

12. Moore, W. D. and R. A. Conover. 1955. Chemical soil treatments for the control of *Rhizoctonia* on snap beans. *Plant Dis. Reptr. 39:* 103–105.

13. Nash, S. M., T. Christou and W. C. Snyder. 1961. Existence of *Fusarium solani* f. *phaseoli* as chlamydospores in soil. *Phytopathology 51:* 308–312.

14. Papavizas, G. C. and C. B. Davey. 1960. *Rhizoctonia* disease of bean as affected by decomposing green plant materials and associated microfloras. *Phytopathology 50:* 516–522.

15. Prasad, K. and J. L. Weigle. 1970. Screening for resistance to *Rhizoctonia solani* in *Phaseolus vulgaris. Plant Dis. Reptr. 54:* 40–44.

16. Richards, B. L. 1923. Soil temperature as a factor affecting the pathogenicity of *Corticium vagum* on the pea and the bean. *J. Agr. Res. 25:* 431–450.

17. Rolfe, F. M. 1903. *Corticium vagum* B. and var. *solani* Burt. A fruiting stage of *Rhizoctonia solani. Science 18:* 729.

18. Sanford, G. B. 1952. Persistence of *Rhizoctonia solani* Kühn in soil. *Can. J. Bot. 30:* 652–664.

19. Tisdale, W. B. and W. D. Moore. 1954. A new suggestion for control of damping-off and root-rot of snap beans. *Florida Agr. Expt. Sta. Circ. S-75.* 6 p.

20. Watson, R. D. 1951. Method for use of soil fumigants to control root rot and seed decay in peas and benas. *Plant Dis. Reptr. 35:* 324–325.

Southern blight (Crown rot)

The disease known as "southern blight," which affects several crops in addition to bean, was first recognized in 1893 in Florida (United States) but occurs now throughout the world in regions with mild to warm environments.

Lower leaves on affected plants develop a slight chlorosis which progresses upward and becomes more severe resulting in eventual defoliation. Water-soaking and discoloration of the cortex of the stem just below the soil line precedes a rapid cortical rot that extends downward involving the underground stem and taproot. A distinctive characteristic of the disease is development around the stem near soil level of a thick collar of mycelium in which are embedded numerous light brown, mostly spherical sclerotia. This collar of mycelium and sclerotia, which may extend above and below the soil line, remains attached to the stem when an affected plant is pulled.

The causal fungus is presently identified as *Botryobasidium rolfsii* but probably is more widely recognized by the name assigned to the asexual stage, *Sclerotium rolfsii.* Sclerotia 0.5–1.5 mm in diameter are light colored early but turn brown later, and hyphae 2–9 μm in width develop clamp connections.

Disease development is largely dependent on environmental conditions and is most severe in seasons with warm temperatures and high moisture. Cardinal temperatures for growth of the fungus are approximately 15, 30 and 37 C. Dissemination of the fungus is presumed to be largely via sclerotia, although mycelial development has been traced from plant to plant in the field. Many vegetables and weed plants are susceptible, but infection of grasses, including small grains and corn, has not been reported.

No completely effective control method has been devised, but planting on well-drained soil in seasons with lower temperatures and low soil and air moisture and in rotation with cereal crops or corn may prove beneficial.

1. Barrett, J. T. 1934. Observations on the basidial stage of *Sclerotium rolfsii. Phytopathology 24:* 1137–1138.

2. Curzi, M. 1931. Studi su lo "*Sclerotium rolfsii.*" *Bol. Roy. Staz. Pat. Veg., n.s. 11:* 306–375.

3. Higgins, B. B. 1922. Notes on the morphology and systematic relationship of *Sclerotium rolfsii* Sacc. *J. Elisha Mitchell Sci. Soc. 37:* 167–172.

4. Higgins, B. B. 1927. Physiology and parasitism of *Sclerotium rolfsii* Sacc. *Phytopathology 17:* 417–448.

5. Leach, L. D. and S. W. Mead. 1936. Viability of sclerotia of *Sclerotium rolfsii* after passage through the digestive tract of cattle and sheep. *J. Agr. Res. 53:* 519–526.

6. Milthorpe, F. L. 1941. Studies on *Corticium rolfsii* (Sacc.) Curzi (*Sclerotium rolfsii* Sacc.) I. Cultural characters and perfect stage. II. Mechanism of parasitism. *Proc. Linn. Soc. New. S. Wales 66:* 65–75.

7. Mordue, J. E. M. 1974. *Corticium rolfsii. Commonwealth Mycol. Inst. Descriptions Pathogenic Fungi Bacteria 410.* 2 p.

8. Taubenhaus, J. J. 1919. Recent studies on *Sclerotium rolfsii* Sacc. *J. Agr. Res. 18:* 127–138.

9. Venkatarayan, S. V. 1950. Notes on some species of *Corticium* and *Pellicularia*. *Indian Phytopathol. 2:* 81–86.

10. West, E. 1947. *Sclerotium rolfsii* Sacc. and its perfect stage on climbing fig. *Phytopathology 37:* 67–69.

11. Zaumeyer, W. J. and H. R. Thomas. 1957. A monographic study of bean diseases and methods for their control. *U.S. Dept. Agr. Tech. Bull. 868.* 255 p.

Sclerotinia wilt

A wilt disease of bean caused by *Sclerotinia sclerotiorum* (Lib.) DBy. (cylindrical asci 8–10 × 125–130 μm and ellipsoidal ascospores 4–6.6 × 9–13 μm borne in stalked, light brown apothecia produced from hard, black irregular sclerotia 1-several mm in diameter) has been reported from throughout the world where beans are grown but usually only following a period of warm, humid weather. Water-soaked spots that gradually enlarge to cause a soft, watery rot often characterized by a sticky, brown exudate develop on the stems, branches and pods of affected plants. Leaves become chlorotic, and affected plants wilt. Superficial mycelium spreads over affected plant parts and sclerotia, first light in color but later dark and eventually black and irregular, develop in the mycelium and characterize the disease.

Infection can originate either from ascospores largely disseminated by wind and rain or from sclerotia that remain in soil or debris. Nearly all vegetable crops losses can be extensive, but disease incidence is dependent on favorable environmental conditions that include warm temperature and high humidity.

Flooding for 3 weeks or more has been shown to hasten deterioration in the field of sclerotia, which may remain viable for 7 years in storage. Broadcast application of calcium cyanamid (920 kg/ha) also has proven successful on marl, but not sandy soils. Reduced seeding rates and wider row spacing on well-drained soil in seasons with low probability of high temperature and humidity may serve to reduce losses from *Sclerotinia* wilt. Foliar sprays with 2,6-dichloro-4-nitroaniline, ziram, Cop-O-Zinc and benomyl have been found to be effective in reducing aerial infections. Differences in susceptibility of dry bean varieties have been reported.

1. Anderson, F. N., J. R. Steadman, D. P. Coyne and H. F. Schwartz. 1974. Tolerance to white mold in *Phaseolus vulgaris* dry edible bean types. *Plant Dis. Reptr. 58:* 782–784.

2. Brien, R. M. 1932. Host range of *Sclerotinia sclerotiorum* in New Zealand. *New Zeal. J. Agr. 44:* 127–129.

3. Dana, B. F. and E. K. Vaughan. 1950. Experiments in control of white mold of beans by fungicides applied as dusts and sprays. *Plant Dis. Reptr. 34:* 8–14.

4. Lumsden, R. D. and R. L. Dow. 1973. Histopathology of *Sclerotinia sclerotiorum* infection of bean. *Phytopathology 63:* 708–715.

5. McClintock, J. A. 1916. *Sclerotinia libertiana* on snap beans. *Phytopathology 6:* 436–441.

6. Moore, W. D. 1948. Weather and the incidence of *Sclerotinia sclerotiorum* on vegetables in south Florida from 1944 to 1948. *Plant Dis. Reptr. 32:* 380–382.

7. Moore, W. D. 1949. Flooding as a means of destroying the sclerotia of *Sclerotinia sclerotiorum. Phytopathology 39:* 920–927.

8. Moore, W. D., R. A. Conover and D. L. Stoddard. 1949. The sclerotiniose disease of vegetable crops in Florida. *Florida Agr. Expt. Sta. Bull. 457.* 20 p.

9. Niedbalski, J. F. and S. F. Rickard. 1969. *Sclerotinia* white mold control in snap and lima beans with 2,6-dichloro-4-nitroaniline. *Plant Dis. Reptr. 53:* 573–575.

10. Vaughan, E. K. and B. F. Dana. 1949. Experimental applications of dusts and sprays to beans for control of *Sclerotinia sclerotiorum. Plant Dis. Reptr. 33:* 12–15.

11. Wallace, M. M. 1944. *Sclerotinia* disease of beans and other crops. *East Afr. Agr. J. 9:* 171–172.

12. Yerkes, W. D., Jr. 1955. *Sclerotinia* wilt of beans in Mexico. *Plant Dis. Reptr. 39:* 47.

Angular leaf spot

Occurrence of angular leaf spot of bean was reported from Italy in 1878 and subsequently from several other European countries, Africa, Australia, South America, the Caribbean area, including Puerto Rico and Trinidad, and from many locations in the United States. Crop losses sustained from angular leaf spot usually have been of little consequence but vary according to local environmental conditions and can be severe. The disease has been one of the most common bean problems in Colombia and the most important disease of crops grown for seed on the south coast of New South Wales (Australia).

Spots that appear on the undersides of leaves are first gray and later brown, without borders and consistently delimited by veins (Fig. 8). Defolia-

Fig. 8. Bean leaf and pod symptoms caused by the angular leaf spot fungus. (Courtesy of G. E. Galvez, Centro Internacional de Agricultura Tropical, Cali.)

tion frequently results from severe infections. Lesions on pods have sharply defined, red-brown centers with black borders and are variable in size and superficial in the beginning but later coalesce to extend across the entire width of pod walls. Stem lesions are dark, elongate and less well defined. Macroscopic, dark, erect coremia develop in all lesions.

Angular leaf spot is caused by *Isariopsis griseola* Sacc. which produces coremia 20–40 μm wide that are darker at the base and lighter toward the tip where 1–3 septate, lightly colored, cylindrical conidia 7–8 × 50–60 μm are borne. Cardinal temperatures for mycelial growth *in vitro* are 8, 24 and 28 C.

Bean, lima bean and cowpea [*Vigna sinensis* (Torner) Savi] are susceptible to infection (through stomata) which can occur during 3 hours' exposure to conditions of high humidity, but extension of the incubation under high humidity to 24 hours increases severity of disease development. The fungus can survive adverse climatic conditions for more than a year in infected debris and also can be transmitted in infected seed. Spores are readily wind-borne and disseminated short distances by splashing water.

Differences in susceptibility of varieties have been reported but offer no possibility for control at this time. Plant debris from infected bean crops should be completely turned under or destroyed, and a rotation cycle of at least 2 years should be followed with crops other than lima bean or cowpea planted between bean crops.

1. Barros, O., C. Cardona and R. Cardeñosa. 1958. The control of angular leaf spot of bean in Colombia. *FAO Plant Protect. Bull. 6:* 97–101.

2. Barros, O., C. Cardona, R. Cardeñosa and R. L. Skiles. 1958. Angular leaf spot of bean in Colombia. *Plant Dis. Reptr. 42:* 420–424.

3. Brock, R. D. 1951. Resistance to angular leaf spot among varieties of beans. *J. Australian Inst. Agr. Sci. 17:* 25–30.

4. Cardona-Alvarez, C. and J. C. Walker. 1956. Angular leaf spot of bean. *Phytopathology 46:* 610–615.

5. Cole, H., Jr. 1966. Angular leaf spot associated with severe defoliation of red kidney beans (*Phaseolus vulgaris*) in Pennsylvania. *Plant Dis. Reptr. 50:* 494.

6. Diaz, P., C. E. de Armas and A. Barros. 1965. La mancha angular de la caraota producida por *Isariopsis griseola* Sacc. en la cuenca del lago de Valencia. *Agron. Trop. (Venezuela) 14:* 261–275.

7. Orozco-Sarria, S. H. and C. Cardona-Alvarez. 1959. Evidence of seed transmission of angular leaf spot of bean. *Phytopathology 49:* 159.

Anthracnose

Anthracnose disease of bean was first recognized in France in 1843 and was reported from other European countries in the next 30 years. The disease now has been discovered throughout Africa, South, Central and North America, the Pacific area, including Australia, New Zealand, Japan and Taiwan, in addition to nearly the whole of Europe. In the last quarter of the nineteenth and the first quarter of the twentieth centuries, anthracnose was quite destructive in Europe and the United States, and severe reductions or complete crop losses were sustained amounting to millions of dollars. Anthracnose was considered the most important disease of bean east of the Mississippi River in the United States until usable control practices were developed.

All aerial parts of plants are susceptible to infection but symptoms on pods are characteristic and more easily identified (Fig. 9). Lesions first appear as light brown spots that enlarge rapidly to a cm in diameter and become almost round with margins lighter than the centers. The margins of lesions are raised, and the color of diseased tissues lightens as affected pods mature, but infection may extend through the pod wall to reach the seed and cause yellow to brown, sunken cankers. Seedlings grown from affected seeds often have similar cankers on cotyledons, and lesions that develop on hypocotyls of young plants are first light brown but gradually elongate and become darker and sunken. The stems of seriously affected plants may break, causing them to topple. Symptoms also develop on the petioles and veins on the undersurfaces of leaves, causing a dark red to purplish discoloration that later turns quite dark. Pinkish spore masses may appear in lesions during humid weather but later dry to gray or brown.

The disease is caused by *Colletotrichum lindemuthianum* (Sacc. & Magn.) Briosi & Cav., characterized by hyaline, septate mycelium that forms stromatoid masses in which are produced acervuli with conidia

Fig. 9. Anthracnose lesions on bean pods and vein discoloration on leaves. (Courtesy of G. E. Galvez, Centro Internacional de Agricultura Tropical, Cali.)

(hyaline, aseptate, cylindrical and 4.4–5.3 × 13–22 μm) that appear pinkish *en masse* and sparse, simple, septate setae 30–100 μm in length.

Infection by direct penetration occurs over a temperature range from 13–27 C but only under conditions of 92 percent, or higher, humidity. Length of the incubation period on pods in transit was found to vary from 5–14 days with decrease in temperature from 22–27 to 7 C, but maximum spore production on pods occurs at 14–18 C. Optimum temperature for spore germination was about 27 C, while best growth in culture occurred at approximately 22 C. The fungus can survive from one season to the next in debris from infected plants as well as in diseased seed, and either source may readily supply primary inoculum, which is disseminated largely by spattering rain. Numerous physiologic races of the fungus have been reported but most have been categorized as belonging to either of 4 pathotypes, i.e. alpha, beta, gamma or delta race groups. Development of simple and satisfactory control procedures, which involve field sanitation, crop rotation of 2 years or more and, especially, planting only disease-free seed (produced in dry western areas), has relegated anthracnose to a position of minor import in the United States. However, in areas where application of these methods is difficult, i.e. in Colombia on beans grown above 1500 m elevation, anthracnose remains a serious disease, but genetic resistance to several pathotypes has been identified.

Control can be achieved by combining crop rotation on a 2–3 year schedule with elimination of debris from infected plants (sanitation) and planting only healthy seed where each of these procedures can be accomplished. Otherwise, foliar application of any of several fungicides, including copper oxychloride, zineb, nabam, ferbam or insoluble copper have been found to effectively control anthracnose. Several varieties with resistance to localized pathotypes have been developed but may not provide adequate control in other locations where different pathotypes predominate.

1. Andrus, C. F. and B. L. Wade. 1942. The factorial interpretation of anthracnose resistance in beans. *U.S. Dept. Agr. Tech. Bull. 810.* 29 p.

2. Ayonoadu, U. W. U. 1974. Races of bean anthracnose in Malawi. *Turrialba 24:* 311-314.

3. Bannerot, H. 1965. Resultats de l'infection d'une collection de haricots par six races physiologiques d'anthracnose. *Ann. Amelior. Plantes 15:* 201–222.

4. Barrus M. F. 1915. An anthracnose-resistant red kidney bean. *Phytopathology 5:* 303-311.

5. Barrus, M. F. 1918. Varietal susceptibility of beans to strains of *Colletotrichum lindemuthianum* (Sacc. and Magn.) B. and C. *Phytopathology 8:* 589–614.

6. Burkholder, W. H. 1918. The production of an anthracnose-resistant White Marrow bean. *Phytopathology 8:* 353–359.

7. Burkholder, W. H. 1923. The gamma strain of *Colletotrichum lindemuthianum* (Sacc. et Magn.) B. et C. *Phytopathology 13:* 316–323.

8. Cardenas, F., M. W. Adams and A. Andersen. 1964. The genetic system for reaction

of field beans (*Phaseolus vulgaris* L.) to infection by three physiologic races of *Colletotrichum lindemuthianum*. *Euphytica 13:* 178–186.

9. Erb, K., M. E. Gallegly and J. G. Leach. 1973. Longevity of mycelium of *Colletotrichum lindemuthianum* in hypocotyl tissue of resistant and susceptible bean cultivars. *Phytopathology 63:* 1334–1335.

10. Goth, R. W. and W. J. Zaumeyer. 1965. Reactions of bean varieties to four races of anthracnose. *Plant Dis. Reptr. 49:* 815–818.

11. Griffey, R. T. and J. G. Leach. 1965. The influence of age of tissue on the development of bean anthracnose lesions. *Phytopathology 55:* 915–918.

12. Leach, J. G. 1923. The parasitism of *Colletotrichum lindemuthianum*. *Minn. Agr. Expt. Sta. Tech. Bull. 14.* 41 p.

13. Leakey, C. L. A. and M. Simbwa-Bunnya. 1972. Races of *Colletotrichum lindemuthianum* and implications for bean breeding in Uganda. *Ann. Appl. Biol. 70:* 25–34.

14. McRostie, G. P. 1919. Inheritance of anthracnose resistance as indicated by a cross between a resistant and a susceptible bean. *Phytopathology 9:* 141–148.

15. Mordue, J. E. M. 1971. *Colletotrichum lindemuthianum*. *Commonwealth Mycol. Inst. Descriptions Pathogenic Fungi Bacteria 316.* 2 p.

16. Oliari, L., C. Vieira and R. E. Wilkinson. 1973. Physiologic races of *Colletotrichum lindemuthianum* in the state of Minas Gerais, Brazil. *Plant Dis. Reptr. 57:* 870–872.

17. Reid, W. D. 1945. Resistance of beans to halo-blight and anthracnose and the occurrence of bean mosaic and bean weevil. *New Zeal. J. Sci. Technol., Sect. A, 27:* 331–335.

18. Saccardo, P. A. 1884. *Gloeosporium lindemuthianum* Sacc. et Magnus. *Sylloge Fungorum 3:* 717.

19. Waterhouse, W. L. 1955. Studies of bean anthracnose in Australia. *Proc. Linn. Soc. New S. Wales 80:* 71–83.

20. Zaumeyer, W. J. and H. R. Thomas. 1957. A monographic study of bean diseases and methods for their control. *U.S. Dept. Agr. Tech. Bull. 868.* 255 p.

Charcoal rot (Ashy stem blight)

Charcoal rot of bean was described from France in 1905 but has been found subsequently throughout the world where beans are grown. Losses vary according to season and locality and have been estimated to reach 65 percent.

Plants infected as seedlings develop conspicuous black, sunken cankers with distinct margins that originate on the stem near the base of the cotyledon. Lesions may extend in both directions and usually develop quickly to involve the entire stem and kill the plant. Infection of older plants progresses more slowly, often in unilateral fashion, and results in yellowing and drooping of foliage. Black pycnidia may develop abundantly in stem lesions, but numerous, small, black sclerotia typically develop in diseased tissues soon after infection.

Sclerotia of *Macrophomina phaseolina* are smooth, black and hard and measure 0.1–1 mm in diameter, while pycnidia are dark brown and 100–200 μm in diameter. Conidia are hyaline, unicellular and obovoid and measure 5–10 $\times$ 14–30 μm.

The causal fungus is infectious to more than 100 host plants and is seed-borne in bean. More serious occurrence of the disease has been found to follow or occur during periods of warm weather (35 C or above) and immature, unthrifty or debilitated plants are particularly susceptible.

Only seed from disease-free plants should be sown in soil not planted recently to a susceptible crop, especially bean. All debris from previous crops where the disease has occurred should be destroyed before planting. Recommended practices for cultivation of bean should be followed and particular attention given to avoid excessive soil moisture.

1. Andrus, C. F. 1938. Seed transmission of *Macrophomina phaseoli*. *Phytopathology 28:* 620–634.

2. Ashby, S. F. 1927. *Macrophomina phaseoli* (Maubl.) comb. nov. the pycnidial stage of *Rhizoctonia bataticola* (Tab.) Butl. *Trans. Brit. Mycol. Soc. 12:* 141–147.

3. Haigh, J. C. 1930. *Macrophomina phaseoli* (Maubl.) Ashby and *Rhizoctonia bataticola* (Taub.) Butler. *Ann. Roy. Bot. Gard. (Peradeniya) 11:* 213–249.

4. Holliday, P. and E. Punithalingam. 1970. *Macrophomina phaseolina*. *Commonwealth Mycol. Inst. Descriptions Pathogenic Fungi Bacteria 275.* 2 p.

5. Kendrick, J. B. 1933. Seedling stem blight of field beans caused by *Rhizoctonia bataticola* at high temperatures. *Phytopathology 23:* 949–963.

6. Ludwig, C. A. 1926. *Macrophoma* stem rot of bean (*Macrophoma phaseoli*). *Plant Dis. Reptr. 10:* 81–82.

7. Luttrell, E. S. 1946. A pycnidial strain of *Macrophomina phaseoli*. *Phytopathology 36:* 978–980.

8. Sarejanni, J. A. and C. B. Cortzas. 1935. Note sur le parasitisme du *Macrophomina phaseoli* (Maubl.) Ashby. *Ann. Inst. Phytopathol. Benaki 1:* 37–44.

9. Small, W. 1924. A *Rhizoctonia* causing root diseases in Uganda. *Trans. Brit. Mycol. Soc. 9:* 152–166.

10. Tompkins, C. M. and M. W. Gardner. 1935. Relation of temperature to infection of bean and cowpea seedlings by *Rhizoctonia bataticola*. *Hilgardia 9:* 219–230.

11. Weimer, J. L. and E. S. Luttrell. 1945. A canker of cowpea and *Macrophomina phaseoli* leaf spot of cowpea and snap bean. *Plant Dis. Reptr. 29:* 127–129.

Fusarium yellows

Progressive, upward yellowing of the foliage on affected plants was observed in California (United States) in 1928. The disease, caused by *Fusarium oxysporum* f. *phaseoli* Kendrick & Snyder, has been reported also from Costa Rica and Australia.

1. Duque, S. L. and L. E. Müller. 1969. Estudio sobre la patogenicidad de *Fusarium oxysporum* f. *phaseoli* en el frijol. I. Patogenésis e histología sintomatológica. *Turrialba 19:* 71–81.

2. Echandi, E. 1967. Amarillamiento del frijol (*Phaselous vulgaris* L.) provocado por *Fusarium oxysporum* f. *phaseoli*. *Turrialba 17:* 409–410.

3. Harter, L. L. 1929. A *Fusarium* disease of beans. *Phytopathology 19:* 84.

4. Kendrick, J. B., Jr. and W. C. Snyder. 1942. *Fusarium* yellows of beans. *Phytopathology 32:* 1010–1014.

Leaf spot and blotch diseases

Alternaria leaf spot (Mancha foliar) (*Alternaria* sp.)

Small, irregular brown spots with darker borders have been reported on bean from the United States, Denmark and Costa Rica, but identity of the causal organism(s) has not been established. Spores of the pathogen in Costa Rica were 16–19 × 120–240 μm.

1. Gonzalez, L. C. 1973. Mancha foliar del frijol (*Phaseolus vulgaris*) causada por *Alternaria* sp. en Costa Rica. *Turrialba 23:* 238–239.

2. Saad, S. and D. J. Hagedorn. 1969. Symptomatology and epidemiology of *Alternaria* leaf spot of bean, *Phaseolus vulgaris*. *Phytopathology 59:* 1530–1533.

3. Saad, S. and D. J. Hagedorn. 1969. Host-parasite relations in the initiation and development of bean *Alternaria* leaf spot. *Phytopathology 59:* 1773–1774.

4. Swank, G. 1951. *Alternaria* leaf spot and dieback of snap bean; a new disease in central Florida. *Plant Dis. Reptr. 35:* 330–332.

Cercospora leaf blotch

Brown spots with dark borders that are usually angular but variable in size and shape and often delimited by veins have been referred to as *Cercospora* leaf blotch caused by *Cercospora cruenta* Sacc. Symptoms develop on mature leaves and centers of lesions dry, crack and often fall. Defoliation frequently follows. (See Cowpea—*Cercospora* leaf spot).

1. Mulder, J. L. and P. Holliday. 1969. *Cercospora cruenta. Commonwealth Mycol. Inst. Descriptions Pathogenic Fungi Bacteria 463.* 2 p.

2. Saccardo, P. A. 1886. *Cercospora cruenta* Sacc. *Sylloge Fungorum 4:* 435.

3. Zaumeyer, W. J. and H. R. Thomas. 1957. A monographic study of bean diseases and methods for their control. *U.S. Dept. Agr. Tech. Bull. 868.* 255 p.

Cercospora leaf spot (Mancha gris)

Mostly circular leaf spots 2–10 mm in diameter with light centers and reddish margins have been observed in the United States, Puerto Rico and China since the disease was first described in 1882 and attributed to *Cercospora canescens* Ell. & Mart. Conidiophores of the causal fungus measure 3–4.5 × 50–100 μm and hyaline, slender to slightly clavate, septate conidia 3–4.5 × 80–220 μm.

Symptoms reported from Brazil and Colombia (Mancha gris) have been caused by a fungus identified as *Cercospora vanderstyi* P. Hennings, while *C. castellanii* and *C. albida* were considered responsible for symptoms in Dominican Republic.

1. Ellis, J. B. and G. B. Martin. 1882. General notes: botany. New species of North American fungi. *Am. Nat. 16:* 1001–1004.

2. Matta, A. and L. Belliard. 1973. Su tre *Cercospora* del fagiolo osservate in Repubblica Dominicana. *Riv. Agr. Subtrop. Trop. 67:* 78–83.

3. Miles, L. E. 1917. Some diseases of economic plants in Porto Rico. *Phytopathology 7:* 345–351.

4. Mulder, J. L. and P. Holliday. 1969. *Cercospora canescens. Commonwealth Mycol. Inst. Descriptions Pathogenic Fungi Bacteria 462.* 2 p.

5. Skiles, R. L. and C. Cardona-Alvarez. 1959. Mancha gris, a new leaf disease of bean in Colombia *Phytopathology 49:* 133–135.

6. Tai, F. L. 1936. Notes on Chinese fungi. VII. *Chinese Bot. Soc. Bull. 2:* 45–66.

7. Vieira, C. and H. L. Shands. 1965. Mancha gris, nova doença do feijoeiro no Brasil. *Rev. Agr. (Brazil) 40:* 3–5.

8. Zaumeyer, W. J. and H. R. Thomas. 1957. A monographic study of bean diseases and methods for their control. *U.S. Dept. Agr. Tech. Bull. 868.* 255 p.

Bean common mosaic

The disease of bean now referred to as common mosaic was observed in Russia in 1894. A similar disease was recognized in the United States in 1908, although it may have been present there earlier. Occurrence of the disease throughout Europe, South and Central America and in Africa was established by 1925. Common mosaic has been identified also from Australia and Japan and probably occurs throughout the world wherever beans are grown. Infection percentage varies according to location, but 100 percent field infection has been reported commonly, and, in a recent study, common mosaic was found to cause 50–60 percent reduction in seed yield.

Plants affected by common mosaic are not often killed, although they do become stunted with mottled leaves and various types of leaf distortion (Fig. 10). Particular symptom expression may vary according to bean variety. Pods often show no symptoms but, on severely affected plants, may

Fig. 10. Leaf symptoms and field appearance of common mosaic on bean. (Courtesy of G. E. Galvez, Centro Internacional de Agricultura Tropical, Cali.)

exhibit small green spots and be smaller with fewer ovules than on healthy plants. Seeds from such pods, which mature late, are frequently smaller than normal and shrivelled.

Another syndrome referred to as "black root" includes loss of green color and wilting of young leaves followed by development of a systemic permanent wilt and frequent death of affected plants. Diseased plants also develop vascular necrosis of leaves, pods, roots and stems, sometimes as unilateral streaking, and external discoloration of petioles and the lower hypocotyl. Necrotic streaking also may extend to both pod sutures with some vascular discoloration throughout affected pods. The occurrence of black root symptoms has been limited to bean varieties with specific parentage that are incubated under particular temperature conditions, but symptom development may vary somewhat even on infected plants grown under controlled (greenhouse) conditions.

Bean common mosaic is caused by a virus, characterized by filamentous particles 15 × 75 nm in length, that is transmitted mechanically and by several aphids, including *Aphis gossypii* Glov., *A. medicaginis* Koch, *Macrosiphum pisi* (Kalt.), *M. solanifolii* (Ashm.) and *Myzus persicae*, in a nonpersistent manner. The virus is inactivated when heated for 10 minutes to about 60 C, diluted more than 10^{-4} or stored (in sap) at room temperature for more than 4 days. Several strains of the virus have been identified.

Transmission of the disease has been demonstrated through seeds produced on plants infected before flowering which may be the primary source of original inoculum and the major means by which the disease is disseminated over long distances. Virus has been recovered from seed stored 30 years, and infection can be transmitted in pollen from infected plants. The virus is infectious on few plants other than bean, but localized infections have been reported on *Chenopodium quinoa* Willd., globe amaranth (*Gomphrena globosa* L.) and *Phaseolus lathyroides* L. *Nicotiana clevelandii* Gray was infected systemically by 2 isolates of the virus.

Response to infection varies among cultivars of bean which have been described as tolerant (slight symptoms only), sensitive (susceptible, rugose mottle, particularly on lower leaves, and misshapen pods) and hypersensitive (necrotic local lesions). Plants with hypersensitive resistance derived from the varieties Corbett Refugee, Kentucky Wonder and Creaseback develop black root symptoms when incubated at a temperature of 32 C or above after inoculation, but plants with resistance derived from Robust or Great Northern U. I. No. 1 are not so affected. Incidence of black root unrelated to temperature has also been reported. Both recessive (Robust) and dominant (Idaho Refugee, U. S. No. 5 Refugee) single-factor resistances have been demonstrated.

High frequency of transmission through infected seed and efficient transmission of the virus by several aphid species make control difficult except by using resistant varieties. Reduction in yield has been found directly correlated with time of infection. The number of virus strains, identified by response of selected bean cultivars to infection, and geographic range of individual virus strains often limit utilization of cultivars with specific disease resistance. Application of insecticide to reduce the aphid population and roguing of diseased plants when they appear may serve to retard spread of the disease in the field. Strict attention to selection of planting seed produced only on virus-free plants is recommended.

1. Ali, M. A. 1950. Genetics of resistance to the common bean mosaic virus (bean virus 1) in the bean (*Phaseolus vulgaris* L.) *Phytopathology 40:* 69–79.

2. Alvarez, A. M. 1965. Determinacion de la resistencia por hipersensibilidad al mosaico comun del fríjol (*Phaseolus* virus 1). *Agr. Téc. 25:* 114–119.

3. Bos, L. 1971. Bean common mosaic virus. *Commonwealth Mycol. Inst. Descriptions Plant Viruses 73.* 4 p.

4. Chamberlain, E. E. 1939. Bean-mosaic (*Phaseolus* virus 1 of Smith, 1937). *New Zeal. J. Sci. Technol., Sec. A, 20:* 381–388.

5. Dean, L. L. and C. W. Hungerford. 1946. A new bean mosaic in Idaho. *Phytopathology 36:* 324–326.

6. Fajardo, T. G. 1930. Studies on the mosaic of the bean (*Phaseolus vulgaris* L.) *Phytopathology 20:* 469–494.

7. Gamez, R. 1973. Los virus del frijol en Centro América. III. Razas del virus del mosaico común del frijol de El Salvador y Nicaragua. *Turrialba 23:* 475–476.

8. Gamez, R., A. Osores and E. Echandi. 1970. Una raza nueva del virus del mosaico común del frijol. *Turrialba 20:* 397–400.

9. Grogan, R. G. and J. C. Walker. 1948. The relation of common mosaic to black root of bean. *J. Agr. Res. 77:* 315–331.

10. Hampton, R. O. 1975. The nature of bean yield reduction by bean yellow and bean common mosaic viruses. *Phytopathology 65:* 1342–1346.

11. Hubbeling, N. 1972. Resistance in beans to strains of bean common mosaic virus. *Wetensch. Gent 37:* 458–466.

12. Jenkins, W. A. 1940. A new virus disease of snap beans. *J. Agr. Res. 60:* 279–288.

13. Jenkins, W. A. 1941. A histological study of snap bean tissues affected with black root. *J. Agr. Res. 62:* 683–690.

14. Kulkarni, H. Y. 1973. Notes on East African plant virus diseases. IV. Bean common mosaic virus. *East Afr. Agr. Forest. J. 38:* 72–76.

15. Lockhart, E. L. and H. U. Fischer. 1974. Chronic infection of seedborne bean common mosaic virus in Morocco. *Plant Dis. Reptr. 58:* 307–308.

16. Medina, A. C. and R. G. Grogan. 1961. Seed transmission of bean mosaic viruses. *Phytopathology 51:* 452–456.

17. Murphy, D. M. and W. H. Pierce. 1938. A mosaic-resistant small red bean. *Phytopathology 28:* 270–273.

18. Nelson, R. 1922. Transference of the bean mosaic virus by *Macrosiphum solanifolii. Science 56:* 342–344.

19. Pierce, W. H. and C. W. Hungerford. 1929. Symptomatology, transmission, infection and control of bean mosaic in Idaho. *Idaho Agr. Expt. Sta. Res. Bull.* 7. 37 p.

20. Rands, R. D. and W. Brotherton, Jr. 1925. Bean varietal tests for disease resistance. *J. Agr. Res. 31:* 101–154.

21. Reddick, D. and V. B. Stewart. 1919. Transmission of the virus of bean mosaic in seed and observations on thermal death-point of seed and virus. *Phytopathology 9:* 445–450.

22. Richards, B. L., Jr. and W. H. Burkholder. 1943. A new mosaic disease of beans. *Phytopathology 33:* 1215–1216.

23. Saettler, A. W. and G. E. Trujillo. 1972. Monroe bean as a local lesion host for bean common mosaic virus. *Phytopathology 62:* 489–490.

24. Silbernagel, M. J. 1969. Mexican strain of bean common mosaic virus. *Phytopathology 59:* 1809–1812.

25. Thomas, H. R. 1954. Factors affecting development of necrosis in some bean varieties inoculated with common bean mosaic virus. *Phytopathology 44:* 508.

26. Thomas, H. R. and H. H. Fisher. 1954. A rapid method of testing snap beans for resistance to common bean mosaic virus. *Plant Dis. Reptr. 38:* 410–411.

27. Zaumeyer, W. J. and R. W. Goth. 1964. A new severe symptom-inducing strain of common bean mosaic virus. *Phytopathology 54:* 1378–1385.

28. Zettler, F. W. 1969. The heterogeneity of bean leaves as sources of bean common mosaic virus for aphids. *Phytopathology 59:* 1109–1110.

Rugose mosaic

A virus disease of bean first observed in Costa Rica in 1964 was described in 1972. The disease has been found subsequently in Guatemala.

Leaflets on affected plants are severely malformed, characteristically with rugose margins (Fig. 11). Mottling and malformation, often with some vein necrosis, persist, while malformation of leaflets may gradually lessen as new leaves appear. Pods on infected plants usually show no symptoms.

Rugose mosaic is caused by a polyhedral virus with particles 29 nm in diameter. Inactivation in sap from infected plants occurs upon exposure for 10 minutes to 65–70 C or dilution of $10^{-4} - 10^{-5}$. Infectivity *in vitro* (at 22 C) was maintained for 2 but not 4 days. The virus is sap transmissible and by at least 3 chrysomelid beetles *Diabrotica adelpha* Herold, *D. balteata* Le Conte and *Cerotoma ruficornis* (Oliv.) which may remain infectious for 2–10 days. Several leguminous species are susceptible to infection, including *Phaseolus acutifolius* Gray, *P. lathyroides*, lima bean, broad bean (*Vicia faba* L.) and soybean [*Glycine max* (L.) Merr.], while others are not, e.g. mung bean (*P. aureus* Roxb.), *P. angularis* (Willd.) Wright, *P. mungo*, hyacinth bean (*Dolichos lablab* L.) and cowpea. *Chenopodium amaran-*

Fig. 11. Rugose mosaic symptoms on bean. (Courtesy of R. Gamez, University of Costa Rica, San Jose.)

ticolor Coste & Reyn responds to inoculation by formation of localized lesions as do some varieties of bean, e.g. Bountiful, Sanilac and Topcrop. The virus is not seed-borne.

Two strains of the virus from Guatemala, A_1 and A_2, have been distinguished, and cytological effects induced by these strains resemble cell alterations caused by bean line pattern mosaic virus (and southern bean mosaic, cowpea strain) and bean pod mottle virus, respectively.

Control includes isolation of new bean plantings from fields where the disease is present and routine application of insecticide to reduce spread of the disease.

1. Gamez, R. 1972. Los virus del frijol en Centroamérica. II. Algunas propiedades y transmisión por crisomélidos del virus del mosaico rugoso del frijol. *Turrialba 22:* 249–257.

2. Kitajima, E. W., A. Tascon, R. Gamez and G. E. Galvez. 1974. Ultrastructural studies on bean leaf tissues infected with two strains of bean rugose mosaic virus. *Turrialba 24:* 393–397.

Southern bean mosaic

A mosaic disease observed in Louisiana (United States) in 1941 was named "southern bean mosaic" and the disease described 2 years later. The disease has been found since then in several more northerly locations in the United States but remains confined largely to warmer temperate and tropical areas of the Americas. It has been found also in Africa and France but generally is not considered of major concern in production.

Infected plants develop a mild systemic mottle with vein banding, rugosity and leaflet malformation, which symptoms reach raximum intensity when plants approach maturity. Pods on affected plants develop irregular, dark, water-soaked or yellow-green areas depending on natural pod coloration.

Some bean cultivars develop, on inoculated leaves only, circular, dark, localized lesions (with no systemic symptoms) in 2–3 days.

Southern bean mosaic is caused by a virus characterized by isometric particles 28 nm in diameter that are rendered noninfectious by exposure for 10 minutes to 90–95 C, or storage *in vitro* (18–22 C) for 20–165 days. Dilution end-point of the virus in sap from infected bean or cowpea is near 10^{-6}. The virus is transmissible mechanically, by the bean leaf beetle [*Cerotoma trifurcata* (Forster)] following a 48-hour acquisition feeding and may be seed-borne in bean and cowpea (3–4 percent in cv. Early Wilt Resistant Ramshorn). Host range of the virus, in addition to bean and cowpea, is limited to lima bean, mung bean and soybean. Strains that differ markedly in host range from the type strain have been described.

Varieties of bean that respond to mechanical inoculation by production of local lesions on inoculated leaves only (viz. no systemic symptoms) are effectively resistant.

1. Lauffer, M. A. and W. C. Price. 1947. Electrophoretic purification of southern bean mosaic virus. *Arch. Biochem. 15:* 115–124.

2. Miller, G. L. and W. C. Price. 1946. Physical and chemical studies on southern bean mosaic virus. I. Size, shape, hydration and elementary composition. *Arch. Biochem. 10:* 467–477.

3. Miller, G. L. and W. C. Price. 1946. Physical and chemical studies on southern bean mosaic virus. II. Crystallization by dialysis. *Arch. Biochem. 11:* 329–335.

4. Miller, G. L. and W. C. Price. 1946. Physical and chemical studies on southern bean mosaic virus. III. Electrophoretic and nucleic acid studies. *Arch. Biochem. 11:* 337–343.

5. Price, W. C. 1945. Crystallization of southern bean mosaic virus. *Science 101:* 515–517.

6. Shepherd, R. J. 1971. Southern bean mosaic virus. *Commonwealth Mycol. Inst. Descriptions Plant Viruses 47.* 4 p.

7. Shepherd, R. J. and R. W. Fulton. 1962. Identity of a seed-borne virus of cowpea. *Phytopathology 52:* 489–493.

8. Zaumeyer, W. J. and L. L. Harter. 1943. Inheritance of symptom expression of bean mosaic virus 4. *J. Agr. Res. 67:* 295–300.

9. Zaumeyer, W. J. and L. L. Harter. 1943. Two new virus diseases of beans. *J. Agr. Res. 67: 305–327.*

10. Zaumeyer, W. J. and L. L. Harter. 1944. A severe necrosis caused by bean-mosaic virus 4 on beans. *Phytopathology 34:* 510–512.

Bean yellow mosaic

Described in 1934 from the United States, bean yellow mosaic is now known to occur throughout the world where legume viruses have been identified. Losses attributed to the disease have been difficult to assess accurately because of similarity between bean yellow mosaic and bean common mosaic symptoms. In a recent study, however, bean yellow mosaic was found to cause more than 30 percent reduction in pods/plant and 40 percent reduction in seed yield.

Infected plants first exhibit epinasty of leaflets by drooping at the pulvini. Chlorotic or necrotic spots 1–3 mm in diameter appear on affected leaves, enlarge and intensify to produce a systemic yellow mosaic. Diseased plants are stunted because internodes are shortened, and maturity, with reduced production, is delayed. Type and severity of symptoms vary according to virus strain and bean cultivar involved.

Particles of the causal virus are filamentous and 15×750 nm, although particles (to 60 nm) greater in length have been observed. Infectivity of the virus in sap is destroyed by exposure to about 60 C for 10 minutes, dilution greater than 10^{-4} or storage *in vitro* for more than 2 days. Yellow mosaic can be transmitted mechanically as well as by several species of aphids, including *Acyrthosiphon pisum* (Harris), *Aphis fabae* Scopoli, *Macrosiphum euphor-*

biae (Thomas) and *Myzus persicae*. Seed transmission is not common but occurs more frequently in lupine (*Lupinus albus* L., *L. luteus* L.), broad bean and white sweet clover (*Melilotus alba* Desr.). The virus is infectious to a number of leguminous plants other than bean, including chickpea (*Cicer arietinum* L.), pigeon pea [*Cajanus cajan* (L.) Millsp.], sweet pea (*Lathyrus odoratus* L.), lentil (*Lens esculenta* Moench.), pea (*Pisum sativum* L.) and clover (*Trifolium pratense* L.), in addition to non-leguminous plants such as gladiolus (*Gladiolus* sp.), fenugreek (*Trigonella foenum-graecum* L.), opium poppy (*Papaver somniferum* L.) and spinach (*Spinacia oleracea* L.). *Chenopodium amaranticolor*, *C. quinoa*, globe amaranth, petunia (*Petunia hybrida* Vilm.) and *Tetragonia expansa* Thunb. are local lesion hosts.

Only seeds harvested from healthy plants should be planted and new plantings isolated from other leguminous crops or non-leguminous crops susceptible to yellow mosaic. Infected plants should be rogued as soon as symptoms appear. Routine application of insecticide may reduce the aphid population and retard in-field spread of the disease.

1. Afanasiev, M. M. 1952. Bean virus 2 (yellow) on Great Northern bean in Montana. *Phytopathology 42:* 101–104.

2. Bagget, J. R. and W. A. Frazier. 1957. The inheritance of resistance to bean yellow mosaic virus in *Phaseolus vulgaris*. *Proc. Am. Soc. Hort. Sci. 70:* 325–333.

3. Bos, L. 1970. Bean yellow mosaic virus. *Commonwealth Mycol. Inst. Descriptions Plant Viruses 40.* 4 p.

4. Evans, I. R. and F. W. Zettler. 1970. Aphid and mechanical transmission properties of bean yellow mosaic virus isolates. *Phytopathology 60:* 1170–1174.

5. Grogan, R. G. and J. C. Walker. 1948. A pod-distorting strain of the yellow mosaic virus of bean. *J. Agr. Res. 77:* 301–314.

6. Hagedorn, D. J. and J. C. Walker. 1950. The relation of bean virus 2 to pea mosaic in Wisconsin. *Phytopathology 40:* 684–698.

7. Hampton, R. O. 1975. The nature of bean yield reduction by bean yellow and bean common mosaic viruses. *Phytopathology 65:* 1342–1346.

8. Hungerford, C. W. and I. G. Hillyer. 1954. Yellow bean mosaic in Idaho. *Plant Dis. Reptr. 38:* 621–627.

9. McCord, R. W. and R. T. Gudauskas. 1968. Properties of a strain of bean yellow mosaic virus isolated from vetch, *Vicia sativa*. *Phytopathology 58:* 1294–1297.

10. McWhorter, F. P. 1949. Susceptibility of selections of Perfection peas to strains of yellow bean mosaic. *Plant Dis. Reptr. 33:* 139–144.

11. McWhorter, F. P., L. Boyle and B. F. Dana. 1947. Production of yellow bean mosaic in beans by virus from mottled gladiolus. *Science 105:* 177–178.

12. Pierce, W. H. 1934. Viroses of the bean. *Phytopathology 24:* 87–115.

13. Schroeder, W. T. and R. Provvidenti. 1966. Further evidence that common pea mosaic virus (PV2) is a strain of bean yellow mosaic virus (BV2). *Plant Dis. Reptr. 50:* 337–340.

14. Thomas, H. R. and W. J. Zaumeyer. 1953. A strain of yellow bean mosaic virus producing local lesions on tobacco. *Phytopathology 43:* 11–15.

15. Vanderveken, J. 1968. Inhibition de la transmission aphidienne du virus de la mosaique jaune du haricot par l'huile. *Bull. Rech. Agron. Gembloux 3:* 748–753.

16. Zaumeyer, W. J. and H. H. Fisher. 1953. A new necrotic-lesions-producing strain of yellow bean mosaic. *Phytopathology 43:* 45–49.

17. Zaumeyer, W. J. and B. L. Wade. 1935. The relationship of certain legume mosaics to bean. *J. Agr. Res. 51:* 715–749.

OTHER DISEASES

Bacterial brown spot (*Pseudomonas syringae* Van Hall) (Fig. 12)

1. Wilson, R. D. 1938. Occurrence of *Bacterium syringae* (Van Hall) E. F. Smith on french beans (*Phaseolus vulgaris* L.) in New South Wales. *J. Australian Inst. Agr. Sci. 4:* 42.

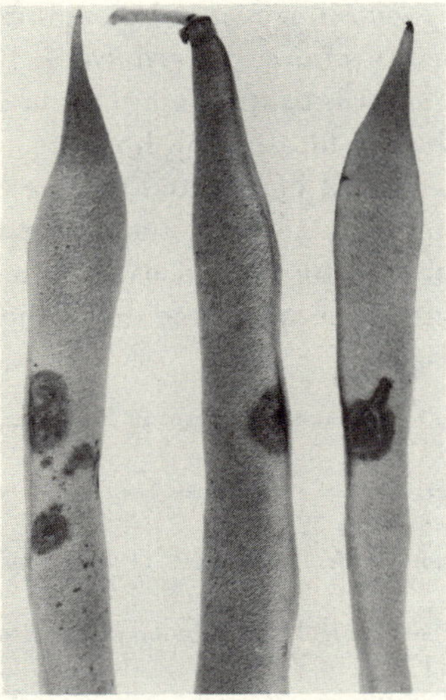

Fig. 12. Bacterial brown spot lesions on bean leaves and pods. (Courtesy of K. G. Pegg, Department of Primary Industries, Indooroopilly.)

Pod twist (*Pseudomonas flectens* Johnson) (Fig. 13)

Unilateral, necrotic lesions that caused characteristic twisting of young pods were described from Australia.

1. Johnson, J. C. 1956. Pod twist: a previously unrecorded bacterial disease of french bean (*Phaseolus vulgaris* L.). *Queensland J. Agr. Sci. 13:* 127–158.

Fig. 13. Pod twist symptoms on bean. (Courtesy of K. G. Pegg, Department of Primary Industries, Indooroopilly.)

Ascochyta disease (*Ascochyta phaseolorum* Sacc.) (Fig. 14)

1. Alcorn, J. L. 1968. Occurrence and host range of *Ascochyta phaseolorum* in Queensland. *Australian J. Biol. Sci. 21:* 1143–1151.

2. Pegg, K. G. and J. L. Alcorn. 1967. *Ascochyta* disease of French beans. *Queensland Agr. J. 93:* 321–323.

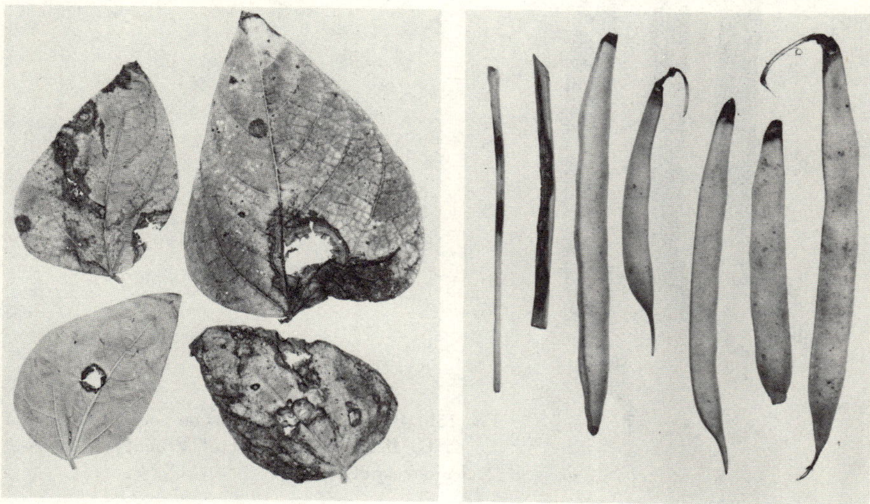

Fig. 14. Leaf lesions and stem-end infections on pods typical of *Ascochyta* disease. (Courtesy of K. G. Pegg, Department of Primary Industries, Indooroopilly.)

Brown spot [*Pleiochaeta setosa* (Kirchn.) Hughes] (Fig. 15)

1. Pegg, K. G. 1968. Brown spot disease of french bean caused by *Pleiochaeta setosa*. *Queensland J. Agr. Animal Sci. 25:* 219–223.

Fig. 15a. Brown spot symptoms on leaves. (Courtesy of K. G. Pegg, Department of Primary Industries, Indooroopilly.)

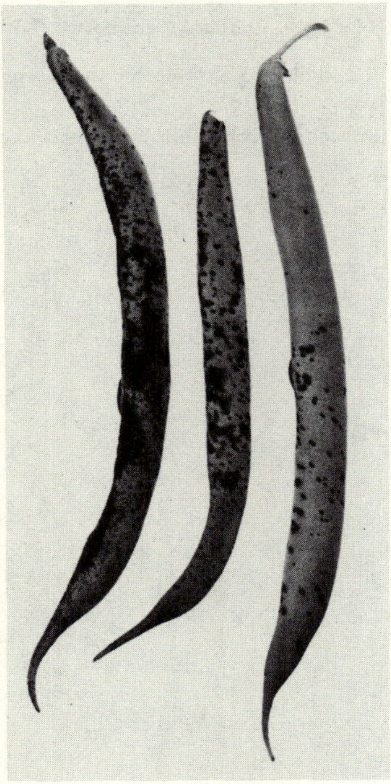

Fig. 15b. Brown spot symptoms on pods. (Courtesy of K. G. Pegg, Department of Primary Industries, Indooroopilly.)

Crumpling (*Euphorbia* mosaic)

Systemic foliar distortion and stunting from the effects on leaf growth of chlorotic or necrotic lesions that developed around feeding punctures of the whitefly vector (*Bemisia tabaci* Genn.) were reported from Brazil (1950). Yield from systemically infected plants was noticeably reduced, but there was no evidence of seed transmission.

1. Costa, A. S. 1965. Three whitefly-transmitted virus diseases of beans in São Paulo, Brazil. *FAO Plant Protect. Bull. 13:* 121–130.

2. Costa, A. S. and C. W. Bennett. 1950. White-fly-transmitted mosaic of *Euphorbia prunifolia*. *Phytopathology 40:* 266–283.

Cucumber mosaic

1. Bird, J., J. Sanchez, R. L. Rodriguez, A. Cortes-Monllor and W. Kaiser. 1974. A mosaic of beans (*Phaseolus vulgaris* L.) caused by a strain of common cucumber mosaic virus. *J. Agr. Univ. Puerto Rico 58:* 151–161.

2. Gibbs, A. J. and B. D. Harrison. 1970. Cucumber mosaic virus. *Commonwealth Mycol. Inst. Descriptions Plant Viruses 1.* 4 p.

Floury leaf spot [*Ramularia phaseoli* (Drummond) Deighton]

Reported to occur in South America, Asia, Africa and the southwest Pacific.

1. Deighton, F. C. 1967. Floury leaf spot of French bean caused by *Ramularia phaseoli* (Drummond) Deighton. *Trans. Brit. Mycol. Soc. 50:* 123–127.

Golden mosaic

Symptoms consisting of a bright mosaic with spots delimited by veins or a yellow net (Fig. 16) appear on young foliage of developing plants little reduced in size. Similar symptoms may be found on pods. The disease, known from Central America, Colombia, Jamaica, Puerto Rico and Brazil, is not seed-borne but has been mechanically transmitted except in Brazil. The whitefly, *Bemisia tabaci* Genn., is a very efficient vector, and the disease has threatened successful production, but no control has been devised other than application of insecticide to reduce whitefly populations.

1. Bird, J., E. Perez, R. Alconero, N. G. Vakili and P. L. Melendez. 1972. A whitefly-transmitted golden-yellow mosaic virus of *Phaseolus lunatus* in Puerto Rico. *J. Agr. Univ. Puerto Rico 56:* 64–74.

2. Costa, A. S. 1965. Three whitefly-transmitted virus diseases of beans in São Paulo, Brazil. *FAO Plant Protect. Bull. 13:* 121–130.

3. Galvez, G. E. and M. Cataño. 1976. Purification of the whitefly-transmitted bean golden mosaic virus. *Turrialba 26:* 205–207.

4. Gamez, R. 1971. Los virus del frijol en Centroamérica. I. Transmision por moscas blancas (*Bemisia tabaci* Gen.) y plantas hospedantes del mosaico dorado. *Turrialba 21:* 22–27.

Fig. 16. Symptoms of golden mosaic disease on bean. (Courtesy of R. Gamez, University of Costa Rica, San Jose.)

Line pattern mosaic

A disease associated with isometric particles 25 nm in diameter has been occasionally found in Brazil.

1. Kitajima, E. W., I. J. B. Camargo, T. Yano and A. S. Costa. 1973. Purificaçao parcial do mosaico em desenho do feijoeiro. *Fitopatologia 8:* 9.

Mottled dwarf (*Abutilon* mosaic)

Bean plants infected by *Abutilon* mosaic in Brazil developed mottling and crinkling symptoms on young leaves with rolling and some rugosity of old leaves. Axillary growth with excessive small leaflets and stems often developed. Plants affected at an early stage were extremely stunted with mottled leaves and crinkled and rolled leaflets. Pod set was severely reduced and pods that did develop were malformed. Transmission was much more effective by the whitefly vector *Bemisia tabaci* than by the mechanical method but was accomplished with difficulty to and from hosts other than bean and from bean to bean. No evidence of seed transmission was noted.

1. Costa, A. S. 1955. Studies on *Abutilon* mosaic in Brazil. *Phytopathol. Z. 24:* 97–112.

2. Costa, A. S. and A. M. B. Carvalho. 1960. Mechanical transmission and properties of the *Abutilon* mosaic virus. *Phytopathol. Z. 37:* 259–272.

3. Costa, A. S. and A. M. B. Carvalho. 1960. Comparative studies between *Abutilon* and *Euphorbia* mosaic viruses. *Phytopathol. Z. 38:* 129–152.

Summer death

Yellowing and downward curling of leaves, vascular discoloration in the roots and wilting and death of plants in Australia has been attributed to summer death disease transmitted by *Orosius argentatus* (Evans).

1. Ballantyne, B. 1968. Summer death—a new disease of beans. *Agr. Gaz. New S. Wales 79:* 486–489.

2. Ballantyne, B. 1969. Transmission of summer death of beans. *Australian J. Sci. 31:* 433–434.

3. Ballantyne, B. 1970. Field reactions of bean varieties to summer death in 1970. *Plant Dis. Reptr. 54:* 903–905.

Swelling mosaic

Symptoms reminiscent of 2-4-D injury, but readily transmitted mechanically, have been observed in El Salvador.

1. Anonymous. 1973. Food legumes production system—Phytopathology. In *Annual Report 1973*, p. 161–170. Centro Internacional de Agricultura Tropical, Cali, Colombia.

Yellow stipple

Mild stippling described from Costa Rica in 1972 was caused by a virus with isometric particles (ca. 30 nm diameter) inactivated by exposure to 74–76 C, 24–48 hours *in vitro* (20 C) or dilution greater than 10^{-3}. The virus was transmitted by chrysomelid beetles *Cerotoma ruficornis* and *Diabrotica balteata* but was not serologically related to the cowpea mosaic virus group.

1. Gamez, R. 1972. Some properties and beetle transmission of bean yellow stipple virus *Phytopathology 62:* 759.

2. Gamez, R. 1976. Bean virus in Central America. IV. Some properties and transmission of bean yellow stipple virus by chrysomelid beetles. *Turrialba 26:* 160–166.

BETEL VINE (PAN): *Piper betle* L.

Bacterial leaf spot

A disease of pan called bacterial leaf spot was recorded in Ceylon in 1896 and the causal bacterium (*Bacterium betle* Rag.) described in 1926. Similar diseases, that typically first appeared as water-soaked spots with or without yellow halos and enlarged to become dark, oily-appearing lesions to 5 mm and coalesced, were described later and attributed to 2 other bacterial pathogens. Colonies of *Bacillus betle*, named by Asthana and Mahmud, produced rod-shaped, motile cells 1.2 × 2.5 μm that formed white colonies, and *Xanthomonas betlicola*, with cells 0.9 × 1.6 μm, motile by 1–2 polar flagella and gram negative, was identified by Patel et al. and produced yellow colonies. Greatest damage by each of the diseases was caused during periods of high rainfall. Early plucking and destroying diseased leaves were recommended.

1. Asthana, R. P. 1945. A new bacterial leaf-spot on *Piper betle. Current Sci. 14:* 73.

2. Asthana, R. P. and K. A. Mahmud. 1944. Bacterial leaf spot of *Piper betle. Indian J. Agr. Sci. 14:* 283–288.

3. Park, M. 1934. Bacterial leaf-spot of betel. *Trop. Agr. (Ceylon) 82:* 393–394.

4. Patel, M. K., Y. S. Kulkarni and G. W. Dhande. 1953. Three bacterial diseases of plants. *Indian J. Agr. Sci. 23:* 41–46.

5. Ragunathan, C. 1926. Bacterial leaf spot of betel. *Trop. Agr. (Ceylon) 67:* 242–243.

Phytophthora wilt (Foot rot) and Leaf spot

Wilt disease of betel vine was reported from India in 1934 and has been found since then in several areas in southeast Asia where this crop is grown, including Burma, Pakistan and Sarawak (Malaysia). Incidence of the different aspects of this complex of symptoms is dependent on environmental conditions and local losses can be heavy.

Early in the development of wilt or foot rot disease, a brown lesion encircling a portion of the underground stem develops. The margins of such lesions are sharply outlined, although irregular, and disintegration of tissues below the lesion is rapid and complete in distinct contrast to the stem tissues above the lesion, which appear little affected with only minor or no vascular discoloration.

Leaves throughout the entire canopy on plants affected by wilt become progressively more flaccid, followed by general chlorosis. Young twigs and branches also become flaccid, and, with the leaves, finally wither and dry to a pale brown color. Leaves do not fall but rather remain attached until the vine dries and breaks at the nodes.

Plants affected by leaf spot only first show on lower leaves circular, dark, wet lesions that may increase rapidly in size, particularly along main veins. Concentric zonations are less pronounced in lesions that expand quickly, and later rot, during periods of high humidity. Defoliation may be quite rapid but infection can extend from the leaves via the petioles into the stem to form lesions which also develop more slowly in dry than in wet weather.

The causal fungus in Indonesia has been identified as *Phytophthora palmivora* (Butl.) Butl. (sporangia 15–36 × 23–68 μm, chlamydospores 19–41 μm, oogonia light brown and 20–45 μm in diameter and antheridia 11–15 × 11–23 μm), while the pathogen in India has been identified as *P. parasitica* Dast. (*P. parasitica* var. *piperina* Dast.) with sporangia 19–69 × 22–91 μm.

Infection by zoospores occurs on both leaf surfaces but is much more frequent on lower surfaces and on mature compared to immature leaves. Incubation varies from 36–72 hours depending on relative humidity. Sporulation also occurs on both surfaces of leaf lesions on sporangiophores that extend in clusters through stomata from the lower leaf surfaces.

Leaf infections result from air-borne sporangia and were found to occur on lower leaves between 0.5–3.5 m in Sarawak.

Protective foliar applications of fungicides, including zineb, thiram, cuprous oxide, copper oxychloride or triphenyl tin hydroxide (each at 0.2 percent), prevented leaf infection but were ineffective in eradicating established infections in laboratory tests.

1. Asthana, R. P. 1947. The role of cuttings in the dissemination of foot-rot of *Piper betle*. *Indian J. Agr. Sci. 18:* 223–225.

2. Chowdhury, S. 1944. Diseases of pan (*Piper betle*) in Sylhet, Assam. II. *Phytophthora* foot-rot and leaf-rot. *Proc. Indian Acad. Sci., Ser. B., 19:* 152–164.

3. Dastur, J. F. 1935. Diseases of pan (*Piper betle*) in the Central Provinces. *Proc. Indian Acad. Sci., Ser. B., 1:* 778–815.

4. Mahmud, K. A. 1952. Reveiw of literature on *Phytophthora* foot rot of pan. *Bull. Bot. Soc. Bengal 6:* 79–88.

5. McRae, W. 1934. Foot rot diseases of *Piper betle* L. in Bengal. *Indian J. Agr. Sci. 4:* 585–617.

6. Mehrotra, R. S. and D. P. Tiwari. 1967. Studies on the foot rot and leaf rot of *Piper betle*. I. Behaviour of *Phytophthora parasitica* var. *piperina* in soil. *Indian Phytopathol. 20:* 161–167.

7. Narasimhan, V. and K. Ramakrishnan. 1969. Studies on wilt disease of betelvine. *Madras Agr. J. 56:* 68–73.

8. Selvaraj, J. C., S. C. Govindaswamy and K. Ramakrishnan. 1973. Effect of soil

temperature on the *Phytophthora* foot rot and wilt disease of *Piper betle*. *Indian Phytopathol. 26:* 626–641.

9. Turner, G. J. 1969. *Phytophthora palmivora* from *Piper betle* in Sarawak. *Trans. Brit. Mycol. Soc. 52:* 411–418.

10. Turner, G. J. 1969. Leaf lesions associated with foot rot of *Piper nigrum* and *P. betle* caused by *Phytophthora palmivora*. *Trans. Brit. Mycol. Soc. 53:* 407–415.

11. Waterhouse, G. M. 1963. Key to the species of *Phytophthora* De Bary. *Commonwealth Mycol. Inst. Mycol. Paper 92.* 22 p.

Anthracnose

A description of anthracnose symptoms appeared in 1935 and again in 1942, both times from India, but the causal organism was not identified until 1948. The disease caused heavy losses in Bengal in 1946 but is not considered an important disease of pan.

Small, dark, round, discolored areas appear on green stems and branches and, under conditions of high humidity, may increase rapidly in size to girdle the stem and cause the plant to wilt. When the humidity is lower, infection progresses more slowly and develops as long, narrow stem lesions that may form concave depressions at infection sites and seriously diseased internodes may become constricted.

Spots on leaves become circular with dark centers surrounded by yellow halos and frequently coalesce to form larger lesions. Necrotic centers of the lesions do not fall, but lesions along the margins of leaf laminae may cause them to curl. Infected leaves often fall prematurely.

The causal fungus is *Colletotrichum dasturii* Roy that produces numerous acervuli with tapering, septate setae 3–4.5 × 110–360 μm and unicellular, hyaline, curved conidia 2–3 × 19–25 μm. (Another species, *C. piperis* Petch, was identified on leaves of pan in Perideniya, Sri Lanka, in 1917 and produced small, pink acervuli with setae 2–4 × 34–100 μm and conidia 3.5–4.5 × 12–19 μm. Also, *C. capsici* (Sydow) Butler & Bisby, with setae 58–348 μm in length and conidia 3.8 × 20 μm, has been reported to infect pan in Pakistan.)

Disease incidence is highest during periods of high rainfall and high temperatures but is of no consequence during dry, cool seasons. Differences in varietal susceptibility have been reported with the variety Kankar offering some promise of resistance.

1. Dastur, J. F. 1935. Diseases of pan (*Piper betle*) in the Central Provinces. *Proc. Indian Acad. Sci., Ser. B., 1:* 778–815.

2. Hafiz, A., K. Akbar and M. Shariff. 1956. *Colletotrichum capsici* (Sydow) Butler and Bisby on *Piper betle*. *Biologica 2:* 216.

3. Hasan, S. and Q. A. Ahmed. 1963. A study of the anthracnose disease of *Piper betle*. *Agr. Pakistan 14:* 204–208.

4. Mordue, J. E. M. 1971. *Colletotrichum capsici. Commonwealth Mycol. Inst. Descriptions Pathogenic Fungi Bacteria 317.* 2 p.

5. Roy, T. C. 1948. Anthracnose disease of *Piper betle* L. caused by *Colletotrichum Dasturii* Roy sp. nov. in Bengal. *J. Indian Bot. Soc. 27:* 96–102.

OTHER DISEASES

Powdery mildew (*Oidium piperis* Uppal & Kamat)

Circular, gray, powdery lesions, first on lower leaf surfaces (later on both leaf surfaces), that increase in size to involve most of the laminae become brown, dry and quite fragile. Shoots may also be affected and growth arrested. The fungus produces septate, hyaline mycelium 5–8 μm in diameter, conidiophores 66–132 μm in length with chains of 3–10 unicellular, hyaline, elliptical conidia 20–75 × 7–24 μm. The disease in India occurs during the cool, dry seasons and severity (20–40 percent losses) is enhanced by close planting and excessive shade. Dusting with sulfur has proven effective for control.

1. Malelu, J. S., M. N. Kamat and M. K. Patel. 1951. Cultivation and diseases of betlevine at Bassein. *Poona Agr. Coll. Mag. 42:* 137–143.

2. Mitra, M. 1930. India: new diseases reported during the year. *Intern. Bull. Plant Protect. 4:* 103–104.

3. Uppal, B. N., M. N. Kamat and M. K. Patel. 1946. Powdery mildew of betel vines. *Proc. Indian Acad. Sci. 24:* 255–259.

Rhizoctonia root rot (*Rhizoctonia solani*)

Incidence of root rot disease, which had been serious near Sylhet, India, was negligible in plots where ulu grass [*Imperata cylindrica* (L). Beauv. = *I. arundinacea* Cyr.] had grown for the previous 4 years.

1. Chowdhury, S. 1944. Diseases of pan (*Piper betle*) in Sylhet, Assam. IV. *Rhizoctonia* root-rot. *Proc. Indian Acad. Sci., Ser. B., 20:* 229–244.

2. Chowdhury, S. 1948. On the control of *Rhizoctonia* root-rot of pan (*Piper betle* L.). *Sci. Cult. 13:* 507–508.

Synchytrium leaf galls (*Synchytrium piperi* Mundkur & Mhatre)

Symptoms reported from India in 1945 included minute, white galls on leaves with one spherical, smooth, brown resting spore (30–38 μm in diameter) in each host cell.

1. Mhatre, J. R. and B. B. Mundkur. 1945. The *Synchytria* of India. *Lloydia 8:* 131–138.

Tip-burn

Marginal or complete necrosis of leaves that remain attached to the stem, occurring late in dry seasons, particularly on plants with root disease problems, has been attributed to physiological conditions, i.e. excessive moisture loss. Proper shading and adequate irrigation have been recommended for control.

1. Asthana, R. P. and K. A. Mahmud. 1944. Tip-burn of *Piper betle* in the central provinces. *Current Sci. 13:* 234.

BLACK GRAM (URD BEAN): *Phaseolus mungo* L.

Downy mildew (*Erysiphe polygoni*)

Rust [*Uromyces appendiculatus* (Pers.) Unger]

Cercospora leaf spot (*Cercospora cruenta*)

1. Bose, R. D. 1932. Studies in Indian pulses. V. Urd or black gram (*Phaseolus mungo* Linn. var *roxburghii* Prain). *Indian J. Agr. Sci. 2:* 625–637.

2. Mulder, J. L. and P. Holliday. 1969. *Cercospora cruenta. Commonwealth Mycol. Inst. Descriptions Pathogenic Fungi Bacteria 463.* 2 p.

3. Williams, F. J., J. S. Grewal and K. S. Amin. 1968. Serious and new diseases of pulse crops in India in 1966. *Plant Dis. Reptr. 52:* 300–304.

Macrophomina phaseolina (=*Rhizoctonia bataticola*)

1. Holliday, P. and E. Punithalingam. 1970. *Macrophomina phaseolina. Commonwealth Mycol. Inst. Descriptions Pathogenic Fungi Bacteria 275.* 2 p.

2. Khare, M. N., N. K. Jain and H. C. Sharma. 1970. Variation among *Rhizoctonia bataticola* isolates from urid bean plant parts and soil. *Phytopathology 60:* 1298.

Bean common mosaic

1. Bos, L. 1971. Bean common mosaic virus. *Commonwealth Mycol. Inst. Descriptions Plant Viruses 73.* 4 p.

2. Nene, Y. L. 1973. Viral diseases of some warm weather pulse crops in India. *Plant Dis. Reptr. 57:* 463–467.

3. Shahare, K. C. and S. P. Raychaudhuri. 1963. Mosaic disease of urid (*Phaseolus mungo* L.). *Indian Phytopathol. 16:* 316–318.

Leaf crinkle

1. Narayanasamy, P. and T. Jaganathan. 1975. Seed transmission of black gram leaf crinkle virus. *Phytopathol. Z. 82:* 107–111.

2. Nene, Y. L. 1972. A survey of viral diseases of pulse crops in Uttar Pradesh. *G. B. Pant Univ. Coll. Agr. Expt. Sta. Res. Bull. 4.* 191 p.

3. Williams, F. J., J. S. Grewal and K. S. Amin. 1968. Serious and new diseases of pulse crops in India in 1966. *Plant Dis. Reptr. 52:* 300–304.

Yellow mosaic

1. Nene, Y. L. 1972. A survey of viral diseases of pulse crops in Uttar Pradesh. *G. B. Pant Univ. Coll. Agr. Expt. Sta. Res. Bull. 4.* 191 p.

2. Nene, Y. L. 1973. Viral diseases of some warm weather pulse crops in India. *Plant Dis. Reptr. 57:* 463–467.

3. Nene, Y. L., S. K. Srivastava and J. S. Naresh. 1972. Evaluation of black gram (*Phaseolus mungo* Roxb.) and green gram (*Phaseolus aureus* Roxb.) varieties and germplasms for resistance to yellow mosaic virus of green gram. *Indian J. Agr. Sci. 42:* 251–254.

BLACK PEPPER: *Piper nigrum* L.

Foot rot (Sudden death)

Notice of a root disease of serious and widespread proportion on pepper in Sarawak appeared in 1955, but there was evidence that the disease had been prevalent there for several years. In fact, a similar disease had been reported from Indonesia in 1936, and a pathogen identified and named. More recently the disease has been reported from Brazil, Jamaica and Puerto Rico. Estimates of losses attributable to foot rot have been based on casual observations only, but 15 percent of the gardens in the main producing district of Sarawak were affected in 1955, and incidence in individual gardens reached 80 percent. Foot rot was reported to be a major problem in establishing pepper culture in Puerto Rico.

Foliage on affected vines, particularly on the lower leaves first, assumes a dull color, and leaves quickly turn chlorotic and develop necrotic apices and petioles. Leaf fall begins before chlorosis is complete, spikes also drop, and lateral branches become necrotic and break at the nodes. Wilt may precede significant leaf fall, but the full cycle of symptoms may take less than 2 weeks. Affected plants are killed.

The underground stems of affected vines typically have deep, dark, sharply demarcated lesions that extend into the cortex. Some of the larger roots also may have dark lesions, and the root system below the necrotic stems or large roots is usually destroyed. Disintegration of the roots exposes the lignified elements which have prompted the name "frayed root-rot" for the disease. Vascular discoloration may extend some distance above the stem lesion.

Foot rot is caused by a species of *Phytophthora (P. palmivora* var. *piperis?*) with simple sporangiophores on which are borne ovoid, papillate sporangia 25–30 × 40–50 μm that germinate by release of zoospores at 20–22 C or below. Hyaline, globose oogonia are terminal or lateral and become brown and 26–45 μm in diameter when mature. Antheridia are amphigynous, terminal or lateral, hyaline and 11–23 × 11–15 μm. Spherical, brown oospores 19–38 μm in diameter have smooth walls 1.8–5.6 μm thick.

Foliar symptoms of foot rot ordinarily do not appear until underground stem lesions and destruction of the root system have become extensive, and the plant can not recover. Once the disease has entered a field, spread of the infection in the field can be fairly rapid and has been found to increase from 8 to 80 percent infected plants in 4 months. The means by which infection spreads have not been reliably established, but apparently dissemination is not entirely dependent on surface water and actually has been enhanced by clean cultivation and weeding. Sporangia of the *Phytophthora* sp. pathogenic on pepper vines in Sarawak have been found to survive passage through the alimentary canal of the giant African snail (*Achatina fulica* Bow.) and to remain infective in moist feces for 3 months.

Control methods have not been established, but cultivation practices that encourage infection, i.e. clean cultivation, should be avoided and plantings should be made only on soils that drain well. Differences in susceptibility of *Piper* species have not been adequately investigated, but the apparent resistance of *P. aduncum* L. may be sufficient to warrant its use as a stock for grafting.

1. Alconero, R., F. Albuquerque, N. Almeyda and A. G. Santiago. 1972. *Phytophthora* root rot of black pepper in Brazil and Puerto Rico. *Phytopathology 62:* 144–148.

2. Albuquerque, F. C. 1961. Podridáo das raizes e do pé da pimento do reino. *Inst. Agron. Norte (Brazil) Circ. 5.* 45 p.

3. Holliday, P. 1961. A root disease of black pepper in Sarawak. *Commonealth Mycol. Conf. Rep. 6:* 156–161.

4. Holliday, P. and W. P. Mowat. 1957. A root disease of *Piper nigrum* L. in Sarawak caused by a species of *Phytophthora. Nature 179:* 543–544.

5. Holliday, P. and W. P. Mowat. 1963. Foot rot of *Piper nigrum* L. *Commonwealth Mycol. Inst. Phytopathol. Paper 5.* 62 p.

6. Leather, R. I. 1967. The occurrence of a *Phytophthora* root and leaf disease of black pepper in Jamaica. *FAO Plant Protect. Bull. 15:* 15–16.

7. Muller, H. R. A. 1936. Het *Phytophthora*—voetrat van Peper (*Piper nigrum* L.) in Nederlandsch-Indie. *Meded. Inst. Plantenziekten (Batavia) 88.* 73 p.

8. Nambiar, E. P., T. J. Nair and N. S. Money. 1965. Preliminary studies on the

incidence of wilt disease of pepper and its relationship to the nitrogen and base status of the soil. *Indian J. Agr. Sci. 35:* 276–281.

9. Raj, J. S. and P. C. Jose. 1966. A *Phytophthora* wilt of pepper, *Piper nigrum. Sci. Cult. 32:* 90–92.

10. Robertson, N. F. 1955. Pepper disease in Sarawak. *Commonwealth Phytopathol. News 1:* 20–23.

11. Ruppel, E. G. and N. Almeyda. 1965. Susceptibility of native *Piper* species to the collar-rot pathogen of black pepper in Puerto Rico. *Plant Dis. Reptr. 49:* 550–551.

12. Turner, G. J. 1962. Production of fusion organs by the species of *Phytophthora* which causes foot rot of *Piper nigrum* L. in Sarawak. *Nature 195:* 201.

13. Turner, G. J. 1967. Snail transmission of species of *Phytophthora* with special reference to foot rot of *Piper nigrum. Trans. Brit. Mycol. Soc. 50:* 251–258.

Anthracnose

Anthracnose was reported to be a minor disease of black pepper in the Philippines in 1940, but occurrence of the disease there apparently had been recognized some years earlier.

Symptoms are confined to the leaves and first appear as small, dark lesions irregular in shape with definite, light colored margins. Coloration of lesions becomes darker and eventually nearly black as they age and enlarge. Lesions may coalesce to form sizeable necrotic areas with distinct, water-soaked margins, and the central portions of such necrotic tissues often drop, leaving holes in the centers of spots. Seriously affected leaves also may fall.

Glomerella cingulata (Stoneman) Spauld. & von Schrenk with oval, hyaline conidia 3.5–7.0 × 12.2–19.2 μm has been identified as the pathogen that causes anthracnose, but only the imperfect stage of the organism has been found.

Seasonal occurrence of the disease has been correlated with onset of the rainy season, and infection is confined to tender, young leaves and older leaves injured or damaged by sunburn. Dissemination of spores occurs primarily by rain and splashing water, although some conidia may be carried short distances by wind.

Partial shade to prevent sunburn damage to leaves has been recommended. Control in the Philippines was achieved by sanitation that included periodic removal and burning of infected leaves, but foliar application of carbamate or other fungicide should prove beneficial.

1. Mordue, J. E. M. 1971. *Glomerella cingulata. Commonwealth Mycol. Inst. Descriptions Pathogenic Fungi Bacteria 315.* 2 p.

2. Vimuktanandana, Y. Y. and M. S. Celino. 1940. Anthracnose of black pepper (*Piper nigrum* Linn.). *Philippine Agr. 29:* 124–141.

OTHER DISEASES

Lampong disease

A "sudden decline" of pepper vines in Sumatra and West Java was attributed to an unidentified bacterium. Vascular discoloration was a diagnostic symptom of the disease that caused leaf fall and premature death of affected vines.

1. Gaumann, E. 1922. Some observations on the Lampong pepper disease. *Teysmannia 33:* 289–293.

Leaf galls (*Elsinoe piperis* Hansford)

Small, dark, corky galls were reported on leaves of *Piper nigrum* in Cameroon in 1956 and attributed to the same fungus found associated with similar galls on *P. capense* L. in Uganda (1941). Galls have been observed on leaves of *Piper* sp. also in Congo and Ghana. Application of copper fungicides gave satisfactory control.

1. Hansford, C. G. 1941. Contributions towards the fungus flora of Uganda. III. Some Uganda ascomycetes. *Proc. Linn. Soc. London, 1940–41:* 4–52.

2. Marticou, M., G. Nonveiller and V. H. Hai. 1966. Détermination des causes qui provoquent l'apparition de galles sur feuilles de poivrier mise en evidence d'un *Elsinoe*. *Agron. Trop. (Paris) 21:* 1407–1414.

Pink disease and Thread blight

Pruning of affected plant parts and spray application of cuprous oxide have been recommended for control.

1. Anonymous. 1962. Pink disease and thread blight of pepper. *Sarawak Dept. Agr. Leafl. 32 A.* 3 p.

Pollu disease (*Colletotrichum* sp.)

Irregular, gray leaf spots and brown discoloration of the rind of individual berries that shrivel have been referred to as "pollu (light berry) disease" in India and considered responsible for 4–13 percent losses. Other forms of "pollu" have been attributed to insect infestation and physiological disorder.

1. Thomas, K. M. and K. K. Menon. 1939. The present position of pollu disease of pepper in Malabar. *Madras Agr. J. 27:* 348–356.

Rhizoctonia blight (*Rhizoctonia solani*)

Dark, irregular leaf lesions that expanded concentrically on pepper plants with a progressive wilt were described from the Congo Republic in 1957. Necrotic tissues often dropped, imparting a ragged appearance to affected leaves, while small, oblong sclerotia (less than 1 mm) developed on completely wilted leaves.

1. Vanderweyen, A. 1957. Une maladie foliaire de poivrier cultivé. *Bull. Agr. Congo Belge 48:* 365–370.

Stunt (virus) disease

In the Philippines in 1961, stunted plants with small, thickened, chlorotic leaves, sometimes with mosaic symptoms, were presumed to be caused by a virus because of graft-transmission of the disease considered similar to a stunting disease of pepper in Sarawak. Yield was noticeably reduced. Similar symptoms have been observed in Laos and Vietnam.

1. Anonymous. 1962. Stunt virus disease of pepper. *Sarawak Dept. Agr. Leafl. 33 A.* 3 p.

2. Del Rosario, M. S. 1961. Note: a new virus disease of black pepper (*Piper nigrum* L.). *Philippine Agr. 45:* 141–143.

3. Holliday, P. 1959. Suspected virus in black pepper. *Commonwealth Phytopathol. News 5:* 49–52.

White root rot

Yellowing and premature dropping of leaves, caused by fine, white rhizomorphs attached to the roots, should not be confused with "foot rot disease." Severely affected plants should be rogued, and the planting site thoroughly cleared of stumps and roots before replanting.

1. Anonymous. 1962. White root rot of pepper. *Sarawak Dept. Agr. Leafl. 31 A.* 3 p.

BROAD BEAN: *Vicia faba* L.

Chocolate spot

The disease of broad beans known as chocolate spot was noted first in 1849 from England and mentioned occasionally in literature for a few years there-

Fig. 17. Chocolate spot lesions on broad bean leaves. (Courtesy of C. O. Gourley, Canadian Department of Agriculture, Kentville, Nova Scotia.)

after. Minor syndrome differences and uncertain identification of the causal agent prompted naming the disease "red-spot" in China. Confusion about the causal nature of the disease was not resolved until 1931 when 2 forms of the disease, based on severity, were described from Spain. Occurrence of the disease has been reported from Australia, Brazil, China, Egypt, England, Japan, Nova Scotia, Spain and Tasmania, but it probably can be found wherever broad beans are grown. Complete crop failures have been caused by chocolate spot, but losses generally have been considerably less, even though of serious consequence.

Symptoms consist of dark brown, necrotic spots from 1–10 mm in diameter on both surfaces of leaflets but often more common on the upper leaflet surfaces (Fig. 17). Lesions may assume a variety of shapes as well as sizes and may either be largely external or extend completely through the laminae. Such lesions may coalesce, and, with necrotic streaking on the stems, cause a blight characterized by some defoliation and death of affected plants. Stem lesions also may be superficial but extensive in length and involve petioles, resulting in death of leaflets. Pods and seed coats may exhibit similar symptoms.

No pathogen was identified in the early accounts of the disease, although it was reported later to be caused by a bacterium. Subsequent study revealed that a fungus was causally related, but some confusion has persisted about species identification of the organism and distinguishable forms of the disease. The causal fungus now is generally considered to be *Botrytis fabae*

Sard., characterized by erect, brown conidiophores 162–350 μm in length and ovoid conidia 11–18 × 15–24 μm borne in heads 39–63 μm in diameter. Black sclerotia measure 1–4 × 1–2 × 0.4–2 μm. Infection of broad bean leaves by *B. cinerea* Pers. has been demonstrated.

Old leaves are much more susceptible than young leaves, and infection can occur over a range of 1–30 C with 20 C optimum temperature. Humidity sufficient to cause formation of a thin film of moisture on leaf surfaces (90–100 percent) is essential to infection. Extended persistence of high humidity conditions after infection enhances development of severe ("aggressive") blight symptoms, while the relatively mild ("non-aggressive") symptoms may result from lowered humidity soon after infection.

Bordeaux mixture (10-10-100) was found to give satisfactory protection under conditions of experimental inoculation in England, and captan (2000 ppm) has been reported to provide systemic protection. However, seasonal planting to avoid heavy rainfall and periods of high humidity, especially when the crop will be maturing, may make spraying unnecessary.

1. Deverall, B. J. and R. K. S. Wood. 1961. Infection of bean plants (*Vicia faba* L.) with *Botrytis cinerea* and *B. fabae*. *Ann. Appl. Biol. 49:* 461–472.

2. Ellis, M. B. and J. M. Waller. 1974. *Botrytis fabae*. *Commonwealth Mycol. Inst. Descriptions Pathogenic Fungi Bacteria 432.* 2 p.

3. Geard, I. D. 1960. Diseases of broad beans and tick beans. *Tasmanian J. Agr. 31:* 68–73.

4. Gourley, C. O. and R. W. Delbridge. 1973. *Botrytis fabae* and *Ascochyta fabae* on broad beans in Nova Scotia. *Can. Plant Dis. Surv. 53:* 79–82.

5. Jauch, C. 1947. La "mancha chocolate" de las habas. *Rev. Invest. Agr. 1:* 65–80.

6. Magee, C. J. 1934. Etiology of the chocolate spot disease of broad beans. *Bull. New S. Wales Dept. Agr. 43.* 8 p.

7. Mansfield, J. W. and B. J. Deverall. 1974. The rates of fungal development and lesion formation in leaves of *Vicia faba* during infection by *Botrytis cinerea* and *Botrytis fabae*. *Ann. Appl. Biol. 76:* 77–89.

8. Napier, E. J., A. Rhodes, D. I. Turner, J. Tootill and A. Dunn. 1957. Systemic action of captan against *Botrytis fabae* (chocolate spot of broad bean). *J. Sci. Food Agr. 8:* 467–474.

9. Purkayastha, R. P. and B. J. Deverall. 1965. The growth of *Botrytis fabae* and *B. cinerea* into leaves of bean (*Vicia faba* L.). *Ann. Appl. Biol. 56:* 139–147.

10. Riker, A. J. and R. S. Riker. 1932. Studies on bacteria associated with the chocolate-spot disease of broad beans. *Ann. Appl. Biol. 19:* 55–64.

11. Sardiña, J. R. 1929. Una nueva especie de "*Botrytis*" que ataca a las habas. *Bol. Patol. Veg. Entomol. Agr. 4:* 93–97.

12. Sardiña, J. R. 1931. Dos nuevas enfermedades de las habas. *Bol. Patol. Veg. Entomol. Agr. 5:* 59–80.

13. Sirry, A. R. 1956. The effect of plant nutrition and soil moisture on *Botrytis fabae* Sard. on broad beans at Shebin El Kom, Egypt. *Ann. Agr. Sci., Fac. Agr. Ain Shams Univ., Cairo, 1:* 319–323.

14. Wilson, A. R. 1937. The chocolate spot disease of beans (*Vicia faba* L.) caused by *Botrytis cinerea* Pers. *Ann. Appl. Biol. 24:* 258–288.

15. Yu, T. F. 1945. The red-spot disease of broad beans (*Vicia faba* L.) caused by *Botrytis fabae* Sardina in China. *Phytopathology 35:* 945–954.

Rust

Rust of broad bean (and several other leguminous crop plants including lentil, pea and vetch) occurs throughout the world. The disease can cause considerable defoliation but is not usually a serious problem except in crops planted late.

Reddish brown pustules frequently surrounded by light colored halos develop mostly on leaves but also on stems and occasionally on pods. Pustules burst, exposing masses of rust-colored spores.

The disease is caused by *Uromyces viciae-fabae* (Pers.) Schroet., characterized by cupulate pycnia 0.3–0.4 mm in diameter often associated with aecia with round aeciospores 18 × 26 μm in diameter. Obovate urediospores with finely echinulate walls which measure 19–22 × 22–28 μm develop in uredia 0.5–1 mm in diameter, while telia are 1–2 mm in diameter and the elliptical teliospores 18–26 × 25–40 μm with smooth walls. The fungus is infectious on lentil, pea, vetch and a number of other legumes in addition to broad bean.

Rust usually is most serious on senescent plants and particularly in late planted crops that are subjected to warm environmental conditions as they mature. Spores are easily wind-blown and germinate readily with sufficient moisture. Pycnial formation is soon followed by the appearance of aecia and aeciospores which are largely responsible for dissemination of the disease. Aecia are formed between 17–22 C, and urediospore production, which continues only for a short while at 25 C, is soon followed by production of teliospores in the same sori. Neither aeciospores or urediospores remain viable for more than 4 months, but teliospores may remain viable for nearly a year and germinate between 12–22 C.

Control may be accomplished by destroying refuse from previous crops, including volunteer plants, and early sowing to avoid premature senescence from unfavorable environmental conditions. Spraying with zineb (0.8 kg/100 ℓ) or lime sulfur (0.7 ℓ 15 percent polysulfide liquid/100 ℓ) alone or combined with colloidal sulfur (0.1 kg 40 percent /100 ℓ) gave satisfactory control in New Zealand. Also, sprays of maneb, thiram or ziram have been used successfully.

1. Geard, I. D. 1960. Diseases of broad beans and tick beans. *Tasmanian J. Agr. 31:* 68–73.

2. Guyot, A. L. 1957. Les Urédinées. *Encycl. Mycol. 29:* 505–554.

3. Jacks, H. 1954. Screening tests with fungicides for control of broad-bean rust. *New Zeal. J. Sci Technol. 36:* 274–279.

4. Jacks, H. and A. J. Webb. 1956. Field tests for control of broad-bean rust. *New Zeal. J. Sci. Technol. 38:* 157–159.

5. Laundon, G. F. and J. M. Waterston. 1965. *Uromyces vicia-fabae. Commonwealth Mycol. Inst. Descriptions Pathogenic Fungi Bacteria 60.* 2 p.

6. Wade, G. C. 1950. Diseases of broad beans and tick beans. *Tasmanian J. Agr. 21:* 28–32.

Ascochyta spot

The disease known as *Ascochyta* blight—or leaf, stem and pod spot of broad bean—was described from Argentina in 1899, but a similar disease of pea had been noted nearly 70 years earlier. It is known to occur now in Europe, Asia, the Pacific, Africa and the Americas and can cause complete crop losses.

Fig. 18. *Ascochyta* spot symptoms on leaves, pods and seeds of broad bean. (Courtesy of C. O. Gourley, Canadian Department of Agriculture, Kentville, Nova Scotia.)

Circular to oval spots that are dark brown at first enlarge up to 2 cm in diameter with white to light brown centers, and dark (red) borders develop on leaves (Fig. 18). Centers of these lesions, in which red-brown pycnidia develop either scattered or in circles, may desiccate and fall to produce a "shot hole" appearance, but lesions may also coalesce and cause severe to complete defoliation.

Stem lesions also are circular to elongate and may extend to 2.5 cm with centers lighter than the red margins but deeply sunken. Pycnidia develop over the entire lesioned areas.

Symptoms on pods are mostly circular and dark colored but with a darker margin, and sunken (Fig. 18). Severely affected pods shrivel, and seeds may either not develop or are markedly reduced in size and shriveled. Black or brown discolored areas that may be sunken (Fig. 18) appear on infected seeds that do not germinate well and may produce infected plants.

The disease is caused by the fungus *Ascochyta pisi* Lib. var. *fabae* Sprague that produces globose, light colored, ostiolate pycnidia 95–270 × 110–300 μm. Hyaline spores 1–30 × 4–8 μm are straight or curved and 1–3 septate.

Conidia germinate readily at temperatures from 14–32 C with an optimum of about 22 C, but optimum temperature for mycelial growth is slightly higher (25 C). Lesion formation requires from 4–7 days and develops first on the lower leaves of plants naturally infected. Seedlings may be killed.

Refuse from previous infected broad bean crops either should be turned under completely some months before replanting or collected and burned. Only seed from disease-free plants or seed lots from which infected seeds have been removed should be planted. Mild infections may be held in check by foliar applications of copper compounds, thiram or captan, but these should be started when symptoms first appear.

1. Beaumont, A. 1950. On the *Ascochyta* spot disease of broad beans. *Trans Brit. Mycol. Soc. 33:* 345–349.

2. Geard, I. D. 1960. Diseases of broad beans and tick beans. *Tasmanian J. Agr. 31:* 68–73.

3. Geard, I. D. 1962. Studies on *Ascochyta, Botrytis* and "seed spot" of *Vicia faba* in Tasmania. *J. Australian Inst. Agr. Sci. 28:* 218–219.

4. Gourley, C. O. and R. W. Delbridge. 1973. *Botrytis fabae* and *Ascochyta fabae* on broad beans in Nova Scotia. *Can. Plant Dis. Surv. 53:* 79–82.

5. Hewett, P. D. 1966. *Ascochyta fabae* Speg. on tick bean seed. *Plant Pathol. 15:* 161–163.

6. Punithalingam, E. and P. Holliday. 1972. *Ascochyta pisi. Commonwealth Mycol. Inst. Descriptions Pathogenic Fungi Bacteria 334.* 2 p.

7. Rathschlag, H. 1930. Zur spezialisierung der auf *Vicia faba* parasitierenden *Ascochyta. Phytopathol. Z. 2:* 493–501.

8. Sprague, R. 1929. Host range and life-history studies of some leguminous *Ascochytae. Phytopathology 19:* 917–932.

9. Yu, T. F. 1947. *Ascochyta* blight and leaf and pod spot of broad bean in China. *Phytopathology 37:* 207–214.

Bean yellow mosaic

Bean yellow mosaic was noted in Wisconsin in 1931 and distinguished from other virus diseases of bean in a report 3 years later. The disease has been found wherever legume viruses have been studied and probably occurs throughout the world. Broad bean is quite susceptible to infection, but the major economic import of the disease results from the wide variety of local strains of the causal entity that exists and the impact of these variants on development of resistant varieties for disease control.

Symptoms in broad bean usually include early vein clearing followed by development of systemic mottle that varies in severity according to virus strain but typically does not include disfiguration of leaves (Fig. 19).

Bean yellow mosaic virus is characterized by flexuous particles 15 × 750 nm usually inactivated by exposure for 10 minutes to 55–60 C, dilution greater than 10^{-4} or storage *in vitro* for more than 2 days. However, different virus strains, use of different test plants or virus source plants may result in divergent physical characters of the virus. (Particles from *Chenopodium amaranticolor* have been found 50 nm or more longer than particles from legumes.) The virus is transmitted by several aphid species, including *Macrosiphum euphorbiae*, *Myzus persicae* and *Aphis fabae*. It is readily transmissible mechanically and is transmitted through up to 6 percent of the seed from infected plants of white and yellow lupine but in lower frequency in pea and white sweet clover seed. Host range includes many leguminous plants, and, also, a number of non-leguminous plants that are infected locally, including *C. quinoa*, globe amaranth and tobacco. Spinach, *N. clevelandii* and *C. amaranticolor* may be infected systemically. The virus is serologically related to pea necrosis virus and to a number of other viruses in the potato Y virus group. Crystalline cytoplasmic inclusions which contain some virus particles are found in epidermal tissues.

Only seeds from disease-free plants should be planted and infected plants rogued from the field when symptoms first appear. Weed sources of the virus, e.g. white sweet clover, should be eliminated or field plantings isolated from such sources of virus. Early sowing to avoid large insect populations when plants are gowing rapidly should serve to reduce surrent season spread of the disease. Development of useful resistant varieties is complicated by differences in local strains of the virus.

1. Anderson, C. W. 1955. *Vigna* and *Crotalaria* viruses in Florida. II. Notations concerning cowpea mosaic virus (*Marmor vignae*). *Plant Dis. Reptr. 39:* 349–353.

2. Barton, D. W., W. T. Schroeder, R. Provvidenti and W. Mishanec. 1964. Clones

Fig. 19. Bean yellow mosaic on broad bean. (Courtesy of I. Evans, Canadian Department of Agriculture, Edmonton, Alberta.)

from segregating progenies of garden pea demonstrate that resistance to BV2 and PV2 is conditioned by the same genotype. *Plant Dis. Reptr. 48:* 353–355.

3. Bos, L. 1969. Inclusion bodies of bean yellow mosaic virus, some less known closely related viruses and beet mosaic virus. *Neth. J. Plant Pathol. 75:* 137–143.

4. Bos, L. 1970. Bean yellow mosaic virus. *Commonwealth Mycol. Inst. Descriptions Plant Viruses 40.* 4 p.

5. Bos, L., C. Kowalska and D. Z. Maat. 1974. The identification of bean mosaic, pea yellow mosaic and pea necrosis strains of bean yellow mosaic virus. *Neth. J. Plant Pathol. 80:* 173–191.

6. Brandes, J. and R. Bercks. 1965. Gross morphology and serology as a basis for classification of elongated plant viruses. *Advances Virus Res. 11:* 1–24.

7. Brierley, P. and F. F. Smith. 1962. Three cowpea mosaic viruses from gladiolus. *Plant Dis. Reptr. 46:* 335–337.

8. Corbett, M. K. 1958. A virus disease of lupines caused by bean yellow mosaic virus. *Phytopathology 48:* 86–91.

9. Evans, I. R. 1973. Seed-borne bean yellow mosaic virus of fababean in Canada. *Can. Plant Dis. Surv. 53:* 123–126.

10. Goodchild, D. J. 1956. Relationships of legume viruses in Australia. I. Strains of bean yellow mosaic virus and pea mosaic virus. *Australian J. Biol. Sci. 9:* 213–230.

11. Goodchild, D. J. 1956. Relationships of legume viruses in Australia. II. Serological relationships of bean yellow mosaic virus and pea mosaic virus. *Australian J. Biol. Sci. 9:* 231–237.

12. Grogan, R. G. and J. C. Walker. 1948. A pod-distorting strain of the yellow mosaic virus of bean. *J. Agr. Res. 77:* 301–314.

13. Hagedorn, D. J. and J. C. Walker. 1950. The relationship of bean virus 2 to pea mosaic in Wisconsin. *Phytopathology 40:* 684–698.

14. Izadpanah, K., A. Dehlavi and A. Saffarian. 1969. The effect of time of mosaic infection on the yield of broad-beans. *Iranian J. Plant Pathol. 5:* 3–4.

15. Kaiser, W. J. 1972. Seed transmission of bean yellow mosaic virus in broad beans in Iran. *Phytopathology 62:* 768.

16. Kaiser, W. J. 1973. Biology of bean yellow mosaic and pea leaf roll viruses affecting *Vicia faba* in Iran. *Phytopathol. Z. 78:* 253–263.

17. Pierce, W. H. 1934. Viroses of the bean. *Phytopathology 24:* 87–115.

18. Schroeder, W. T. and R. Provvidenti. 1966. Further evidence that common pea mosaic virus (PV2) is a strain of bean yellow mosaic virus (BV2). *Plant Dis. Reptr. 50:* 337–340.

19. Taylor, R. H. and P. R. Smith. 1968. The relationship between bean yellow mosaic virus and pea mosaic virus. *Australian J. Biol. Sci. 21:* 429–437.

20. Thomas, H. R. and W. J. Zaumeyer. 1953. A strain of yellow bean mosaic virus producing local lesions on tobacco. *Phytopathology 43:* 11–15.

21. Von Der Pahlen, A. 1962. El "mosaico amarillo del poroto" [*Phaseolus* virus 2 (Pierce) Smith] en cultivos de haba, arveja y poroto de los alrededores de Buenos Aires. *Rev. Invest. Agr. 16:* 87–92.

22. Zaumeyer, W. J. and H. H. Fisher. 1953. A new necrotic-lesion-producing strain of yellow bean mosaic. *Phytopathology 43:* 45–49.

Broad bean leaf roll

A disease of broad beans described from Germany in 1954 now also occurs in England, the Netherlands and Belgium. Yield from infected plants has been estimated at 50 percent, mostly from reduction in number of pods formed.

Fig. 20. Leaf roll virus disease symptoms in broad bean (healthy plant on left). (Courtesy of W. J. Kaiser, East African Agriculture and Forestry Research Organization, Nairobi.)

The top-most leaves of infected plants first roll as they become lighter in color. These symptoms extend until the upper halves of diseased plants exhibit rolled, bright yellow leaves that are thickened and brittle. Affected plants are stunted, and there is considerable defoliation, while the growing points of affected plants are often killed (Fig. 20).

The virus disease is not transmitted mechanically but by extended feeding periods for *Myzus persicae* and *Macrosiphum pisi*. Incubation for 2 to 4 weeks is required for symptom development.

Varietal differences in percentage of diseased plants have been attributed to differences in susceptibility to infection, but numbers of plants infected were consistently lower in late-sown plants.

1. Aldrich, D. T. A., A. J. Gibbs and L. R. Taylor. 1965. The incidence of bean leaf roll virus in some varieties of field beans (*Vicia faba* L.). *Plant Pathol. 14:* 11–14.

2. Heathcote, G. D. and A. J. Gibbs. 1962. Virus diseases in British crops of field beans (*Vicia faba* L.). *Plant Pathol. 11:* 69–73.

3. Quantz, L. and J. Völk. 1954. Die blattrollkrankheit der ackerbohne und erbse, eine neue viruskrankheit bei leguminosen. Nachrichtenbl. *Deut. Pflanzenschutzdienst (Braunschweig) 6:* 177–182.

4. Tinsley, T. W. 1959. Pea leaf roll, a new virus disease of legumes in England. *Plant Pathol. 8:* 17–18.

Broad bean true mosaic

A number of mottling diseases of broad bean were observed in Europe and China in the first quarter of this century, but the similarity of field symptoms and causal agents served to complicate distinction. Broad bean true mosaic (echten ackerbohnen mosaik) was described from Germany in 1953 but is now known to occur in several other European countries and northern Africa. Losses from mosaic have been assessed experimentally by bean weight at less than 15 percent/plant, but epidemiological factors cause the disease to be of potential consequence.

Infected plants exhibit systemic mottling (Fig. 21) that is much less severe at temperatures of 25 C or above. Symptoms on plants grown below this temperature often develop a systemic wilt and stem necrosis. Development of symptoms in the youngest leaves varies by cycles, but no evidence of infection can be detected by appearance of the seeds.

Particles of the causal virus, which is readily sap transmissible, are isometric (25 nm) but sediment as two components of 98 S and 119 S. The virus is inactivated by heating for 10 minutes at 65–75 C, storage at room temperature for more than 6–7 days or dilution of sap from infected plants 10^{-4} or more. Infectivity of the virus is limited to the *Leguminosae*, and the disease in pea consists of severe, systemic chlorosis in stunted plants with malformed leaves. Infection of bean varieties is limited to chlorotic flecks.

Fig. 21. Symptoms of broad bean true mosaic virus disease. (Courtesy of A. J. Cockbain, Rothamsted Experiment Station, Harpenden, Herts.)

Chenopodium amaranticolor, globe amaranth, tobacco and *N. clevelandii* are not infected.

No resistant varieties of broad bean have been identified, and seed transmission of the disease may result in 15 percent infected progeny plants. Although the disease is known to spread over sizeable distances within a field of broad beans, no insect vector has been identified. Despite morphological similarities of the particles with other viruses in the cowpea mosaic group, no serological relations have been established with other viruses in this group. Similarity of foliage symptoms caused by mosaic, mottle and stain virus diseases of broad bean precludes positive visual identification of these diseases in naturally infected field plants.

Planting of seed harvested from disease-free plants is imperative and roguing of infected plants when symptoms appear recommended. New fields should not be planted near old broad bean fields in which mosaic occurred.

1. Gibbs, A. J. and H. L. Paul. 1970. Echtes ackerbohnenmosaik-virus. *Commonwealth Mycol. Inst. Descriptions Plant Viruses 20.* 4 p.

2. Gibbs, A. J., G. Giussani-Belli and H. G. Smith. 1968. Broad-bean stain and true broad-bean mosaic viruses. *Ann. Appl. Biol. 61:* 99–107.

3. Paul, H. L. 1963. Untersuchungen über das echte ackerbohnenmosaik-virus. *Phytopathol. Z. 49:* 161–176.

4. Paul, H. L., J. Brandes and L. Quantz. 1958. Versuche zur reinigung und darstellung des echten ackerbohnenmosaik-virus. *Phytopathol. Z. 31:* 441–443.

5. Quantz, L. 1953. Untersuchungen über ein samenübertragbares mosaikvirus der ackerbohne (*Vicia faba*). *Phytopathol. Z. 20:* 421–448.

6. Wetter, C., H. L. Paul, J. Brandes and L. Quantz. 1960. Vergleich zwischen eigenschaften des echten ackerbohnenmosaik-virus und des broad bean mottle-virus. *Z. Naturforsch. 15 B:* 444–447.

Broad bean mottle

A disease of broad bean was observed in England in 1947 and reported as "mottle" disease 4 years later. It has not been reported to occur naturally elsewhere.

Inoculated plants develop vein clearing in the youngest leaves 8–10 days after inoculation, which is followed by a bright, interveinal systemic mottle and extensive necrosis. The margins of leaf laminae first become necrotic, and soon thereafter axillary shoots and the growing tip are killed.

The causal agent, a spherical virus with nominal diameter of 26 nm, is readily transmissible mechanically, but it is not transmitted by 5 species of aphids, including *Myzus persicae*, *Acyrthosiphon pisum*, *Macrosiphum euphorbiae*, *Aphis fabae* and *Hyperomyzus tulipaellus* (Theobald). The striped cucumber beetle [*Acalymma vittata* (Fabricius)], spotted cucumber beetle (*Diabrotica undecimpunctata howardi* Barber) and the grape colaspis [*Colaspis brunnea* (Fabricius)] can transmit the virus but are not very efficient vectors. Sap from infected broad bean remains infectious after dilution to 10^{-3} or 10 minutes exposure to 95 C (but not 100 C) but will not withstand 20 days *in vitro* at 15 C. The virus infects sweet pea, red, crimson and subterranean clovers (*Trifolium pratense*, *T. incarnatum* L., and *T. subterraneum* L., respectively), *Nicotiana clevelandii* and soybean, produces necrotic local lesions on several varieties of bean and kills pea, but does not infect vetch (*Vicia sativa* L.), jimson weed (*Datura stramonium* L.), tomato (*Lycopersicum esculentum* Mil.) or tobacco. Globe amaranth is susceptible but remains symptomless, while *Chenopodium album* L., *C. amaranticolor* and *C. quinoa* all develop chlorotic local lesions on inoculated leaves. X-bodies in the cells of infected broad bean plants are first granular, later vacuolate, and composed primarily of spherical virus particles.

Procedures for control have not been devised, but routine application of insecticide may prove beneficial.

1. Bawden, F. C., R. P. Chaudhuri and B. Kassanis. 1951. Some properties of broad-bean mottle virus. *Ann. Appl. Biol. 38:* 774–784.

2. Gibbs, A. J. 1972. Broad bean mottle virus. *Commonwealth Mycol. Inst. Descriptions Plant Viruses 101.* 4 p.

3. Hollings, M. 1959. *Nicotiana clevelandii* Gray as a test plant for plant viruses. *Plant Pathol. 8:* 133–137.

4. Hollings, M. 1959. Host range studies with fifty-two plant viruses. *Ann. Appl. Biol. 47:* 98–108.

5. Paul, H. L. 1961. Physikalische und chemische untersuchungen am broad bean mottle-virus. *Z. Naturforsch. 16 B:* 786–791.

6. Rubio, M. and D. H. M. van Slogteren. 1956. Light and electron microscopy of X-bodies associated with broad-bean mottle virus and *Phaseolus* virus 2. *Phytopathology 46:* 401–402.

7. Walters, H. J. and P. Surin. 1973. Transmission and host range studies of broad bean mottle virus. *Plant Dis. Reptr. 57:* 833–836.

8. Yamazaki, H., J. B. Bancroft and P. Kaesberg. 1961. Biophysical studies of broad bean mottle virus. *Proc. Nat. Acad. Sci. 47:* 979–983.

Broad bean stain (Evesham stain)

A disease of broad bean characterized by disfiguration of seeds was observed in England in 1960, reported from there in 1965 and further studied and described in 1968. It has been postulated that the disease was imported into England from North Africa. The disease has recently been identified in Australia in plants grown from seed commercially available there. Losses of 50 percent in harvested bean weight have been noted.

Infected plants typically exhibit diseased foliage followed by production of younger, apparently healthy leaves followed in turn by still younger leaves, again with symptoms that consist of a mottle pattern of dark green and chlorotic leaf areas. Growth of leaves may be affected, resulting in irregular leaf margins (Fig. 22).

A characteristic symptom of this disease is discoloration of the seed testa (Evesham stain), often as an irregular, necrotic band around the periphery of the seed. Discoloration occurs more frequently in infected seed, but not all infected seed exhibit discoloration.

The causal virus is isometric (25 nm) but sediments as 3 components with coefficients of 60 S, 100 S and 127 S. It is seed-borne (up to 10 percent) and readily transmissible mechanically, but no insect vector has been identified. Sap from infected plants is not infectious after being diluted 10^{-4}, heated for 10 minutes at 65C or stored for 17 days *in vitro* at room temperature. Broad bean stain virus is serologically related to some viruses in the cowpea mosaic group but not to broad bean mosaic nor broad bean mottle viruses. Infectivity is limited to 7 leguminous species, including bean varieties Tendergreen (Fig. 22) and Canadian Wonder, which develop systemic symptoms.

Control measures have not been determined, but only seed from healthy plants should be harvested for future plantings. New plantings should not be near old fields where the broad bean stain disease has occurred, and infected plants should be rogued as soon as symptoms appear.

1. Gibbs, A. J., G. Giussani-Belli and H. G. Smith. 1968. Broad-bean stain and true broad-bean mosaic viruses. *Ann. Appl. Biol. 61:* 99–107.

Fig. 22. Symptoms on broad bean (left) and bean cv. Tendergreen (right) inoculated with broad bean stain virus. (Courtesy of R. Francki and S. M. Moghal, Waite Agricultural Research Institute, Glen Osmond, South Australia.)

2. Gibbs, A. J. and H. G. Smith. 1970. Broad bean stain virus. *Commonwealth Mycol. Inst. Descriptions Plant Viruses 29.* 3 p.

3. Lloyd, A. T. E., H. G. Smith and L. H. Jones. 1965. Evesham stain—a virus disease of broad beans (*Vicia faba* L.). *Hort. Res. 5:* 13–18.

4. Moghal, S. M. and R. I. B. Francki. 1974. Occurrence and properties of broad bean stain virus in South Australia. *Australian J. Biol. Sci. 27:* 341–348.

Broad bean (virus) wilt

A wilt of broad beans in home gardens near Melbourne, Australia, had been recognized for 15 years and attributed to *Pythium* disease before it was shown to be virus-induced in 1947. The disease is now known to occur in Europe (Spain), the United States (New York) and Japan, as well as Australia.

The growing tips of infected plants bend laterally, and veins in the youngest leaves become blackened. As the disease progresses, the tops of diseased plants become flattened and assume a rosette formation. Affected plants develop a progressive wilt and often die but may survive and recover from severe symptoms. If they survive, distorted leaves with a blister-type, dark green mottle appear. Leaves at the top of the plant later are small, stiff and erect, feel leathery to the touch and become rolled inward. Vascular discoloration, more noticeable in the collar region in transverse section, and disintegration of the root system are characteristic.

The causal virus is mechanically and aphid transmissible, remains infectious *in vitro* 54 but not 72 hours, is inactivated by exposure to 58–60 C for 10 minutes or by dilution of juice from infected plants more than 10^{-4}. Virus particles are isometric and about 25 nm in diameter. Systemic symptoms develop in pea, sweet pea, blue lupine (*Lupinus angustifolius* L.), jimson weed, tomato, petunia, pepper (*Capsicum annuum* L.), tobacco, *Nicotiana glutinosa* L., beet (*Beta vulgaris* L.), *Chenopodium album*, spinach and snapdragon (*Antirrhinum majus* L.). No symptoms appear on zinnia (*Zinnia elegans* Jacq.), marigold (*Calendula officinalis* L.), lettuce (*Lactuca sativa* L.), bean, cucumber (*Cucumis sativus* L.) or nasturtium (*Tropaeolum majus* L.).

The disease develops most rapidly in spring or early winter but is not transmitted through seed from infected plants. *Aphis craccivora*, although it reproduces prolifically on broad bean, is an inefficient vector as is the more infrequent vector *Macrosiphum euphorbiae*. *Myzus persicae* is an efficient vector of the virus but does not reproduce well on broad bean.

Late planting to avoid plant growth during periods of high incidence of *M. persicae* will serve to reduce losses. Infected plants should be rogued at the first sign of symptoms.

1. Anonymous. 1944. Diseases of broad beans. *Agr. Gaz. New S. Wales 55:* 426–429.

2. Bruckhart, W. L. and J. W. Lorbeer. 1975. Recent occurrences of cucumber mosaic, lettuce mosaic and broad bean wilt viruses in lettuce and celery fields in New York. *Plant Dis. Reptr. 59:* 203–206.

3. Doel, T. R. 1975. Comparative properties of type, nasturtium ringspot and petunia ringspot strains of broad bean wilt virus. *J. Gen. Virol. 26:* 95–108.

4. Frowd, J. A. and J. A. Tomlinson. 1972. Relationships between a parsley virus, nasturtium ringspot virus and broad bean wilt virus. *Ann. Appl. Biol. 72:* 189–195.

5. Stubbs, L. L. 1947. A destructive vascular wilt virus disease of broad bean (*Vicia faba* L.) in Victoria. *J. Dept. Agr. Victoria 45:* 323–332.

6. Stubbs, L. L. 1960. Aphid transmission of broad bean wilt virus and comparative transmission efficiency of three vector species. *Australian J. Agr. Res. 11:* 734–741.

7. Taylor, R. H., P. R. Smith, C. Reinganum and A. J. Gibbs. 1968. Purification and properties of broad bean wilt virus. *Australian J. Biol. Sci. 21:* 929–935.

8. Taylor, R. H. and L. L. Stubbs. 1972. Broad bean wilt virus. *Commonwealth Mycol. Inst. Descriptions Plant Viruses 81.* 4 p.

Broad bean necrosis

Particles of the soil-borne virus in Japan that causes mild systemic spotting and mottling of broad bean are rod-shaped, 25 × 150–250 nm (Fig. 23), inactivated by exposure to 50–55 C for 10 minutes or aging 8–17 days at 20

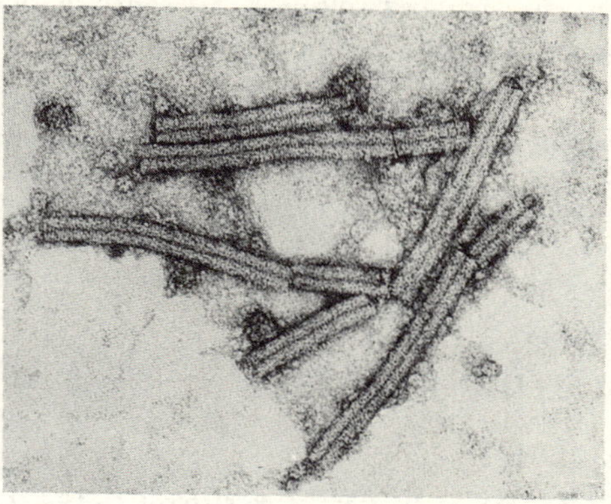

Fig. 23. Lesions on leaves of broad bean (top left) and *Chenopodium amaranticolor* (lower left) caused by inoculation with broad bean necrosis virus (particles × 200,000). (Courtesy of T. Inouye, University of Osaka Prefecture, Sakai, Osaka.)

C. The virus is mechanically transmitted and infectious to several leguminous species, including pea and sweet pea, and causes local, chlorotic lesions on *Nicotiana rustica* L., *N. clevelandii*, *Chenopodium amaranticolor* (Fig. 23) and *Tetragonia expansa*.

1. Inouye, T. and M. Asatani. 1968. Broadbean necrosis virus. *Ann. Phytopathol. Soc. Jap. 34:* 317–322.

Pea early browning

No visible symptoms are seen on broad bean plants systemically infected with the pea early browning disease, which is caused by short, rigid, thick virus particles with a bimodal length distribution of about 105 and 210 nm. The virus is readily transmissible mechanically. Infectious sap is inactivated by exposure to 65–70 C for 10 minutes or by storage *in vitro* for 6–8 months. Snapdragon, globe amaranth, *Nicotiana rustica*, nasturtium and zinnia, in addition to pea, are susceptible to infection. The virus is seed-borne in pea and transmitted by the nematodes *Trichodorus teres* Hooper and *T.*

pachydermus Seinhorst. Susceptible varieties of broad bean or other crops should not be grown on infested soil.

1. Bos, L. and J. P. H. van der Want. 1962. Early browning of pea, a disease caused by a soil- and seed-borne virus. *Tijdschr. Planteziekten 68:* 368–390.

2. Harrison, B. D. 1973. Pea early-browning virus. *Commonwealth Mycol. Inst. Descriptions Plant Viruses 120.* 4 p.

3. Van Hoof, H. A. 1962. *Trichodorus pachydermus* and *T. teres*, vectors of the early browning virus of peas. *Tijdschr. Planteziekten 68:* 391–396.

OTHER DISEASES

Downy mildew [*Peronospora viciae* (Berk.) Casp.]

1. Glasscock, H. H. 1963. New or uncommon plant diseases and pests. Downy mildew of broad bean. *Plant Pathol. 12:* 92.

Pythium blight (*Pythium graminicola*)

1. Sen, B. and S. P. Lal. 1969. Blight of broad bean—a *Pythium* disease. *Plant Dis. Reptr. 53:* 470–472.

Pythium root rot (*Pythium fabae* Cheney)

1. Cheney, G. M. 1932. *Pythium* root-rot of broad beans in Victoria. *Australian J. Expt. Biol. Med. Sci. 10:* 143–155.

Rhizoctonia damping-off (*Rhizoctonia solani*)

1. Fahim, M. M., M. M. Ragab and M. I. Shawky. 1967. Relative efficiency of different fungicides as protectants against *Rhizoctonia* damping-off of broad-bean seedlings. *Phytopathol. Mediter. 6:* 149–153.

Fusarium disease (*Fusarium* sp.)

1. Casals, C. T. 1951. *Fusarium* sp. sobre *Vicia faba*. *Microbiol. Espan. 4:* 15–22.

Stemphylium leaf spot (*Stemphylium botryosum* Wallr.)

1. Abdou, Y. A. and M. M. Fahim. 1969. A *Stemphylium* leaf spot of *Vicia faba* in the United Arab Republic. *Plant Dis. Reptr. 53:* 157–159.

Alfalfa mosaic

1. Nitzany, F. E. and S. Cohen. 1964. Virus affecting broad beans in Israel. *Phytopathol. Mediter. 3:* 1–8.

Mild mosaic (Rosette)

1. Yu, T. F. 1939. Mild-mosaic virus of broad bean. *Phytopathology 29:* 448–455.

2. Yu, T. F. 1939. A list of plant viroses observed in China. *Phytopathology 29:* 459–461.

Mosaic virus A/B

1. Klesser, P. J. 1960. Virus diseases of broad beans. *Bothalia 7:* 283–294.

Pea enation mosaic

1. Musil, M. and O. Leskova. 1969. Some properties of pea enation mosaic virus isolated from field pea and broad bean plants in Bohemia. *Biol. Plant. (Praha) 11:* 319–323.

Pea mosaic

1. Nour, M. A. and J. J. Nour. 1962. Broad bean mosaic caused by pea mosaic virus in the Sudan. *Phytopathology 52:* 398–403.

Spotted wilt

1. Yu, T. F. 1947. Spotted wilt of broad bean. *Phytopathology 37:* 191–192.

C

CACAO: *Theobroma cacao* L.

Black pod

The disease known as black pod or black rot is the most widespread major disease of cacao and occurs wherever the crop is grown. First recorded occurrence of the disease has been associated with the disaster that struck the cacao industry in Trinidad in 1727. Losses from black pod are more

Fig. 24. Black pod symptoms on cacao fruits. (Courtesy of J. Lawrence, Centro Agronomico Tropical de Investigacion y Enseñanza, Turrialba.)

consistent than with other cacao diseases and have been assessed as high as 80 percent. Annual financial loss from black pod has been estimated to be nearly 50 million dollars and to contribute to the unstable world price of cocoa.

Fruit infections consist of brown lesions that may originate anywhere on the fruit but more often first appear at either end (Fig. 24). Affected tissue is clearly and abruptly demarcated, but lesions extend rapidly and eventually cover whole pods, which turn black in a few days. Scant mycelial growth may appear on the surface of lesions under humid conditions. Internal invasion of fruits is more rapid than external, but infection throughout finally is complete and the beans destroyed.

Bark cankers in the trunk or branches also are caused by the same fungus and develop from dark, moist lesions in which the bark often splits and a reddish colored gum exudes. Infections often extend to encircle the affected branch or trunk.

A wilt and dieback of chupons and seedlings may be caused by the same fungus which can infect leaves as well as cause irregular, necrotic stem lesions. Root infection under natural conditions has been demonstrated.

The causal fungus of black pod is *Phytophthora palmivora*, characterized by ovoid sporangia 31–35 × 50–60 μm, hyphae no more than 5 μm in width and thin-walled terminal or intercalary chlamydospores usually 30–35 μm in diameter. Oospores formed *in vitro* by pairing opposite mating strains of *P. palmivora* or by culturing together with *P. nicotianae* average 30 μm in diameter, and the oval to spherical antheridia measure 14 × 15 μm. Some morphological and pathological differences noted among cultures of *P. palmivora* may be sufficient to warrant distinct designations.

The causal fungus was found indigenous to forest soils in Ghana, where prevalence in soils was variable and unrelated to atmospheric temperature. Sporangia, produced most abundantly in lesions on infected pods, are readily disseminated in the field by rain and germinate either directly or indirectly under conditions of high relative humidity (above 95 percent). Pod infection results from penetration either through the cuticle or stomata within hours after germination and more often occurs at either end of the fruit. Sporangia and zoospores may lodge in the depression around the peduncle or be washed in droplets down to the distal end of hanging fruits. Incidence of black pod in Ceylon has been directly correlated with minimum ambient temperature a month earlier, and disease severity was estimated to increase progressively with decrease in temperature from 20 to 15 C. Black pod developed faster under shade in Ghana and Brazil.

No single control measure has proven completely adequate for control, but a combination of one or more practices can serve to reduce losses to a minimum. Plant spacing and sanitation to provide maximum air and water drainage, combined with removal of infected pods as early as possible and frequent harvest of healthy pods, have been recommended. Diseased pods should be removed from the field and the hulls destroyed after opening. Sprays, particularly during cool periods with high rainfall, are necessary to provide fruit protection. Copper-containing materials, viz. copper oxide and copper oxychloride, have given promising results. No varietal resistance sufficient to control pod rot has been found, but available resistance has permitted reduction in frequency and expense of fungicide applications. Seedlings of the Catongo cultivar have been considered resistant to stem canker development but no resistance has been reported to root infection.

1. Ashby, S. F. 1929. Strains and taxonomy of *Phytophthora palmivora* Butler (*P. faberi* Maubl.). *Trans. Brit. Mycol. Soc. 14:* 18–33.

2. Asomaning, E. J. A. 1964. Varietal resistance of young clones and seedlings of cocoa (*Theobroma cacao* L.) to root infection by *Phytophthora palmivora*. *Trop. Agr. (Trinidad) 41:* 251–256.

3. Chant, S. R. 1959. Investigations on a seedling die-back of *Theobroma cacao* L. in Nigeria. I. Description of the disease and its spread in the nursery. *Trop. Agr. (Trinidad) 36:* 138–144.

4. Chant, S. R. 1959. Investigations on a seedling die-back of *Theobroma cacao* L. in Nigeria. II. Factors affecting the incidence of the disease and its control. *Trop. Agr. (Trinidad) 36:* 145–149.

5. Crowdy, S. H. and R. S. Elias. 1956. Cacao diseases. *Outlook Agr. 1:* 64–69.

6. Dakwa, J. T. 1973. The relationship between blackpod incidence and the weather in Ghana. *Ghana J. Agr. Sci. 6:* 93–102.

7. Dakwa, J. T. 1974. The development of blackpod disease (*Phytophthora palmivora*) in Ghana. *Turrialba 24:* 367–372.

8. Dakwa, J. T. 1974. The occurrence of *Phytophthora palmivora* (Butl.) Butl. in soil in Ghana. *Ghana J. Agr. Sci. 7:* 37–41.

9. Desrosiers, R. and J. Diaz M. 1957. The world distribution of diseases of cacao. *VI Reunião Comité Téc. Interam. Cacau (1956):* 331–341.

10. Evans, E., E. F. Collingwood, J. W. H. Taylor and E. P. Whitaker. 1962. Field trials with Bordeaux mixture and various formulations of copper oxychloride for control of late blight of potatoes and black pod of cacao. *Trans. Brit. Mycol. Soc. 45:* 321–334.

11. Evans, H. C. 1971. Transmission of *Phytophthora* pod rot of cocoa by invertebrates. *Nature 232:* 346–347.

12. Hansen, A. J. 1961. Resistencia y susceptibilidad de tipos de cacao a *P. palmivora*. *Cacao 6:* 9–10.

13. Hardy, F. 1960. *Cacao manual.* Inter-Am. Inst. Agr. Sci., Turrialba. 395 p.

14. Hislop, E. C. 1963. Studies on the chemical control of *Phytophthora palmivora* (Butl.) Butl. on *Theobroma cacao* L. in Nigeria. IV. Further laboratory and field trails of fungicides. *Ann. Appl. Biol. 52:* 465–480.

15. Hislop, E. C. 1963. Studies on the chemical control of *Phytophthora palmivora* (Butl.) Butl. on *Theobroma cacao* L. in Nigeria. V. Comparisons of three spraying machines for applying fungicides. *Ann. Appl. Biol. 52:* 481–492.

16. Hislop, E. C. and P. O. Park. 1962. Studies on the chemical control of *Phytophthora palmivora* (Butl.) Butl. on *Theobroma cacao* L. in Nigeria. *Ann. Appl. Biol. 50:* 57–65; 67–76; 77–88.

17. Holliday, P. 1960. A spraying and harvesting trial in Trinidad against cocoa pod diseases. *Trop. Agr. (Trinidad) 37:* 235–242.

18. Okaisabor, E. K. 1965. Preliminary studies on the epidemiology of *Phytophthora*

palmivora. I. Outbreak of blackpod disease of cocoa. *Nigeria Agr. J. 2:* 67–70.

19. Orellana, R. G. 1954. Les maladies du cacaoyer au Venezuela, en Colombia, en Equateur et à la Trinité. *FAO Bull. Phytosanitaire 2:* 49–52.

20. Orellana, R. G. 1955. Cacao diseases in Mexico, Nicaragua, Costa Rica and Jamaica. *FAO Plant Protect. Bull. 4:* 35–37.

21. Orellana, R. G. 1959. Variation in *Phytophthora palmivora* isolated from cacao and rubber. *Phytopathology 49:* 210–213.

22. Orellana, R. G. and R. K. Som. 1957. Correlation between low temperature and incidence of *Phytophthora* pod rot of cacao in Ceylon. *FAO Plant Protect. Bull. 6:* 6–8.

23. Owen, H. 1951. Cocoa pod diseases in West Africa. *Ann. Appl. Biol. 38:* 715–718.

24. Rocha, H. M. 1965. Cacao varieties resistant to *Phytophthora palmivora* (Butl.) Butl. *Cacao 10:* 1–9.

25. Rocha, H. M. and A. D. Machado. 1972. Factores ambientais associados com a "podridão parda" do cacaoeiro. *Rev. Theobroma 2:* 26–34.

26. Sansome, E., C. M. Brasier and M. J. Griffin. 1975. Chromosome size differences in *Phytophthora palmivora*, a pathogen of cocoa. *Nature 255:* 704–705.

27. Shaw, D. E. 1963. Diseases of cacao in Papua and New Guinea. *Papua New Guinea Agr. J. 15:* 79–90.

28. Spence, J. A. 1961. Probable mechanism of resistance of varieties of cocoa to black pod disease caused by *Phytophthora palmivora* (Butl.) Butl. *Nature 192:* 278.

29. Spence, J. A. 1961. Black-pod disease of cocoa. *Ann. Appl. Biol. 49:* 717–722; 723–734.

30. Thorold, C. A. 1955. Observations on black-pod disease (*Phytophthora palmivora*) of cacao in Nigeria. *Trans. Brit. Mycol. Soc. 38:* 435–452.

31. Thorold, C. A. 1959. Methods of controlling black-pod disease (caused by *Phytophthora palmivora*) of *Theobroma cacao* in Nigeria. *Ann. Appl. Biol. 47:* 708–715.

32. Tollenaar, D. 1958. *Phytophthora palmivora* of cocoa and its control. *Neth. J. Agr. Sci. 6:* 24–38.

33. Turner, P. D. 1960. Distribution of strains of *Phytophthora palmivora* from *Theobroma cacao* in West Africa. *FAO Plant Protect. Bull. 8:* 53–54.

34. Turner, P. D. 1960. A world survey of *Phytophthora palmivora* on cocoa. *Plant Dis. Reptr. 44:* 221.

35. Turner, P. D. 1960. Variation in *Phytophthora palmivora* (Butl.) Butl. on *Theobroma cacao* L. in West Africa. *Nature 186:* 495–496.

36. Turner, P. D. 1960. Strains of *Phytophthora palmivora* (Butl.) Butl. from *Theobroma cacao* L. I. Isolates from West Africa. *Trans. Brit. Mycol. Soc. 43:* 665–672.

37. Turner, P. D. 1961. Strains of *Phytophthora palmivora* (Butl.) Butl. from *Theobroma cacao* L. II. Isolates from non-African countries. *Trans. Brit. Mycol. Soc. 44:* 409–416.

38. Turner, P. D. 1961. Complementary isolates of *Phytophthora palmivora* from cacao and rubber, and their taxonomy. *Phytopathology 51:* 161–164.

39. Turner, P. D. 1962. Distribución de las razas de *Phytophthora palmivora* sobre *Theobroma cacao* en el Africa Occidental. *Ager 13:* 12–14.

40. Turner, P. D. 1963. Influence of root exudates of cacao and other plants on spore development of *Phytophthora palmivora. Phytopathology 53:* 1337–1339.

41. Turner, P. D. 1965. Comparative sporulation as an indication of resistance by cacao to *Phytophthora* pod rot. *Trop. Agr. (Trinidad) 42:* 305–309.

42. Turner, P. D. 1965. Effect of spacing and pruning on *Phytophthora* pod rot of cacao. *Plant Dis. Reptr. 49:* 257.

43. Turner, P. D. 1965. Polyphenoloxidase activity in cacao selections showing variable resistance to *Phytophthora* pod rot. *Plant Dis. Reptr. 49:* 319–321.

44. Turner, P. D. and E. J. A. Asomaning. 1962. Root infection of *Theobroma cacao* by *Phytophthora palmivora. Trop. Agr. (Trinidad) 39:* 339–343.

45. Wharton, A. L. 1960. Black pod disease of cocoa in Ghana. *Ghana Farmer 4:* 18–23.

46. Zaiger, D. and G. A. Zentmyer. 1965. *Phytophthora* canker of cacao in the Caroline Islands. *Plant Dis. Reptr. 49:* 565–567.

47. Zentmyer, G. A., S. M. Mircetich and D. M. Mitchell. 1968. Tests for resistance of cacao to *Phytophthora palmivora. Plant Dis. Reptr. 52:* 790–791.

Green point cushion gall

A disease of cacao, which is now recognized as green point cushion gall, was described from Guyana (British Guiana) in 1905. An assortment of names, including "agallas leñosas," "verruga del cojin floral" and "la buba" or "agallas del cacao," have been applied to this disease in the various areas where it occurs, including most of Latin America, Ghana and Sri Lanka. The disease does affect pod set with losses in succeeding years which become progressively greater and often result finally in complete absence of production.

Flower cushions primarily are affected by the disease, and symptoms include hypertrophy of cushion tissue and excessive floral bud production, but very few buds ever open. Cushion tissue assumes a gall-like, approximately half-spherical shape and enlarges rapidly (Fig. 25). Galls several cm in diameter and thickness have been reported and are easily separated from the tree and broken apart while still green. They turn dark when dead and crumble readily. Galls are more often seen in flower cushions but can develop near a leaf scar or near the bases of chupons.

Descriptive names for what are presently considered different manifestations of the same disease include "disc gall," "fan gall," "flowery gall" and "knob gall." The first and last descriptions imply only variations in hypertrophied shape of the cushion gall, while fan gall is characterized by branched, twig-like outgrowths. An excessive number of normal-appearing flowers, some of which open and set normal fruit if hand-pollinated, are typical of flowery gall.

Calonectria rigidiuscula (Berk. & Br.) Sacc. (*Fusarium decemcellulare* Brick) is considered the primary cause of green point cushion gall, although galls have been induced by *F. roseum* under experimental conditions. Hyaline, oval, occasionally uniseptate microconidia 3–5 × 10–15 μm of *F. decemcellulare* are formed in chains from phialides 4–5 × 28–36 μm. Cylindrical, curved 7–10 septate macroconidia 6–10 × 55–130 μm are borne on phialides 5–6 × 30–40 μm. Globose, light colored perithecia 190–300 × 200–320μm contain clavate, 4-spored asci 12–14 × 70–100 μm with hyaline, ellipsoid 4-celled ascospores 7–10 × 22–28 μm.

It has been established that *C. rigidiuscula* is often associated with green point cushion gall of cacao. However, experimental production of

Fig. 25. Green point cushion gall on cacao. (Courtesy of R. G. Orellana, U. S. Department of Agriculture, Beltsville.)

similar galls in cacao by *F. roseum* and transmission of gall diseases from mango and pigeon pea to cacao leave some possibility for doubt of an absolute causal relationship. The same fungus has been identified as a cause of dieback in cacao.

Satisfactory control measures can be developed only after more information about the etiology and causal relations of the disease has been established. Until then, infected trees should be rogued and destroyed and care taken to select propagation and replacement materials free of gall diseases. A few resistant cacao cultivars have been identified.

1. Alexander, J. V. and R. J. Cook. 1965. *Fusarium* species and *Calonectria* associated with cushion gall of cacao in Central America. *Cacao 10:* 15–16.

2. Archibald, J. F. 1961. Transmission of gall-diseases of cacao, mango, and pigeon pea. *Nature 190:* 284.

3. Bates, J. F. 1963. Early records of cushion gall diseases of cocoa from British Guiana. *FAO Plant Protect. Bull. 11:* 34–35.

4. Booth, C. 1971. *The genus* Fusarium. Commonwealth Mycological Institute, Kew, Surrey. 237 p.

5. Booth, C. and J. M. Waterston. 1964. *Calonectria rigidiuscula. Commonwealth Mycol. Inst. Descriptions Pathogenic Fungi Bacteria 21.* 2 p.

6. Bourret, J. A. and E. J. Ford. 1965. Distribution of *Calonectria rigidiuscula* and other *Fusarium* species in relation to cushion gall of cacao. *Cacao 10:* 13–15.

7. Brunt, A. A. F. and A. L. Wharton. 1960. Transmission of a gall disease of cocoa. *Nature 187:* 80–81.

8. Brunt, A. A. F. and A. L. Wharton. 1962. Etiology of a gall disease of cocoa in Ghana caused by *Calonectria rigidiuscula* (Berk. & Br.) Sacc. *Ann. Appl. Biol. 50:* 283–289.

9. Brunt, A. A. and A. L. Wharton. 1963. *Calonectria rigidiuscula* (Berk. and Br.) Sacc. the cause of a gall disease of cocoa in Ghana. *Nature 193:* 903–904.

10. Crowdy, S. H. 1947. Observations on the pathogenicity of *Calonectria rigidiuscula* (Berk. & Br.) Sacc. on *Theobroma cacao* L. *Ann. Appl. Biol. 34:* 45–59.

11. Ford, E. J., J. A. Bourret and W. C. Snyder. 1967. Biologic specialization in *Calonectria (Fusarium) rigidiuscula* in relation to green point gall of cocoa. *Phytopathology 57:* 710–712.

12. Hansen, A. J. 1963. The role of *Fusarium decemcellulare* and *Fusarium roseum* in the greenpoint cushion gall complex of cacao. Turrialba 13: 80–87.

13. Hansen, A. J. 1966. Fusaria as agents of cacao greenpoint cushion gall in the Caribbean and in Latin America. *Plant Dis. Reptr. 50:* 229–233.

14. Hansen, A. J. and L. de Reyes. 1963. Agalla de punto verdes en el cacao de Venezuela y Costa Rica. *Turrialba 13:* 128–129.

15. Hardy, F. 1960. *Cacao manual.* Inter-Am. Inst. Agr. Sci., Turrialba. 395 p.

16. Hutchins, L. M. 1959. Current surveys for cushion gall. *J. Agr. Soc. Trinidad Tobago 59:* 325–340.

17. Hutchins, L. M. 1959. Transmission of cushion gall of cacao by means of tissue transplantation. *Comunicaciones Turrialba 66.* 6 p.

18. Hutchins, L. M. 1965. Loss of gall-inducing capacity of cacao, when *Calonectria rigidiuscula* passes from the conidial (*Fusarium*) stage through the perfect (ascospore) stage. *Plant Dis. Reptr. 49:* 564–565.

19. Hutchins, L. M. 1966. Loss of gall-inducing capacity of *Fusarium rigidiuscula* on cacao by interference from another *Fusarium* clone. *Plant Dis. Reptr. 50:* 219–222.

20. Hutchins, L. M. and L. Siller. 1960. Cushion gall types in cacao. *Proc. VIII Inter-Am. Cacao Conf. (Trinidad Tobago 1960):* 381–389.

21. Kevorkian, A. G. 1951. The cushion gall disease of cacao. *Phytopathology 41:* 562.

22. Malaguti, G. 1959. La buba o "agallas" del cacao. *Agr. Venezolana 23:* 8–9.

23. Malaguti, G. and L. C. de Reyes. 1964. A gall disease of cacao and mango in Venezuela caused by *Calonectria rigidiuscula. Phytopathology 54:* 499.

24. Orellana, R. G. 1959. Cushion gall of cacao in Ceylon. *FAO Plant Protect. Bull. 7:* 53–54.

25. Owen, H. 1956. Further observations on the pathogenicity of *Calonectria rigidiuscula* (Berk. & Br.) Sacc. to *Theobroma cacao* L. *Ann. Appl. Biol. 44:* 307–321.

26. Reichle, R. E. and W. C. Snyder. 1964. Heterothallism and ascospore number in *Calonectria rigidiuscula. Phytopathology 54:* 1297–1299.

27. Shaw, D. E. and W. M. Burnett. 1969. Galls of cacao in Papua and New Guinea. *Papua New Guinea Agr. J. 21:* 25–48.

28. Wellman, F. L. and R. G. Orellana. 1955. Buba or cushion gall of cacao in Nicaragua. *FAO Plant Protect. Bull. 3:* 71–73.

Mal del machete (Canker, Necrosis del tronco, La polilla, Llaga macana)

The disease of cacao commonly known as "mal del machete" was described from Ecuador in 1918 but, since 1950, has caused serious losses in Venezuela, Colombia and Mexico. It has been found in Trinidad and remains a

Fig. 26. Mal del machete disease of cacao. Note dead leaves that remain hanging. (Courtesy of J. Lawrence, Centro Agronomico Tropical de Investigacion y Ensenañza, Turrialba.)

factor in cacao production in Latin America.

Infection is restricted to the aerial parts of the tree where affected areas on the trunk or branches are slightly sunken and colored dark red or purple. The portion of the tree above or distal to the infected area dies, but the leaves remain attached for a considerable time (Fig. 26). A wilt of chupons has been attributed to the same disease. Insects that bore holes into dead wood, particularly those of the genus *Xyleborus*, are usually associated with the disease.

Infected pods exhibit shallow, dark lesions that enlarge slowly as a pod rot. Scant mycelial growth can be seen sometimes in the centers of lesions, and the seeds beneath such lesions develop a dry, brown rot. Pod infections often develop from wounds, and fructifications of the causal fungus frequently can be found in fruit lesions, particularly on the inside wall of pods with symptoms.

The causal fungus for mal del machete disease is *Ceratocystis fimbriata*, characterized by slender, septate, hyaline conidiophores up to 160 μm in length on which are borne cylindrical, hyaline conidia 4–5.5 × 11–25 μm. Thick-walled, colored, oval chlamydospores 6–13 × 9–18 μm are produced terminally. Hyaline, elliptical, aseptate, "hat-shaped" ascospores 2.5–5.5 × 4.5–8 μm are produced in dark, globose, partially immersed perithecia 140–220 μm in diameter with long necks up to 900 μm in length.

Wounding often precedes infection but has not been established as a prerequisite for infection. *Xyleborus* beetles, while consistently associated with the disease, have not been demonstrated to predispose plants to infection by creating wounds in otherwise healthy tissue and actually show a preference for trees already infected. However, they may play an indirect role in dissemination of the fungus whereby chlamydospores that form in the insect frass in the galleries are forced out of the holes with the frass and are air-borne to new infection sites.

Care should be taken to minimize injury to trees during pruning that includes removal of chupons. Tools used for pruning, or any other operation that might involve injury to healthy tissues, should be sterilized after each cut if possible. Formalin has been recommended for this purpose. Certainly all dead and diseased trees and branches should be removed and destroyed and a regular spray program for insect control followed. Varietal resistance has been reported in various species and hybrids of cacao relatives, including *Theobroma angustifolia* Moc. & Sesse ex DC, *T. mammosa* Cuatr. & Leon, *T. mammosa* × *T. simiarum* Donn.-Sm. and *T. cacao* clones IMC 67, Pound 12, SPA 9, UF 29, 296, 613, CC 38, 41, 42 and PA 121.

1. Delgado, J. C. and E. Echandi. 1965. Evaluación de la resistencia de especies y clones de cacao al mal del machete provocado por *Ceratocystis fimbriata*. *Turrialba 15:* 286–289.

2. Desrosiers, R. 1956. Diferenciación entre variedades de cacao con base en su susceptibilidad a la infección con *Ceratostomella fimbriata* (E. & H.) Elliot, en el Ecuador. *Turrialba 6:* 48–52.

3. Desrosiers, R. 1961. La *Ceratostomella* o el mal del machete. *Cacaotero 4:* 5–6.

4. Desrosiers, R. and A. von Buchwald. 1955. Fungicidal treatments for control of cacao cutting disease in Ecuador. *FAO Plant Protect. Bull. 4:* 17–21.

5. Giraldo, E. A. 1957. La llaga macana del tronco del cacao. *Acta Agron. (Palmira) 7:* 71–103.

6. Idrobo, S. 1958. El complejo Xyleborus-Ceratostomella en Colombia. *Proc. VII Conf. Interam. Cacao (Palmira, 1958):* 73–79.

7. Iton, E. F. 1960. *Ceratostomella* wilt in cacao in Trinidad. *J. Agr. Soc. Trinidad Tobago 60:* 461–464.

8. Malaguti, G. 1956. La necrosis del tronco del cacao en Venezuela. *Agron. Trop. (Venezuela) 5:* 207–226.

9. Morgan-Jones, G. 1967. *Ceratocystis fimbriata. Commonwealth Mycol. Inst. Descriptions Pathogenic Fungi Bacteria 141.* 2 p.

10. Orellana, R. G. 1954. Enfermedades del cacao en Venezuela, Colombia, Ecuador y Trinidad. *FAO Boletín Fitosanit. 2:* 49–52.

11. Wallenius, K. E. 1958. Observaciones sobre *Xyleborus* y el combate del mismo en el cacao. *Proc. VII Conf. Inter-Am. Cacao (Palmira 1958):* 270–273.

Watery pod rot (Quevedo disease)

A watery rot of cacao pods was reported from Ecuador in 1915 and later found in all countries on the northwestern coast of South America. It was not found in eastern Panama until 1956 but apparently remains confined to this general area. The disease involves only the pods and incidence varies, but losses of more than 50 percent have been reported. Watery pod rot is considered a major disease of cacao where it does occur.

Symptoms of infection often remain masked until pods approach maturity, when colored spots similar to premature ripening appear. Such lesions enlarge into brown spots that may involve a large part of the fruit and are typically surrounded by yellow borders. A white to light brown mycelial growth with abundant sporulation may cover the lesion in humid weather. Seeds in infected pods are destroyed by a wet rot (Fig. 27), and a considerable amount of liquid may accumulate inside the pods.

The disease is caused by *Monilia roreri* Cif. that produces simple, hyaline conidiophores 9–12 × 30–50 μm on which are borne globose, hyaline, catenate conidia 7–10 × 9–14 μm.

Disease incidence is directly related to rainfall during the period of early pod growth, and fruit wounds caused by insects increase the probability of infection. Spores of the fungus are easily dislodged and readily carried by wind as well as by insects.

Fig. 27. External sporulation on infected cacao pod and wet, internal rotting of fruit caused by watery pod rot (Quevedo disease). (Courtesy of J. Lawrence, Centro Agronomico Tropical de Investigacion y Enseñanza, Turrialba.)

Criollo cacao has been reported more resistant than Forastero or Trinitario, but complete susceptibility of the former type to black pod and mal del machete diseases precludes successful use of such resistance. Control and significant increase in production have been achieved in Ecuador with either zineb (0.2 kg/100 ℓ water) or copper hydroxide, provided applications were begun at flowering and continued through fruit formation. Fruits should be harvested at frequent intervals and all infected fruits picked and destroyed. Beans should not be removed in the field nor husk piles left in the field.

1. Bastidas, R. A. 1953. Patogenidad de *Monilia* sp. en *Theobroma cacao* L. *Cacao Colombia 2:* 139–152.

2. Desrosiers, R., A. von Buchwald and C. W. Bolaños. 1955. Effect of rainfall on the incidence of *Monilia* pod rot of cacao in Ecuador. *FAO Plant Protect. Bull. 3:* 161–164.

3. Desrosiers, R. and J. Diaz M. 1956. Efecto de diversos fungicidas en el control de la *Monilia*. *Turrialba 6:* 19–22.

4. Hardy, F. 1960. *Cacao manual.* Inter-Am. Inst. Agr. Sci., Turrialba. 395 p.

5. Holliday, P. 1970. *Monilia roreri. Commonwealth Mycol. Inst. Descriptions Pathogenic Fungi Bacteria 226.* 2 p.

6. Naundorf, G. 1954. Contribuciones al problema de la Moniliasis en cacao. *Cacao Columbia 3:* 35–61.

7. Orellana, R. G. 1956. Occurrence of *Monilia* pod rot and other cacao diseases in eastern Panama. *FAO Plant Protect. Bull. 4:* 168–169.

8. Rorer, J. B. 1926. *Monilia* and witches' broom diseases in cacao plantations in Ecuador. *Trop. Agr. (Trinidad) 3:* 46–47, 68–69.

Witches' broom (Escoba de bruja)

The disease of cacao known as witches' broom was originally described from Surinam in 1895 and since then has been found in all the countries of South America where cacao is grown. It is considered endemic in Trinidad, Tobago and Grenada, also, but has not been reported from Central America, including Panama, or other islands in the Caribbean area. Escoba de bruja can exact considerable losses in the field and has been considered largely responsible for failure of the cacao industry in Guyana and Surinam early in the twentieth century. Fruit losses of 50 percent or more in recent years necessitate that witches' broom still be considered a major disease of cacao, even though incidence does vary according to climatic conditions.

Symptoms of witches' broom disease may be found on twigs, flower cushions and fruits where infection is confined to meristematic or developing tissues (Fig. 28). Visible evidence of infection may not be apparent until after several days, and then in various forms, although some degree of malformation usually can be detected. Infection of vegetative shoots results in abnormal proliferation of axillary buds for a short distance from the growing tip. This growth imparts a "broom" effect accompanied by a noticeable thicken-

Fig. 28. Twig malformation, fruit symptoms and fructifications of the causal fungus (*Marasmius perniciosus*) of cacao witches' broom (escoba de bruja). (Courtesy of Instituto Nacional de Investigaciones Agropecuarias, Pichilingue.)

ing of the affected part of the twig. On occasion, similar proliferated twig growth may originate from a single bud to form a "lateral broom." Brooms may retain a compact appearance or assume an elongated shape as a result of increase in internodal length. On occasion, normal growth may again arise distal to terminal brooms formed either in twigs or chupons, which has been described as "grown-through" brooms. Neither leaf production nor size are seriously affected immediately, but leaves on brooms begin to die soon after broom formation and are quickly followed by complete necrosis of the entire broom. Dead brooms often remain attached to the tree for some time but are easily broken when dry.

Infection of flower cushions often results in production of vegetative shoots, which assume broom configuration, or masses of abnormal flowers. Some of the flowers in a "cushion broom" may have thickened pedicels or distinctive conformation known as "star bloom."

Pod infection may cause a number of different symptoms, but all include some form of hypertrophy or malformation. In all cases, infection of pods presumably occurs while they are still developing and a considerable time before any necrosis is apparent. The different pod symptoms include a thickening of the pedicel, broadening of the shoulders of the fruit, asymmetrical distortion and premature ripening, as well as various degrees of necrosis from a slight speckling to extensive necrotic areas on the pod wall. Necrotic tissues remain firm, and damage to beans varies from little to more serious losses where beans may remain undeveloped or be completely destroyed. Diseased beans often are closely adherent to the pod wall and to each other.

Witches' broom is caused by *Marasmius perniciosus* Stahel [*Crinipellis perniciosa* (Stahel) Singer], identified by pinkish basidiocarps with a dark spot in the center of the pileus 0.5–1.5 cm in diameter (Fig. 28). The lamellae are whitish, as is the stipe, and the spore print composed of basidiospores 4–5 × 7–11 μm. Arthrospore production *in vitro* has been reported, but the role of this spore form in the disease cycle has not been established.

Sporophores are produced only on necrotic brooms or pods, often still hanging in the tree, and infection is known to result only from basidiospores, which are readily air-borne and germinate almost immediately. Basidiospores are produced most abundantly at night and do not withstand desiccation well. Infection presumably is confined to meristematic or young, developing tissues and occurs most readily under conditions of high humidity. The intercellular infection usually becomes apparent after 3–6 weeks, but sporophore production may not develop thereon until 5–6 months after infection and is directly related to the advent of high rainfall seasons. The existence of pathotypes of the fungus has been suggested but has not been adequately demonstrated.

Physical removal of brooms and infected pods at 3-month intervals has been recommended for control because the minimum time for first sporophore formation from green brooms has been reported to exceed this period. The labor required, the necessity for widespread application and the need for extensive pruning in some cases followed by reduction in subsequent yields make this theoretically successful practice less than adequate in application. Foliar sprays, including zineb, Bordeaux mixture or other copper-containing formulations, have provided some improvement in disease control but have been far from completely satisfactory. Selections with apparent resistance have been found in Forastero cacao, and variation in response to inoculation has been noted in Trinitario. Hybridization and selection of disease-resistant types acceptable for commercial use gave considerable promise of success in disease control. However, the recent appearance of witches' broom in materials previously considered resistant has necessitated reassessment of programs to develop resistant plant types. Genetic resistance remains more promising than any other method for eventual successful control of witches' broom disease.

1. Baker, R. E. D. 1943. Witches' broom disease investigations. *Trop. Agr. (Trinidad) 20:* 5–12; 156–158.

2. Baker, R. E. D. and S. H. Crowdy. 1942. Witches' broom disease investigations. II. Notes on the susceptibility of ICS selections at River Estate to witches' broom disease of cacao. *Trop. Agr. (Trinidad) 19:* 207–209.

3. Baker, R. E. D. and S. H. Crowdy. 1943. Studies in the witches' broom disease of cacao caused by *Marasmius perniciosus* Stahel. I. Introduction, symptoms and etiology. *Mem. Imp. Coll. Trop. Agr., Trinidad 7.* 28 p.

4. Baker, R. E. D. and S. H. Crowdy. 1944. Studies in the witches' broom disease of cacao caused by *Marasmius perniciosus* Stahel. II. Field studies and control methods. *Mem. Imp. Coll. Trop. Agr., Trinidad 8.* 28 p.

5. Baker, R. E. D., S. H. Crowdy and C. A. Thorold. 1941. Witches' broom disease investigations. I. Seasonal variations in intensity of infection and their effect on control methods. *Trop. Agr. (Trinidad) 18:* 107–116.

6. Baker, R. E. D. and W. T. Dale. 1944. Witches' broom investigations. *Trop. Agr. (Trinidad) 21:* 170–175; 198–199.

7. Baker, R. E. D. and P. Holliday. Witches' broom disease of cacao (*Marasmius perniciosus* Stahel). *Commonwealth Mycol. Inst. Phytopathol. Paper 2.* 42 p.

8. Baker, R. E. D. and R. K. McKee. 1943. Witches' broom disease investigations. VI. The infection of flower cushions and pods of cacao by *Marasmius perniciosus* Stahel. *Trop. Agr. (Trinidad) 20:* 188–194.

9. Briton-Jones, H. R. 1929. Witch broom disease. *Trop. Agr. (Trinidad) 6:* 20–22.

10. Briton-Jones, H. R. and E. E. Cheesman. 1931. Witch broom control. *Trop. Agr. (Trinidad) 8:* 79–89.

11. Dale, W. T. 1946. Witches' broom disease investigations. XII. Further studies on the infection of cacao pods. *Trop. Agr. (Trinidad) 23:* 217–221.

12. Delgado, J. C. and A. A. Cook. 1976. Nuclear condition of the basidia, basidiospores, and mycelium of *Marasmius perniciosus. Can. J. Bot. 54:* 66–72.

13. Delgado, J. C. and A. A. Cook. 1976. Arthroconidial formation in cultures of *Marasmius perniciosus* Stahel. *Phytopathology 66:* 717–718.

14. Desrosiers, R. and C. W. Bolaños. 1955. Inhibition of spore development in *Marasmius perniciosus* Stahel on cacao. *Turrialba 5:* 28–32.

15. Desrosiers, R., C. W. Bolaños and J. Vargas. 1955. Evaluación de clones de cacao en relación con su resistancia a la escoba de bruja (*Marasmius perniciosus* Stahel). *Turrialba 5:* 78–82.

16. Desrosiers, R., A. von Buchwald and C. W. Bolaños. 1955. Selección en masa de plántulas de cacao con base a su resistancia a la escoba de bruja. *Turrialba 5:* 134–138.

17. Freeman, W. G. 1928. Witch broom in Trinidad. *Trop. Agr. (Trinidad) 5:* 287–288.

18. Garcia, C. 1956. La escoba de bruja en Colombia. *Cacao Colombia 5:* 19–23.

19. Goncalves, J. R. C. 1965. *Theobroma grandiflorum* (Spreng.) Schum as a source of inoculum of witch-broom disease of *Theobroma cacao* L. *Trop. Agr. (Trinidad) 42:* 261–263.

20. Hardy, F. 1960. *Cacao manual.* Inter-Am. Inst. Agr. Sci., Turrialba. 395 p.

21. Holliday, P. 1954. Control of witches' broom disease of cacao in Trinidad. *Trop. Agr. (Trinidad) 31:* 312–317.

22. Holliday, P. 1970. *Crinipellis perniciosa. Commonwealth Mycol. Inst. Descriptions Pathogenic Fungi Bacteria 223.* 2 p.

23. Lindberg, G. and K. Molin. 1949. Notes on the physiology of the cocoa parasite *Marasmius perniciosus. Physiol. Plant 2:* 138–144.

24. Montserin, B. G. 1932. Contributions to the knowledge of witch broom disease. *Trop. Agr. (Trinidad) 9:* 167–176.

25. Pound, F. J. 1940. Witches' broom resistance in cacao. *Trop. Agr. (Trinidad) 17:* 6–8.

26. Pound, F. J. 1944. La escoba de bruja del cacao. *Hacienda 30:* 158–159.

27. Rorer, J. B. 1910. The Surinam witch-broom disease of cacao. *Trinidad Tobago Dept. Agr. Circ. 10.* 13 p.

28. Singer, R. 1942. A monographic study of the genera *Crinipellis* and *Chaetocalathus. Lilloa 8:* 441–534.

29. Stell, F. 1928. Witch broom disease of cacao and its control. *Bull. Dept. Agr. Trinidad Tobago 21:* 3–14.

30. Thorold, C. A. 1943. Witches' broom disease investigations. V. Large scale experiments on direct control. *Trop. Agr. (Trinidad) 20:* 176–181.

31. Thorold, C. A. 1953. Observations on fungicide control of witches' broom, black pod, and pink disease of *Theobroma cacao* L. *Ann. Appl. Biol. 40:* 362–376.

32. Tollenaar, D. 1959. Increased cocoa production by foliar copper applications as an effect additional to witches' broom disease control. *Trop. Agr. (Trinidad) 36:* 177–188.

33. Wardlaw, C. W. 1929. Witch broom in Surinam. *Trop. Agr. (Trinidad) 6:* 348–349.

Swollen shoot

A disease of cacao that included swellings on branches and dieback symptoms was observed and described from the Gold Coast (Ghana) in 1936. Leaf symptoms similar to those induced by swollen shoot were noted in Trinidad in 1943, and different but still typical leaf symptoms were observed in Ceylon 10 years later. The disease is now known to occur in the important cacao areas of West Africa, including Nigeria, Ghana, Togoland, Ivory Coast and Sierra Leone but is not reported to occur in the Caribbean area or South America. (Transmission of cacao viruses by mealybugs in Trinidad has been reported.) Early losses to the disease were quite heavy, and, despite concerted efforts to control the disease, it has been estimated that annual losses

Fig. 29. Swollen shoot of cacao. (Courtesy of R. G. Orellana, U. S. Department of Agriculture, Beltsville.)

in eastern Ghana alone still exceed 18,000 metric tons. Consequently, swollen shoot remains the major disease of cacao in Africa.

Symptoms of the disease when it was first discovered were not the same in the different areas of Ghana where it occurred. Accordingly, letter designations were applied to these symptom types, but later letter and name designations were applied to still different syndromes observed in Trinidad.

Strain A (New Juaben or virulent strain) of the disease, which occurs in Ghana, is characterized by swellings that develop on fan and chupon branches as well as on roots (Fig. 29). These swellings form anywhere along the branch, including nodes, but often appear terminal, because growth is severely retarded in that portion of the branch distal to the swelling, and the apical bud is often killed. Leaf symptoms consist of vein clearing and banding in the smallest veins and opaque, yellowish lesions that may become necrotic as the leaf matures. Symptoms are seen only in leaves formed after infection, which also may develop a fine mosaic or "fern pattern" (see "oak leaf" reference below in this section) that may vary in intensity. Vein banding seen in leaves just after emergence from the bud may be reddish, which color fades as the leaf develops green coloration. Developing pods assume a dark green mottle with red blotching later and become more smooth and rounded than usual. Pods that are yellow when mature retain a green mottle. Affected pods are smaller, the beans weigh considerably less than those from healthy fruits and are flatter and lighter in color than normal. Diseased trees are noticeably less vigorous, and leaves with symptoms fall prematurely. Leaves from successive flushes are progressively smaller, and dieback begins as the tree continually weakens. Some axillary bud growth may be stimulated but soon aborted before the tree is killed within 2 years or less of infection. An attenuated form of strain A has been described which causes infrequent, small internodal swellings on fan branches but inconspicuous leaf or pod symptoms and no evident reduction in growth or yield.

The swellings caused by strain B (Bisa or mild strain) on branches and roots often are more elongated than those caused by strain A. Pod and leaf symptoms are absent in natural infections, although chlorosis may appear in the second or third leaf flush only after experimental inoculation. Neither yield nor growth is affected.

Swellings on branches, which may appear infrequently between nodes near the ends of fan branches, are not typical of strain D (Nkawkaw strain), which is distinguished by a fine mosaic on leaves. Small yellow spots seen as first symptoms are succeeded in later flushes by a uniform mosaic of yellow spots and transparent flecks which may later become necrotic. Pod symptoms are absent, but leaf size is reduced and growth retarded.

The "red-mottle" type of the disease as it occurs in Trinidad is characterized by a reddish banding of a few to several main leaf veins that may fade as the leaf matures. Some form of mosaic frequently accompanies vein banding, but these symptoms are not found on all leaves of a single flush. Red blotches on pods are found on cacao types with normally yellow fruits, and affected trees may develop dieback.

A similar but distinguishable form of the disease known as the "vein clearing" type also has been described from Trinidad and is notable for persistence of pronounced vein clearing even of small veins in leaves after they reach maturity.

Both stem swellings and mottle or "oak leaf" patterns along major leaf veins and vein clearing have been observed on cacao in Ceylon.

Transmission of the swollen shoot disease by grafting and by insects (mealybugs) was the basis for assuming the causal agents of the various syndromes to be virus strains until mechanical transmission was accomplished in 1960. Partial purification of 5 virus preparations revealed particles with an average dimension of 28 × 121 nm that were rendered noninfectious by exposure to 45–50 C for 10 minutes or dilution to 10^{-2}. Longevity *in vitro* is greater than 96 hours. Transmission of the virus types in Africa has been accomplished by mealybugs, including *Planococcoides njalensis* Laing, *Planococcus citri* Risso and *Ferrisia virgata* Cockerell. Host range of swollen shoot disease includes *Adansonia digitata* L., *Bombax brevicuspe* Sprague, and more than 20 other species in the *Tiliales* but apparently is confined to families in this order.

All of the manifestations included in the swollen shoot disease complex of cacao are virus-induced (*Theobroma* virus 1), but exact relations between the various components have not been established. Cross-protection between the virus strains is not uniformly consistent, but absence of synergistic effect from double infections is indicative of some relation. The irregular growth pattern of cacao characterized by flushes of growth serves to complicate accurate timing of incubation periods for comparison purposes. Incubation has been estimated to persist for as long as a year. Symptoms of the different disease manifestations in other than identical plant types have precluded ideal comparisons, but successful purification procedures may permit determination of more definitive strain relations by serological study.

Control of swollen shoot has proven exceptionally difficult and has been abandoned altogether in some areas of severe infection. Vector control and roguing of infected and all neighboring trees within a given distance have been either ineffective as a control or too deleterious to yield to be economic. No cacao types immune to infection have been found, but some forms with apparent resistance to field infection have been identified, and further selection within crosses among these types offers promise for satisfactory control. Until resistant selections are available, only frequent inspection and roguing of infected trees can be recommended.

1. Adegbola, M. O. K. 1972. Peculiarity of a Nigerian isolate of the cocoa swollen shoot virus (CSSV). *Plant Dis. Reptr. 56:* 755–758.

2. Adomako, D. and G. K. Owusu. 1974. Studies on the mechanical transmission of cocoa swollen shoot virus: some factors affecting virus multiplication and symptom development in cocoa. *Ghana J. Agr. Sci. 7:* 7–15.

3. Albert, H. 1946. Note preliminaire sur une nouvelle maladie du cacaoyer le "swollen shoot." *Agron. Trop. (Paris) 1:* 34–43.

4. Attafuah, A., J. W. Blencowe and A. A. Brunt. 1963. Swollen-shoot disease of cocoa in Sierra Leone. *Trop. Agr. (Trinidad) 40:* 229–232.

5. Attafuah, A. and D. R. Glendinning. 1965. Studies on resistance and tolerance to cocoa viruses in Ghana. I. A survey of the T17 progeny. *Ann. Appl. Biol. 56:* 219–225.

6. Attafuah, A. and T. W. Tinsley. 1958. Virus diseases of *Adansonia digitata* L. *Bombacaceae*) and their relation to cacao in Ghana. *Ann. Appl. Biol. 46:* 20–22.

7. Baker, R. E. D. and W. T. Dale. 1947. Virus diseases of cacao in Trinidad. II. *Trop. Agr. (Trinidad) 24:* 127–130.

8. Baker, R. E. D. and W. T. Dale. 1947. Notes on a virus disease of cacao. *Ann. Appl. Biol. 34:* 60–65.

9. Bigger, M. 1972. Recent work on the mealybug vectors of cocoa swollen shoot disease in Ghana. *PANS 18:* 61–70.

10. Box, H. E. 1945. Insect transmission of the "swollen shoot" virus in West African cocoa. *Nature 153:* 608–609.

11. Brunt, A. A. 1970. Cacao swollen shoot virus. *Commonwealth Mycol. Inst. Descriptions Plant Viruses 10.* 4 p.

12. Brunt, A. A. 1975. The effects of cocoa swollen-shoot virus on the growth and yield of Amelonado and Amazon cocoa (*Theobroma cacao*) in Ghana. *Ann. Appl. Biol. 80:* 169–180.

13. Brunt, A. A. and R. H. Kenten. 1960. Mechanical transmission of cocoa swollen-shoot virus. *Virology 12:* 328–330.

14. Brunt, A. A. and R. H. Kenten. 1962. Mechanical transmission of cocoa mottle leaf

virus and three distinct isolates of the cocoa swollen-shoot virus. *Virology 16:* 199–201.

15. Brunt, A. A. and R. H. Kenten. 1962. Mechanical transmission of cocoa swollen-shoot virus to and from cocoa and other hosts. *Ann. Appl. Biol. 50:* 749–754.

16. Brunt, A. A. and R. H. Kenten. 1963. The use of protein in the extraction of cocoa swollen-shoot virus from cocoa leaves. *Virology 19:* 388–392.

17. Brunt, A. A., R. H. Kenten and H. L. Nixon. 1964. Some properties of cocoa swollen-shoot virus. *J. Gen. Microbiol. 36:* 303–309.

18. Cornwell, P. B. 1956. Some aspects of mealybug behaviour in relation to the efficiency of measures for the control of virus diseases of cacao in the Gold Coast. *Bull. Entomol. Res. 47:* 137–166.

19. Crowdy, S. H. and A, F. Posnette. 1947. Virus diseases of cacao in West Africa. II. Cross-immunity experiments with virus 1A, 1B and 1C. *Ann. Appl. Biol. 34:* 403–411.

20. Glendinning, D. R., J. T. Legg, N. K. Lovi and V. A. Martinson. 1966. A field experiment in Ghana on the tolerance of cocoa seedlings to cocoa swollen-shoot and cocoa mottle-leaf viruses. *Ann. Appl. Biol. 57:* 389–396.

21. Hancock, B. L. 1949. A laboratory colour test for the diagnosis of swollen shoot of *Theobroma cacao. Trop. Agr. (Trinidad) 26:* 54–56.

22. Hanna, A. D. and W. Heatherington. 1957. Arrest of the swollen-shoot virus disease of cacao in the Gold Coast by controlling the mealybug vectors with the systemic insecticide, dimefox. *Ann. Appl. Biol. 45:* 473–480.

23. Kenten, R. H. and J. T. Legg. 1971. Varietal resistance of cocoa to swollen shoot disease in West Africa. *FAO Plant Protect. Bull. 19:* 1–12.

24. Kirkpatrick, T. W. 1950. Insect transmission of cacao virus disease in Trinidad. *Bull. Entomol. Res. 41:* 99–117.

25. Knight, R. and T. W. Tinsley. 1958. Some histological observations on virus-infected *Theobroma cacao* L. *Ann. Appl. Biol. 46:* 7–10.

26. Legg, J. T. 1972. Measures to control spread of cocoa swollen shoot disease in Ghana. *PANS 18:* 57–60.

27. Lockard, R. G. and E. J. A. Asomaning. 1965. Mineral nutrition of cacao *(Theobroma cacao* L.). II. Effects of swollen shoot virus on the growth and nutrient content of plants grown under nutrient deficient, excess and control conditions in sand culture. *Trop. Agr. (Trinidad) 42:* 39–53.

28. Lockwood, G. 1973. Effects of Ghanaian cocoa swollen-shoot virus on young bearing trees of a range of varieties. *Expt. Agr. 9:* 31–39.

29. Longworth, J. F. and J. M. Thresh. 1963. The reaction of different cacao types to infection with swollen-shoot virus. *Ann. Appl. Biol. 52:* 117–124.

30. Meiffren, M. 1949. Swollen shoot, maladie du cacaoyer. *Agron. Trop. 4:* 563–578.

31. Nicol, J., H. Owen and A. H. Strickland. 1950. Biological control of the mealybug vectors of swollen shoot virus of cacao. *Nature 165:* 490.

32. Orellana, R. G. and J. W. L. Peiris. 1957. The swollen shoot phase of the virus disease of cacao in Ceylon. *FAO Plant Protect. Bull. 5:* 165–168.

33. Peiris, J. W. L. 1953. A virus disease of cacao in Ceylon. *Trop. Agr. (Ceylon) 109:* 135–138.

34. Posnette, A. F. 1940. Transmission of "swollen shoot" disease of cacao. *Trop. Agr. (Trinidad) 17:* 98.

35. Posnette, A. F. 1941. Swollen-shoot virus disease of cacao. *Trop. Agr. (Trinidad) 18:* 87–90.

36. Posnette, A. F. 1943. Control measures against swollen shoot virus disease of cacao. *Trop. Agr. (Trinidad) 20:* 116–123.

37. Posnette, A. F. 1944. The diagnosis of swollen-shoot disease of cacao. *Trop. Agr. (Trinidad) 21:* 56–58.

38. Posnette, A. F. 1944. Virus diseases of cacao in Trinidad. *Trop. Agr. (Trinidad) 21:* 105–106.

39. Posnette, A. F. 1947. Virus diseases of cacao in West Africa. I. Cacao viruses 1A, 1B, 1C and 1D. *Ann. Appl. Biol. 34:* 388–402.

40. Posnette, A. F. 1950. Virus diseases of cacao in West Africa. VII. Virus transmission by different vector species. *Ann. Appl. Biol. 37:* 378–384.

41. Posnette, A. F., N. F. Robertson and J. M. Todd. 1950. Virus diseases of cacao in West Africa. V. Alternative host plants. *Ann. Appl. Biol. 37:* 229–240.

42. Posnette, A. F. and J. M. Todd. 1951. Virus diseases of cacao in West Africa. VIII. The search for virus-resistant cacao. *Ann. Appl. Biol. 38:* 785–800.

43. Posnette, A. F. and J. M. Todd. 1955. Virus diseases of cacao in West Africa. IX. Strain variation and interference in virus 1A. *Ann. Appl. Biol. 43:* 433–453.

44. Rojter, S., J. K. Bonney and J. T. Legg. 1966. Investigations into the use of a pathogenic fungus (*Cephalosporium* sp.) as a means of controlling the mealybug (*Pseudococcidae*) vectors of swollen shoot virus in Ghana. *Ghana J. Sci. 6:* 110–114.

45. Thresh, J. M. 1959. The control of cacao swollen shoot disease in Nigeria. *Trop. Agr. (Trinidad) 36:* 35–44.

46. Thresh, J. M. 1961. Some isolates of virus causing swollen-shoot disease of cacao in Nigeria and their interrelationships. *Ann. Appl. Biol. 49:* 340–346.

47. Thresh, J. M. and R. M. Lister. 1960. Coppicing experiments on the spread and control of cacao swollen-shoot disease in Nigeria. *Ann. Appl. Biol. 48:* 65–74.

48. Thresh, J. M. and T. W. Tinsley. 1960. The virus complex causing swollen shoot disease of cacao in West Africa. *FAO Plant Protect. Bull. 8:* 65–69.

49. Tinsley, T. W. and A. L. Wharton. 1958. Studies on the host ranges of viruses from *Theobroma cacao* L. *Ann. Appl. Biol. 46:* 1–6.

50. Voelcker, O. J. and J. West. 1942. Swollen shoot and die-back of cacao. *Trop. Agr. (Trinidad) 19:* 83.

51. Wessel-Riemens, P. C. 1965. The use of proteins and alkaloids in mechanical transmission of Nigerian isolates of cocoa swollen-shoot virus. *Virology 27:* 566–570.

Mottle leaf

One of the diseases with distinctive symptoms (Kpeve isolate) originally included among the variant syndrome forms caused by isolates of *Theobroma* virus 1 was designated "cacao mottle leaf" in 1947. This particular disease was first observed near Kpeve (Togoland), Ghana, but a disease with many similar and some dissimilar characteristics has been found in Nigeria (Alaparun disease). The apparent restriction of the disease to the Ghana-Nigeria area of Africa and relatively infrequent occurrence there make mottle leaf of little economic import now.

Leaf symptoms with no necrosis are prominent in both acute and chronic stages of infection, but there are no swellings on roots or stems. Early vein clearing is followed by conspicuous vein banding with red coloration in flush leaves. The mottle also seen on these leaves may persist as clearing and banding associated only with smaller leaf veins between the larger veins, which contributes to a characteristic chlorotic mottle on mature leaves.

Mottle leaf disease is caused by a virus which is transmissible mechanically by rubbing and by mealybugs, not including *Ferrisia virgata*. Mechanical transmission rates achieved have been relatively low despite infiltration of infected leaves used for inoculum with reducing agents prior to maceration. Host range of mottle leaf disease is apparently restricted to the *Tiliales* and *Malvales*, including *Hildegardia barteri* (Mast.) Kosterm., *Theobroma bicolor* Humb. & Bonpl., *Adansonia digitata* and *Bombax brevicuspe*. The causal virus is unrelated to other viruses infectious to cacao based on cross-protection tests, although minor differences have been detected in isolates of mottle leaf from different hosts.

Symptoms of mottle leaf appeared earlier than those of swollen shoot in graft-inoculated trees in Ghana. Both growth and first-crop yield were reduced, but infected trees were not killed.

No satisfactory control has been devised.

1. Brunt, A. A. and R. H. Kenten. 1962. Mechanical transmission of cocoa mottle leaf virus and three distinct isolates of the cocoa swollen-shoot virus. *Virology 16:* 199–201.

2. Glendinning, D. R., J. T. Legg, N. K. Lovi and V. A. Martinson. 1966. A field experiment in Ghana on the tolerance of cocoa seedlings to cocoa swollen-shoot and cocoa mottle-leaf viruses. *Ann. Appl. Biol. 57:* 389–396.

3. Posnette, A. F. 1947. The use of seeds in the insect transmission of some plant viruses. *Nature 159:* 500–501.

4. Thresh, J. M. and T. W. Tinsley. 1959. The viruses of cacao. *West Afr. Cocoa Res. Inst. Tech. Bull.* 7. 32 p.

Cocoa necrosis (virus) disease

Necrosis of small leaf veins in spots that enlarge to form irregular necrotic areas and eventually involve the entire leaf laminae (Fig. 30) characterizes the necrosis disease of cocoa, first noted in 1948 and now known to occur only in Nigeria and Ghana. Shoot tips of infected seedlings are killed, as are axillary buds subsequently produced. The disease has been transmitted by grafting and by mechanical inoculation of whole seeds with a partially purified preparation infectious at 10^{-3} but not 10^{-4}, but neither *Planococcoides njalensis* nor *Ferrisia virgata* transmit the infection. Eradication of infected plants has provided effective control.

1. Lister, R. M. and J. M. Thresh. 1957. Virus research. Nigerian isolates of cacao viruses. *Repub. West Afr. Cocoa Res. Inst. 1955–56:* 81.

2. Martini, C. H. K. 1960. Virus research. Cocoa necrosis virus. *Repub. West Afr. Cocoa Res. Inst. 1958–59:* 67–68.

3. Owusu, G. K. 1971. Cocoa necrosis virus in Ghana. *Trop. Agr. (Trinidad) 48:* 133–139.

4. Thresh, J. M. 1958. Virus research. Nigerian isolates of cacao viruses. *Repub. West Afr. Cocoa Res. Inst. 1956–57:* 71–73.

Fig. 30. Cocoa necrosis virus disease symptom on leaves (malformation on left and transparent lesions along major veins, middle) and hexagonal virus particles (24–28 nm diameter). Occasional particles 11–12 nm are found, see arrow. Bar = 100 nm. (Courtesy of G. K. Owusu, Cocoa Research Institute, Tafo.)

Yellow mosaic

Plants with distinctive, virus-like symptoms were observed in Sierra Leone in 1958, and studies reported in 1963–65 established yellow mosaic as an identifiable disease of cacao. It remains a disease of minor importance in the field and presently is restricted to Sierra Leone.

Symptoms on infected plants consist only of chlorotic blotches or mosaic on the leaves with vein necrosis sometimes apparent in the chlorotic areas. Some plants may become completely chlorotic, but growth of diseased plants is not noticeably affected.

Particles of the causal virus measure about 25 nm in diameter and leaf extracts are rendered non-infectious by exposure to 65 C for 10 minutes or storage at room temperature (20–30 C) for more than 16 days, while infectivity may persist for more than 75 days following storage at 0–4 C. The virus, which is serologically related to wild cucumber mosaic virus, is readily transmitted by sap inoculation to species in several plant families, including *Apocynaceae* (*Vinca rosea* L.), *Bombacaceae* [*Adansonia digitata* and *Ceiba pentandra* (L.) Gaertn.], *Chenopodiaceae* (*Chenopodium amaranticolor* and *C. quinoa*), *Cucurbitaceae* (*Cucurbita melo-pepo* L., cucumber and *Momordica charantia* L.), *Leguminosae* (cowpea), *Solanaceae* (*Nicotiana clevelandii*) and *Sterculiaceae* (*Cola laterita* K. Schum., *Sterculia foetida* and *Theobroma* sp.). The virus has been recovered from naturally infected plants of *Cola nitida* (Vent.) Schott & Endl. and *Culcasia scandens* P. Beauv. but is not transmitted by mealybugs.

1. Blencowe, J. W., A. A. Brunt, R. H. Kenten and N. K. Lovi. 1963. A new virus disease of cocoa in Sierra Leone. *Trop. Agr. (Trinidad) 40:* 233–236.

2. Brunt, A. A. 1970. Cacao yellow mosaic virus. *Commonwealth Mycol. Inst. Descriptions Plant Viruses 11.* 4 p.

3. Brunt, A. A., R. H. Kenten, A. J. Gibbs and H. L. Nixon. 1965. Further studies on cocoa yellow mosaic virus. *J. Gen. Microbiol. 38:* 81–90.

Cherelle wilt

Attention was first given to the condition in cacao known as cherelle wilt in Trinidad some time after 1930 and has continued sporadically because of its general occurrence wherever cacao is grown. The effect on yield has not been accurately assessed because no control measure(s) have been established, but experiments involving simulated flower/fruit removal did not result in serious influence on production.

Cessation of growth followed rapidly by yellowing, desiccation and necrosis (blackening) of young cacao fruits up to about 5 cm in length has been termed "cherelle wilt." Fruits 50–70 days old are particularly susceptible, usually mummify and often remain hanging but may eventually fall.

Various theories have been advanced concerning the causal nature of cherelle wilt, including physiological irregularities, unfavorable environmental conditions and deleterious genetic relations, but none has provided a completely satisfactory explanation of cause. Consequently, control measures remain to be established.

1. Alvim, P. de T. 1954. Studies on the cause of cherelle wilt of cacao. *Turrialba 4:* 72–78.

2. Alvim, P. de T. 1967. Causas del marchitamiento prematuro de los frutos del cacao. *Vida Agr. 44:* 421–424.

3. Bartolome, R. 1954. A study on the effect of fertilizer application on the incidence of cherelle wilt of cacao. *Philippine J. Agr. 17:* 55–88.

4. Hardy, F. 1960. *Cacao manual.* Inter-Am. Inst. Agr. Sci., Turrialba. 395 p.

5. Humphries, E. C. 1943. Wilt of cacao fruits (*Theobroma cacao*). I. An investigation into the causes. *Ann. Bot., n. s. 7:* 31–44.

6. Jaramillo, R. A. 1953. La urea en el marchitamiento de las frutas jovenes de cacao. *Cacao Colombia 2:* 63–76.

7. Krekule, J. 1969. Cherelle wilting in cocoa induced by TIBA. *Trop. Agr. (Trinidad) 46:* 69–72.

8. McKelvie, A. D. 1956. Cherelle wilt of cacao. I. Pod development and its relation to wilt. *J. Expt. Bot. 7:* 252–263.

9. McKelvie, A. D. 1960. Cherelle wilt of cacao. II. Wilt in relation to yield. *J. Expt. Bot. 11:* 413–424.

10. Nichols, R. 1965. Studies of fruit development of cacao (*Theobroma cacao*) in relation to cherelle wilt. *Ann. Bot., n.s. 29:* 197–203.

11. Thrower, L. B. 1960. Observations on the diseases of cacao pods in Papua-New Guinea. II. Cherelle wilt. *Trop. Agr. (Trinidad) 37:* 121–124.

Mineral deficiency symptoms

Young leaves on plants deficient in *nitrogen* are smaller in size and somewhat lighter in color, but loss of green coloration and yellowing are most readily evident in older leaves. Some light brown necrosis, beginning at the leaf apex and spreading toward the base, may appear and slowly enlarge before severly affected leaves fall. The angles formed by leaf petioles and stems of deficient plants are acute and plant growth noticeably retarded.

Phosphorus deficiency results in smaller plants and leaves which are light green in color with irregular yellowish areas. Severe deficiency is characterized by unilateral development of necrosis within the yellow areas. The blotchy coloration extends to involve entire leaves with islands that remain green as the leaves later turn brown.

Deficiency of *potassium* causes older leaves to develop yellow coloration between secondary veins that may not extend completely to the leaf margin. Necrosis usually follows chlorosis and also develops from the midrib outward in interveinal areas and from the leaf base toward the apical end.

Affected leaves drop soon after necrosis becomes complete, and defoliation may be extensive.

Leaves on *calcium*-deficient plants remain small and often become distorted at the apical ends. Interveinal areas of immature leaves on such plants assume a light green coloration in which develop small, brown, necrotic spots. Entire leaves become brown and necrotic later with interveinal areas thin and fragile. Severe deficiency results in shoot dieback and stunting of plants.

Dark green vein banding, first apparent in the oldest leaves, characterizes *magnesium* deficiency. The width of the dark green banding is narrower along smaller veins, and interveinal tissues gradually assume a lighter green coloration in which develop brown, necrotic spots along the leaf margins near the apical ends of affected leaves. The chlorosis extends, and necrotic spots enlarge gradually to involve entire leaves.

A general yellow-white chlorosis of young leaves with veins that remain green typifies *iron* deficiency.

Mature leaves on plants deficient in *manganese* exhibit interveinal chlorosis, while young leaves may curl downward and develop light colored spots that become necrotic along the veins. Interveinal tissues later turn brown and become necrotic.

Zinc deficiency causes narrowing of laminae, particularly near the leaf bases, and twisting and distortion of young leaves. Coloration of laminae changes to light green or white but veins remain dark green.

Leaves on plants deficient in *boron* curl downward and the leaf edges roll under. No major color change is apparent, although young leaves may be lighter green than normal. Leaf texture becomes tough and leathery, and the leaves remain attached to the plant.

Immature leaves become necrotic and drop from plants deficient in *copper*, and new leaves fail to develop as the condition progresses down the shoot. Internodes become shortened, and lateral buds arise from loss of apical dominance, but they, too, fail to develop. One or more stem swellings may develop in the area from which leaves drop.

Molybdenum deficiency appears as a mild interveinal chlorosis of older leaves accompanied by a marginal necrosis that originates at leaf apices. Veins on affected leaves remain green, and leaves remain attached for some time but eventually fall.

1. Acquaye, D. K., R. W. Smith and R. G. Lockard. 1965. Potassium deficiency in unshaded Amazon cocoa (*Theobroma cacao* L.) in Ghana. *J. Hort. Sci. 40:* 100–108.

2. Ahenkorah, Y. 1969. A note on zinc deficiency in cacao (*Theobroma cacao* L.). *Ghana J. Agr. Sci. 2:* 3–6.

3. Greenwood, M. and R. K. Djokoto. 1952. Symptoms of mineral deficiency in cacao. *J. Hort. Sci. 27:* 223–236.

4. Greenwood, M. and R. J. Hayfron. 1951. Iron and zinc deficiencies in cacao in the

Gold Coast. *Empire J. Expt. Agr. 9:* 73–86.

5. Lockard, R. G. and E. J. A. Asomaning. 1964. Mineral nutrition of cacao (*Theobroma cacao* L.). I. Deficiency symptoms and nutrient levels in plants grown in sand culture. *Plant Soil 21:* 142–152.

6. Lockard, R. G. and E. J. A. Asomaning. 1965. Mineral nutrition of cacao (*Theobroma cacao* L.). IV. Effects of nitrate and urea nitrogen on deficiency symptoms of phosphorus, potassium, calcium and magnesium in sand culture. *Trop. Agr. (Trinidad) 42:* 55–62.

7. Lockard, R. G., P. Vamathevan and S. Thamboo. 1959. Mineral deficiency symptoms of cacao grown in sand-culture. *Malaya Dept. Agr. Bull. 107.* 20 p.

8. Lockard, R. G., P. Vamathaven and S. Thamboo. 1959. Swellings on the shoots of cacao plants deficient in copper. *Nature 184:* 75–76.

9. Machiado, M. and P. de T. Alvim. 1954. Sintomatologia de las deficiencias minerales de cacao. *Turrialba 4:* 155–163.

10. Maskell, E. J., H. Evans and D. B. Murray. 1953. The symptoms of nutritional deficiencies in cacao produced in sand and water cultures. *Rep. Cacao Res. Trinidad, 1949–51:* 53–64.

11. Schroo, H. 1959. Acute zinc deficiency observed in cacao on certain soil types in Netherlands New Guinea. *Neth. J. Agr. Sci. 7:* 309–316.

OTHER DISEASES

Mealypod (*Trachysphaera fructigena* Tabor & Bunting)

Pod lesions are very similar to black pod symptoms (Fig 31).

1. Dakwa, J. T. 1972. Occurrence of mealypod disease of *Theobroma cacao* in Ghana. *Plant Dis. Reptr. 56:* 1011–1013.

2. Holliday, P. 1970. *Trachysphaera fructigena. Commonwealth Mycol. Inst. Descriptions Pathogenic Fungi Bacteria 229.* 2 p.

3. Tabor, R. J. and R. H. Bunting. 1923. On a disease of cocoa and coffee fruits caused by a fungus hitherto undescribed. *Ann. Bot. 37:* 153–157.

Ceratocystis pod rot [*Ceratocystis paradoxa* (Dade) C. Moreau]

Soft, brown lesions rapidly enlarge to result in complete destruction of pods. Infection occurs only through deep wounds on nearly mature pods in Ecuador and Costa Rica.

1. Desrosiers, R. 1955. *Thielaviopsis* pod rot of cacao in Ecuador. *FAO Plant Protect. Bull. 3:* 154–155.

2. Hardy, F. 1960. *Cacao manual.* Inter-Am. Inst. Agr. Sci., Turrialba. 395 p.

3. Morgan-Jones, G. 1967. *Ceratocystis paradoxa. Commonwealth Mycol. Inst. Descriptions Pathogenic Fungi Bacteria 143.* 2 p.

Fig. 31. Mealypod symptoms on cacao fruit. (Courtesy of J. T. Dakwa, Cocoa Research Institute, Tafo.)

Thread blight (Mal de hilachas) [*Ceratobasidium koleroga* (Cke.) Venkat.]

Defoliation and twig dieback with dead leaves held to the tree by strong, dark mycelial threads are characteristic. Foliar applications of Bordeaux mixture (4–4–50) or Crag 531 fungicide provide effective control.

1. Orellana, R. and C. Bianchini. 1961. Control con fungicidas de *Pellicularia* "mal de hilachas" en cacao. *Cacaotero 3:* 11–12.

Pink Disease [*Botryobasidium salmonicolor* (Berk. & Br.) Venkat.]

Pink disease is similar to "thread blight" except that leaves fall and fungus strands are lighter in color. (See Cacao—Thread blight for control.)

1. Thorold, C. A. 1953. Observations on fungicide control of witches' broom, black pod and pink disease of *Theobroma cacao*. *Ann. Appl. Biol. 40:* 362–376.

Root rots

White root rot (*Fomes lignosus* Klotz)

Brown root rot (*Fomes noxius* Corner)

Black root rot [*Rosellinia bunodes* (Berk. & Br.) Sacc.] (Fig. 32)

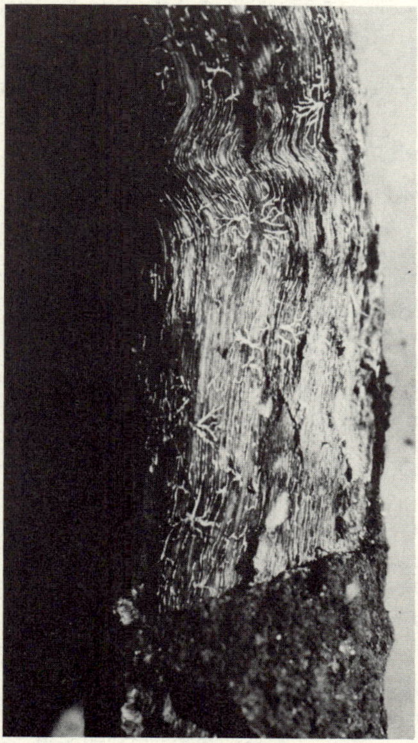

Fig. 32. White mycelial fans formed under the cortex in root rot of cacao caused by *Rosellinia* sp. (Courtesy of J. Lawrence, Centro Agronomico Tropical de Investigacion y Enseñanza, Turrialba.)

Wet root rot [*Ganoderma pseudoferreum* (Wakef.) Ov. & Stein]

Dry root rot [*Ustilina deusta* (Hoffm. ex Fr.) Petr.]

Infection of roots and subsequently trunks to a height of 0.3–1 m usually originates from rhizomorphs from diseased roots of neighboring or dead (and removed) trees. Identification of the pathogen involved is achieved by symptoms produced or, more commonly, by fructifications on trunk or stump of the dead tree. Complete removal of diseased plant material, i.e. stump, trunk and roots, before replanting reduces probability of reinfection.

1. Corner, E. J. H. 1932. The identification of the brown-rot fungus. *Gardners' Bull. 5:* 317–350.

2. Dadant, R. 1954. Contribution à l'étude des maladies du cocotier, du cacaoyer et du cafeier aux Nouvelles-Hébrides. *Agron. Trop. (Nogent) 9:* 41–48.

3. Hardy, F. 1960. *Cacao manual.* Inter-Am. Inst. Agr. Sci., Turrialba. 395 p.

4. Sivanesan, A. and P. Holliday. 1972. *Rosellinia bunodes. Commonwealth Mycol. Inst. Descriptions Pathogenic Fungi Bacteria 351.* 2 p.

5. Thrower, L. B. 1955. The root diseases of cacao in Papua and New Guinea. *Papua New Guinea Agr. J. 10:* 1–14.

6. Thrower, L. B. 1965. Parasitism of cacao by *Fomes noxius* in Papua-New Guinea. *Trop. Agr. (Trinidad) 42:* 63–67.

Charcoal rot (*Diplodia theobromae* Nowell)

1. Desrosiers, R. and A. von Buchwald. 1955. Fungicidal treatments for control of caco cutting disease in Ecuador. *FAO Plant Protect. Bull. 4:* 17–21.

Anthracnose [Colletotrichum gloeosporioides (Penz.) Sacc.] (Fig. 33)

1. Hardy, F. 1960. *Cacao manual.* Inter-Am. Inst. Agr. Sci., Turrialba. 395 p.

2. Mordue, J. E. M. 1971. *Glomerella cingulata. Commonwealth Mycol. Inst. Descriptions Pathogenic Fungi Bacteria 315.* 2 p.

3. Orellana, R. G. 1956. Cacao diseases in Haiti. *FAO Plant Protect. Bull. 4:* 148–151.

Mata-cacau (Cacao-killer)

Parasitism of cacao roots by *Acanthosyris pauloalvimii* Barr. (*Santalaceae*) has been reported in Brazil.

1. Alvim, P. de T. and K. W. Seeschaaf. 1968. Die-back and death of cacao trees caused by a new species of parasitic tree. *Nature 219:* 1386–1387.

Fig. 33. Anthracnose lesions on cacao pod. (Courtesy of J. Lawrence, Centro Agronomico Tropical de Investigacion y Enseñanza, Turrialba.)

Vascular-streak dieback

Chlorosis with islands of green color develops in a single leaf which falls in 2–3 days and is followed by dropping of successive leaves both up and down the branch or seedling until the growing tip is reached and dies (Fig. 34). Axillary bud growth is often stimulated by the causal fungus *Oncobasidium theobromae* Talbot & Keane which also causes brown streaks in the xylem of affected shoots.

1. Bridgeland, L. A., J. M. Richardson and I. L. Edward. 1966/1967. Dieback diseases of cacao. I. II. III. *S. Pacific Plants 1:* 13–20, 28; 3–6, 22; 9–11, 16–17.

2. Keane, P. J., N. T. Flentje and K. P. Lamb. 1972. Investigations of vascular-streak dieback of cocoa in Papua New Guinea. *Australian J. Biol. Sci. 25:* 553–564.

3. Shaw, D. E. 1962. Diseases of cacao in Papua and New Guinea. *Papua New Guinea Agr. J. 15:* 79–90.

4. Talbot, P. H. B. and P. J. Keane. 1971. *Oncobasidium*, a new genus of tulasnelloid fungi. *Australian J. Bot. 19:* 203–206.

Fig. 34. Chlorotic leaf in typical position some distance below growing tip (top left) and green islets (top right) caused by vascular streak dieback of cacao. Fruit bodies of *Oncobasidium theobromae* around leaf scars where mycelium has emerged (bottom). Note swollen lenticels. (Courtesy of P. J. Keane, La Trobe University, Bundoora, Victoria.)

Dieback (Seca de ramas)

Defoliation and death of terminal ends of twigs and small branches results in repeated branching and unthrifty appearance of affected trees. Severely affected trees may be killed, but the causal nature has not been definitely established, although variously attributed to insect damage or unfavorable environmental conditions, especially water relations.

1. Hardy, F. 1960. *Cacao manual.* Inter-Am. Inst. Agr. Sci., Turrialba. 395 p.

2. Hardy, F. and R. Bazan. 1966. Note on cacao disback in the Turrialba plantings. *Cacao 11:* 3–7.

3. Rubies Trias, J. M. 1962. Principales causas de la seca de ramas de los cacaos de Fernando Poo. *Ager 13:* 12–13.

4. Rubies Trias, J. M. 1963. Aspectos de la "seca de ramas" en Fernando Poo. *Ager 14:* 17–18.

5. Voelcker, O. J. and J. West. 1940. Cacao die-back. *Trop. Agr. (Trinidad) 17:* 27–30.

Verticillium wilt (*Verticillium dahliae* Kleb.) (See Sage—*Verticillium* wilt)

Sudden wilting of leaves starting from the top of ends of fan branches and developing unilaterally, on occasion, followed by chlorosis and desiccation of leaves, which remain attached for some time before falling, are the most obvious symptoms of *Verticillium* wilt. Dark brown or black discontinuous streaking in vascular tissues throughout the plant is the most prominent internal symptom. The disease is known to occur presently only in Uganda.

1. Emechebe, A. M. and C. L. A. Leaky. 1968. *Verticillium dahliae* established as the cause of sudden death of cocoa. *FAO Plant Protect. Bull. 16:* 13.

2. Emechebe, A. M., C. L. A. Leaky and W. B. Banage. 1971. *Verticillium* wilt of cacao in Uganda: symptoms and establishment of pathogenicity. *Ann. Appl. Biol. 69:* 223–227.

3. Emechebe, A. M., C. L. A. Leakey and W. B. Banage. 1973. *Verticillium* wilt of cacao in Uganda: the relationship between *Verticillium dahliae* and cacao roots. *Ann. Appl. Biol. 70:* 157–162.

4. Emechebe, A. M., C. L. A. Leakey and W. B. Banage. 1974. *Verticillium* wilt of cacao in Uganda: wilt induction by mechanical vessel blockage and mode of recovery of diseased plants. *East Afr. Agr. Forest. J. 39:* 337–343.

5. Hawksworth, D. L. and P. W. Talboys. 1970. *Verticillium dahliae. Commonwealth Mycol. Inst. Descriptions Pathogenic Fungi Bacteria 256.* 2 p.

6. Leaky, C. L. A. 1965. Sudden death disease of cocoa in Uganda associated with *Verticillium dahliae* Kleb. *East Afr. Agr. Forest. J. 31:* 21–24.

Hoja estrecha abollada (Narrow, rugose leaf)

A disease reported to occur in Colombia and Dominican Republic (1943) was characterized by rugosity of abnormally narrow leaves and presumed to be virus-induced because of apparent graft transmissibility. Shoot dieback symptoms were included in the syndrome (see Calcium deficiency symptoms).

1. Ciferri, R. 1948. Una virosis del cacao en Colombia y en la Republica Dominicana. *Trop. Agr. (Venezuela) 21:* 36.

Corynespora cassiicola (Berk. & Curt.) Wei. (Fig. 35)

1. Duarte, M. L. R., F. C. Albuquerque and A. S. Prabhu. 1974. A new disease of cocoa (*Theobroma cacau* L.) induced by the fungus *Corynespora cassiicola* (Berk. & Curt.) Wei. *Fitopatologia 9:* 52.

2. Ellis, M. M. and P. Holliday. 1971. *Corynespora cassiicola. Commonwealth Mycol. Inst. Descriptions Pathogenic Fungi Bacteria 303.* 2 p.

Lysigenous canals

1. Shaw, D. E. 1968. Traumatic vertical lysigenous canals in cacao in Papua and New Guinea. *Papua New Guinea Agr. J. 20:* 65–70.

CAMPHOR: *Cinnamomum camphora* (L.) Nees & Eberm.

White rot (*Polystictus leoninus* (Klotzsch)

1. Chattopadhyay, S. B. 1967. *Diseases of plants yielding drugs, dyes and spices.* Indian Council of Agricultural Research, New Delhi. 100 p.

Fig. 35. Spots on cacao leaves caused by *Corynespora cassiicola*. (Courtesy of M. de L. R. Duarte, Instituto de Pesquisa Agropecuaria do Norte, Belém).

(Camphor) Virus

1. Hino, I. and T. Hino. 1953. Suspected virus diseases of camphor trees. *Yamaguti Univ. Fac. Agr. Bull. 4:* 185–197.

SMALL CARDAMOM: *Elettaria cardamomum* Maton

LARGE CARDAMOM: *Ammomum subulatum* Roxb.

Chirke

The disease of large cardamom, now known as chirke, was first observed in India in 1956 with infection ranging to 10 percent. Losses of more than 90 percent in number and weight of fruits and seeds harvested have been determined under experimental conditions.

Chlorotic streaks that later become necrotic and coalesce to cause death of the leaves develop on affected plants, which are nearly the same size as healthy plants. The major practical effect of the disease is severe reduction in number of flowers developed and seed production.

Cause of chirke disease has been established as a virus that is readily transmitted mechanically as well as by aphids, including the corn aphid (*Rhopalosiphum maidis* Fitch.) and other unidentified species. Aphids can transmit the virus after an acquisition feeding of 5 minutes and transmission feeding of 15–30 minutes. Virus in sap from infected plants is inactivated by exposure for 10 minutes to 60 C, dilution to 5×10^{-3} and storage *in vitro* (at laboratory temperature) for 8 days. Wheat (*Triticum aestivum* L.), arrowroot (*Curcuma angustifolia* Roxb.) and wild cardamom (*A. aromaticum* Roxb.) also are susceptible to infection, as are ginger (*Zingiber officinale* Roscoe) and a perennial (in India) weed, *Acorus calamus* L. Maize and oats (*Avena* sp.) are not known to be susceptible.

Neither heat treatment of infected rhizomes (variable to 40 C for 12 hours) nor soil drenches with various materials, except hydroquinone (0.075–0.1 percent concentrations), provided any therapeutic effect. Differences in varietal susceptibility have been noted.

The most promising method for control presently is use of the tolerant varieties Kopringe and Sawney. Infected plants should be rogued when symptoms first appear and regular application of insecticide to reduce aphid populations may serve to slow spread of the disease.

1. Ganguly, B. 1965. A rapid test for detecting chirke affected large cardamom (*Ammomum subulatum* Roxb.) plants in field. *Sci. Cult. 32:* 95–96.

2. Raychaudhuri, S. P. and S. N. Chatterjee. 1955. Transmission of chirke disease of large cardamom by aphid species. *Indian J. Entomol. 27:* 272–276.

3. Raychaudhuri, S. P. and S. N. Chatterjee. 1961. Chirke a new virus threat to cardamom. *Indian Farming 11:* 11–12.

4. Raychaudhuri, S. P. and S. N. Chatterjee. 1964. Aphid transmission of a new mosaic streak disease of large cardamom. *Phytopathology 54:* 904.

5. Raychaudhuri, S. P. and B. Ganguly. 1965. Further studies on chirke disease of large cardamom (*Ammomum subulatum* Roxb.). *Indian Phytopathol. 18:* 373–377.

Foorkey

A disease of large cardamom noted in India in 1936 threatened the industry near Darjeeling 20 years later. Means to reduce losses have been devised, but the disease remains one of concern to growers because of potential economic significance.

Diseased plants become lighter green in color, develop broad, slightly chlorotic streaks in leaves and produce numerous short, vegetative shoots at the bases. Growth of plants is retarded, and they become dwarfed as leaves and stalks are greatly reduced and vegetative growth from the rhizomes becomes profuse, imparting a characteristic appearance to diseased plants. Flowering is gradually suppressed and finally ceases within 3–4 years after symptoms first appear. Rhizomes become progressively reduced in size, and, while diseased plants may survive for a few years, they remain unproductive and sterile.

Foorkey disease is caused by a virus readily transmitted by the banana aphid, *Pentalonia nigronervosa* Coq., which can remain infectious for 12 days after feeding on diseased plants, but the virus is not sap transmissible. The virus is infectious to large and small cardamom, but banana (*Musa sapientum* L.), ginger, wheat, sorghum (*Sorghum vulgare* Pers.) and maize are not susceptible. Viruses causing foorkey and "katte" diseases (of small cardamom) are not related (i.e. afford no cross protection), and both may be found in the same plant. The 2 viruses cause distinguishable symptoms on a wild species of *Ammomum.*

There is no means by which infected plants can be rehabilitated and made productive again, so roguing of diseased plants is recommended. Diseased plants must be destroyed *in situ*, which can be achieved by treating plants and rhizomes with herbicide (40 percent Agroxone) and replacing them with healthy plants grown from seed. The disease is not seed transmitted nor is there evidence of soil transmission. Application of insecticide to reduce populations of banana aphids may serve to reduce spread of the disease.

1. Chattopadhyay, S. B. and T. P. Bhowmik. 1965. Control of 'foorkey' disease of larger cardamom in West Bengal. *Indian J. Agr. Sci. 35:* 272–275.

2. Mukherji, D. K. 1973. Large cardamom. *World Crops 25:* 31–33.

3. Varma, P. M. and S. P. Capoor. 1964. 'Foorkey' disease of large cardamom. *Indian J. Agr. Sci. 34:* 56–62.

Mosaic (Katte, Marble)

Infection of small cardamom with mosaic was first noted in 1900 in India and attributed to various causes until studies 40 years later established its infectious nature. The disease remains confined to India, where it has occurred in serious proportions locally.

Diseased plants exhibit general chlorosis as an early symptom, accompanied by discontinuous streaks of green parallel to the veins. Streaking continues to develop, becoming more obvious and finally resulting in a characteristic mosaic pattern. Affected plant clumps degenerate quickly with progressive reduction in productivity, and, finally, they wither and die.

The disease is caused by a virus which is not sap transmissible but is readily transmitted by the banana aphid *Pentalonia nigronervosa*. No plant other than cardamom is known to be susceptible.

Satisfactory control has been achieved by roguing of infected plants when they appear and replacing diseased plants with healthy clumps.

1. Uppal, B. N., P. M. Varma and S. P. Capoor. 1945. A mosaic disease of cardamom. *Current Sci. 14:* 208–209.

2. Varma, P. M. 1956. Relation of mosaic diseases of cardamom to banana aphis (*Pentalonia nigronervosa* Coq.). *Proc. 43 Indian Sci. Cong. 4:* 48.

3. Varma, P. M. and S. P. Capoor. 1953. 'Marble' disease of cardamom. *Indian Farming 3:* 22–23.

OTHER DISEASES

Fruit and leaf rot (*Phytophthora nicotianae* var. *nicotianae* B. de Haan)

Small, brown lesions on fruit that fall in 3–6 days and large, water-soaked, dark leaf lesions were found during the monsoon season in India.

1. Thankamma, L. and P. N. R. Pillai. 1973. Fruit rot and leaf rot diseases of cardamom in India. *FAO Plant Protect. Bull. 21:* 83–84.

2. Waterhouse, G. M. and J. M. Waterston. 1964. *Phytophthora nicotianae* var. *nicotianae*. *Commonwealth Mycol. Inst. Descriptions Pathogenic Fungi Bacteria 34.* 2 p.

Leaf spots

Sphaceloma sp.

Brown blotches with red margins along the edges of leaf laminae were observed in India.

1. Muthappa, B. N. 1965. A new leaf spot of cardamom from India. *Plant Dis. Reptr. 49:* 917.

Phaeotrichoconis crotalariae (Salam & Rao) Subramaniam

Irregular, white leaf spots with brown margins in which the centers sometimes dried and fell have been described from India.

1. Dhanalakshmy, C. and K. M. Leelavathy. 1976. Leaf spot of cardamon caused by *Phaeotrichoconis crotalariae. Plant Dis. Reptr. 60:* 188.
2. Subramaniam, C. V. 1956. *Phaeotrichoconis* a new genus of the *Dematiaceae. Proc. Indian Acad. Sci., Sect. B, 44:* 1–2.

Rust (*Uredo elettariae* Thirumalachar)

Powdery pustules of white, echinulate urediospores (19–27 × 26–33 μm) are characteristic.

1. Thirumalachar, M. J. 1943. A new rust disease of cardamoms. *Current Sci. 12:* 231–232.

CASSAVA: *Manihot esculenta* Crantz

Bacterial blight

A bacterial disease of cassava was reported from Brazil in 1912 and a similar disease, although attributed to another pathogen, from Tasmania in 1926. Bacterial diseases of cassava in various manifestations have been described since from several South American and African countries. Accounts of these diseases, i.e. blight, wilt and leaf spot (mancha angular), often have included different names for the associated causal agents. However, only one bacterium has been positively identified as the cause of extensive losses, but geographic limitations for this disease have not been established. There is a great need for further study to identify and resolve relationships, if any, between all of these reported pathogens of cassava.

Syndromes of the described diseases have included all of the symptoms typically ascribed to bacterial diseases. Leaf spot symptoms appear as small (up to 2 mm), angular, water-soaked spots that soon enlarge (Fig. 36). As the centers turn brown, the lesions become translucent but retain sizeable yellow halos. Bacterial exudate appears as fragile scales after drying on the lower leaf surfaces. Veins in affected leaf areas assume a dark brown, necrotic coloration that may extend for a few mm beyond the margins of the lesions. Under conditions of high humidity, lesions may coalesce to affect considerable areas of the laminae, resulting in defoliation.

Plants affected by the wilt disease exhibit flagging of lower leaves, followed by wilting of all leaves, dieback and eventual death of the plants (Fig. 36). Blight symptoms may include one or more of the above manifestations of infection and vascular discoloration and disruption in the stems of affected plants.

Confusion concerning identity of the causal bacteria for the various diseases described will remain until comparative study of pathogens can be accomplished. The recent explanation that the Brazilian pathogen is an atypical (white) strain of the (yellow) pigmented bacterium that occurs in Africa and elsewhere, named *Xanthomonas manihotis* (Arthaud-Berthet) Starr, has been accepted herein. The bacteria have been described as gram negative rods 0.6–0.9 × 1.6–2 μm with single, polar flagella.

The bacteria gain entrance either through stomata or wounds and cause extensive internal damage to parenchymatous tissue before invading vascular elements. Infection in mature, lignified stems is confined to vascular tissues (Fig. 36). Severity of the diseases, particularly leaf spotting, increases during periods of high humidity and abundant rainfall, which has been shown to be important in local spread of the bacteria. Leaf spot and blight symptoms result from stomatal infections with inoculum from exudate or other "external" sources, while wilt symptoms result from use of planting material with "internal" vascular infection. Dissemination and persistence of the pathogen occurs largely through infected propagating materials, although the bacteria do remain viable for 4 months in dry exudate or debris.

Care should be taken in selection and use made only of healthy planting stocks, which should not be planted near other (diseased) cassava plantings. Infected plants should be rogued immediately, and all debris destroyed. Resistant cultivars have been reported but are not presently used extensively.

A technique has been developed to produce rooted cuttings free of bacterial blight infection. Shoots that develop from axillary buds below pruning cuts on (infected) mature, lignified stems remain free of infection for a month or more, while bacteria remain confined to vascular elements not

Fig. 36. Systemic (top) and localized leaflet (middle) infections of bacterial blight of cassava. Discoloration of vascular elements from upper-lower nodes (bottom). (Courtesy of J. C. Lozano, Centro Internacional de Agricultura Tropical, Palmira.)

directly connected to axillary buds. Axillary shoots excised during this time readily root under mist, and, if infected, quickly develop symptoms under such optimum conditions for disease development and can be eliminated. Rooted axillary shoots that remain free of bacterial blight infection can be used, thereafter, to propagate more healthy cuttings to be rooted.

1. Anonymous. 1972. Cassava production systems. *Centro Internacional de Agricultura Tropical Annual Report, 1972,* p. 60–64, Cali, Colombia.

2. Bondar, G. 1912. Uma nova molestia bacteriana das hastes da mandioca. *Chacaras Quintaes 5:* 15–18.

3. Bondar, G. 1915. Molestia bacteriana da mandioca. *Bol. Agr. (São Paulo) 16:* 513–524.

4. Bradbury, J. F. 1975. Bacterial diseases of cassava. *PANS 21:* 44.

5. Burkholder, W. H. 1942. Three bacterial plant pathogens: *Phytomonas caryophylli* sp. n., *Phytomonas alliicola* sp. n., and *Phytomonas manihotis* (Arthaud-Berthet et Bondar) Viegas. *Phytopathology 32:* 141–149.

6. Drummond, O. A. and O. Hipolito. 1941. Notas sobre a bacteriose da mandioca. *Ceres 2:* 280–307.

7. Drummond-Goncalves, R. 1953. A bacteriose e a mandioca Guaxupé. *Biológico 19:* 114–117.

8. Franco do Amaral, J. 1942. Estudo do organismo causador de bacteriose da mandioca. *Arq. Inst. Biol. 13:* 119–126.

9. Franco do Amaral, J. and L. G. Vasconcellos. 1945. Novos estudos do agente etiologico da bacteriose da mandioca. *Arq. Inst. Biol. 16:* 361–368.

10. Freire, J. R. J. 1951. Considerações sôbre o problema da bacteriose da mandioca. *Rev. Agron. 15:* 103–104.

11. Lozano, J. C. 1974. El añublo bacterial de la yuca (*Manihot esculenta* Crantz) en America: etiología, epidemiología y control. *Fitopatologia 9:* 110–119.

12. Lozano, J. C. 1975. Bacterial blight of cassava. *PANS 21:* 38–43.

13. Lozano, J. C. and L. Sequeira. 1974. Bacterial blight of cassava in Colombia: etiology. *Phytopathology 64:* 74–82.

14. Lozano, J. C. and L. Sequeira. 1974. Bacterial blight of cassava in Colombia: epidemiology and control. *Phytopathology 64:* 83–88.

15. Lozano, J. C. and D. W. Wholey. 1974. The production of bacteria-free planting stock of cassava. *World Crops 26:* 114–118.

16. Maraite, H. and J. A. Meyer. 1975. *Xanthomonas manihotis* (Arthaud-Berthet) Starr causal agent of bacterial wilt, blight and leaf spots of cassava in Zaire. *PANS 21:* 27–37.

17. Pereira, A. L. G. and A. G. Zagatto. 1967. Etiologia da "mancha angular" na folha da mandioca, (*Manihot utilissima*). *Arq. Inst. Biol. 34:* 153–160.

18. Robbs, C. F., R. de L. D. Ribeiro, O. Kimura and F. Akiba. 1972. Variações em *Xanthomonas manihotis* (Arthaud Berthet) Starr. *Rev. Soc. Brasileira Fitopatol. 5:* 67–75.

19. Robbs, C. F., R. de L. D. Ribeiro, O. Kimura and F. Akiba. 1973. Studies on the white xanthomonads in Brazil. *Abstracts of Papers, Second International Congress of Plant Pathology, No. 1056.*

20. Takatsu, A. and J. C. Lozano. 1975. Translocacion del agente causal del añublo bacterial de la yuca (*Manihot esculenta* Crantz) en los tejidos del hospedero. *Fitopatologia 10:* 13–22.

21. Wiehe, P. O. and W. J. Dowson. 1953. A bacterial disease of cassava (*Manihot utilissima*) in Nyasaland. *Empire J. Expt. Agr. 21:* 141–143.

22. Williams, R. J., S. D. Agboola and R. W. Schneider. 1973. Bacterial wilt of cassava in Nigeria. *Plant Dis. Reptr. 57:* 824–827.

Cercospora leaf spots

A number of leaf spot diseases of cassava caused by species of *Cercospora* have been reported, but 2 are more important because of wider geographic distribution and greater symptom severity. Brown leaf spot is found wherever cassava is grown in Africa, Asia, North and South America and is one of the most important leaf diseases of cassava. White leaf spot has been reported from all of the above areas, but incidence is limited to cooler locations where cassava is grown. Two additional *Cercospora* leaf spot diseases, both relatively minor in importance, have been reported. One (*C. viscosae* Muller & Chupp) is presently known to occur in Brazil. Economic losses from these diseases have not been accurately assessed, but some reports have described serious seasonal damages.

Typical symptoms of brown leaf spot include numerous, well-defined, circular, brown lesions 3–12 mm in diameter that extend through the leaf laminae (Fig. 37). Lesions along the margins of laminae or bordered by main veins may be irregular or angular in shape, while smaller veins in the lesions are noticeably blackened. A narrow, darker margin delimits each lesion but is somewhat less noticeable from the lower leaf surface where the presence of conidiophores imparts a lighter color to the center of each spot. Severely affected leaves turn yellow and drop.

Lesions of white leaf spot are mostly circular in shape but smaller in size (1–7 mm diameter) and white rather than brown in color (Fig. 37). A dark colored border, sometimes diffuse on the lower leaf surface and with a yellowish halo, surrounds each lesion, which is visibly sunken from both leaf surfaces. Sporulation develops mostly on the lower leaf surfaces and gives the lesions on this leaf surface a lighter, velvety appearance.

The leaf spot that occurs in Brazil and Colombia (*C. viscosae*) is characterized by large brown lesions with poorly defined margins, and each lesion may involve a sizeable portion of a leaflet (Fig. 37). Symptoms are mostly confined to older, lower leaves and are not usually of serious consequence.

C. manihobae has been reported to cause snow-white lesions on *M. esculenta* in Brazil.

Brown leaf spot is caused by *Cercospora henningsii* Allescher characterized by brown conidiophores 3–5 × 10–50 μm on which are borne amphygenous conidia 4–6 × 30–60 μm with 2–8 septae. A perfect stage, *Mycosphaerella manihotis* Sydow, has been reported for the fungus and includes black perithecia (100 μm in diameter), clavate asci 10–13 × 55–72 μm and ovoid, 2-celled ascospores 5–7 × 17–22 μm.

Conidiophores of *C. caribaea* Chupp & Ciferri (white leaf spot) measure 3–5 × 50–200 μm and the hypophyllous, hyaline, 1–6 spetate conidia 4–8 × 20–90 μm.

Fig. 37. Leaf spots of cassava caused by *Cercospora henningsii* (top left), *C. caribaea* (top right) and *C. viscosae* (bottom). (Courtesy of J. C. Lozano, Central Internacional de Agricultura Tropical, Palmira.)

C. viscosae produces dark, coremoid fascicles of conidiophores 4–6 × 50–150 μm and conidia 4–6 × 25–100 μm.

The dark colored conidiophores of *C. manihobae* are 3–5 × 50–200 μm and the nearly hyaline conidia 4–8 × 20–90 μm.

Conidia from debris or mature leaf lesions are carried by wind or rain to leaves where germination and infection by indirect penetration may occur within 12 hours in warm, humid weather. Older leaves usually are more susceptible than younger leaves, but petioles and even fruits of some varieties may be infected. *C. henningsii* infects *M. glaziovii* Muell. (cera rubber), *M. piauhynsis* Ule. and sweet potato [*Ipomea batata* (L.) Lam.] as well as *M. esculenta*, but only the latter species is a known suscept of *C. caribaea* and *C. manihobae*.

Different geographic distributions of the 2 more important diseases, brown leaf spot and white leaf spot, may result from the influence of environmental conditions. Optimum and maximum temperatures for conidial germination of *C. henningsii* and *C. caribaea* are 39–43 and 33–35 C, respectively. Conidia of the former will germinate at 50 percent minimum relative humidity (optimum 90 percent), while conidia of the latter will not germinate below 65 percent (optimum 100 percent) relative humidity. Conidial production and disease development for all species is greater during seasons of high humidity, but occurrence of *M. manihotis* is nearly as common in dry as in wet seasons and more frequent in Africa than in the Americas. Incidence of *C. viscosae* is similar to that of brown leaf spot.

A number of methods have been recommended for *Cercospora* leaf spot control, including planting of cassava on a 3–5 year crop rotation schedule. Topping of plants during the dormant season to 6 inches above the soil and burning the debris, frequent weeding and wider spacing of plants serve to reduce early inoculum and lower humidity during the production season. Application of copper oxychloride suspended in mineral oil at 12 ℓ/ha has provided good control in Africa, and differences in varietal susceptibility have been reported.

1. Chevaugeon, J. 1956. Les maladies cryptogamiques du manioc en Afrique Occidentale. *Encycl. Mycol. 28:* 1–205.

2. Chupp, C. 1953. *A monograph of Cercospora.* Ithaca, N.Y. 667 p.

3. Ghesquiere, J. 1932. Sur la "Mycosphaerellose" des feuilles du manioc. *Bull. Inst. Roy. Col. Belge 3:* 160–178.

4. Golato, C. 1963. *Cercospora henningsii* sulla manioca in Nigeria. *Riv. Agr. Subtrop. Trop. 57:* 60–66.

5. Golato, C. and E. Meossi. 1966. Una nuova malattia fogliare della manioc in Somalia. *Riv. Agr. Subtrop. Trop. 60:* 182–186.

6. Lozano, J. C. and R. H. Booth. 1974. Diseases of cassava (*Manihot esculenta* Crantz). *PANS 20:* 30–54.

7. Springensguth, W. 1940. Die kultur des manioks, siene krankheiten und schadlinge im litoral des staates St. Catharine (Brasilien). *Tropenpflanzer 43:* 286–306.

8. Sydow, C. 1901. *Mycosphaerella manihotis* Syd. n. sp. Fungi Novi Brasilienses. *Bull. l'Herbier Boissier 1:* 78.

9. Viegas, A. 1941. Manchas das folhas da mandioca produzidas por Cercosporas. *Bragantia 1:* 233–248.

10. Viennot-Bourgin, G. and J. Grimald. 1950. Les *Cercospora* parasites des feuilles de manioc. *Rev. Bot. Appl. Agr. Trop. 30:* 138–146.

11. Welles, C. 1924. Observations on taxonomic factors used in the genus *Cercospora. Science, n. s. 59:* 216–218.

12. Welles, C. 1925. Taxonomic studies on the genus *Cercospora* in the Philippine Islands. *Am. J. Bot. 12:* 195–218.

Ash disease

Originally reported from Africa (1913), ash disease now also occurs commonly in South America and Asia but is not usually of serious consequence.

The white mycelium visible on the leaf surfaces causes chlorotic spots which develop into poorly defined, yellow lesions that include angular, necrotic areas (Fig. 38). Symptoms are more often seen on mature leaves, but young leaves are also susceptible.

The causal fungus, *Oidium manihotis* P. Henn., produces unicellular, hyaline, cylindrical conidia 12–20 × 20–40 μm in basipetal chains. A perfect stage, *Erysiphe manihotis*, has been described.

Occurrence of the disease in warmer areas, where cassava is grown

Fig. 38. Ash disease of cassava. (Courtesy of J. C. Lozano, Centro Internacional de Agricultura Tropical, Palmira.)

more often during the dry seasons, usually does not warrant control measures, although foliar application of sulfur materials has been suggested. Varietal resistance has been reported.

1. Ferdinando, G., H. Tokeshi, P. C. T. Carvalho, E. Balmer, H. Kimati, C. O. N. Cardoso and C. L. Salgado. 1968. *Manual de fitopatologia. Doenças das plantas e seu controle.* Bibl. Agron. Ceres, São Paulo. 640 p.

2. Lozano, J. C. and R. H. Booth. 1974. Diseases of cassava (*Manihot esculenta* Crantz). *PANS 20:* 30–54.

3. Saccardo, P. A. 1913. *Sylloge fungorum 22:* 1250.

4. Viegas, A. P. 1943. Alguns fungos da mandioca. *Bragantia 3:* 1–19.

Phyllosticta leaf spot

Similar leaf spot diseases of cassava have been reported from tropical Africa and Colombia after first having been reported from the Philippines in 1913. The disease occurs only under particular environmental conditions. This has precluded wide distribution of the disease, but susceptible varieties may be seriously defoliated.

Large, brown leaf spots, often with irregular margins, develop along the margins of leaf lobes, near the midrib or large veins (Fig. 39). From the under surface of leaves, lesions are uniformly brown, but concentric rings of pycnidia develop on the upper surfaces of lesions, which continue to enlarge and eventually cause the entire leaf and petiole to become necrotic and fall. Infection can extend into young shoots, which die back and on which abundant pycnidia are produced.

Various names have been proposed for the causal agent(s) herein designated *Phyllosticta manihoticola* Sydow that produces unicellular, ovoid conidia 15 × 20 μm in dark, globose pycnidia 100–170 μm in diameter.

The disease is restricted to the cooler areas (higher altitudes) where cassava is grown or to (cooler) rainy seasons at lower altitudes. Optimum temperature range for spore germination is 20–25 C, and only experimental inoculations at temperatures below 25 C have been successful.

Control measures have not been devised, but foliar application of fungicide should be beneficial. Varietal resistance has been observed in Colombia.

1. Lozano, J. C. and R. H. Booth. 1974. Diseases of cassava (*Manihot esculenta* Crantz). *PANS 20:* 30–54.

2. Reinking, O. A. 1919. Philippines plant disease. *Phytopathology 9:* 114–140.

3. Saccardo, P. A. 1931. *Sylloge fungorum 25:* 36, 773.

4. Sydow, H. P. 1913. Enumeration of Philippine fungi, with notes and description of new species. I. Micromycetes. *Philippines J. Sci. 8:* 265–285.

5. Viegas, A. P. 1943. Alguns fungos da mandioca. *Bragantia 3:* 1–19; 20–29.

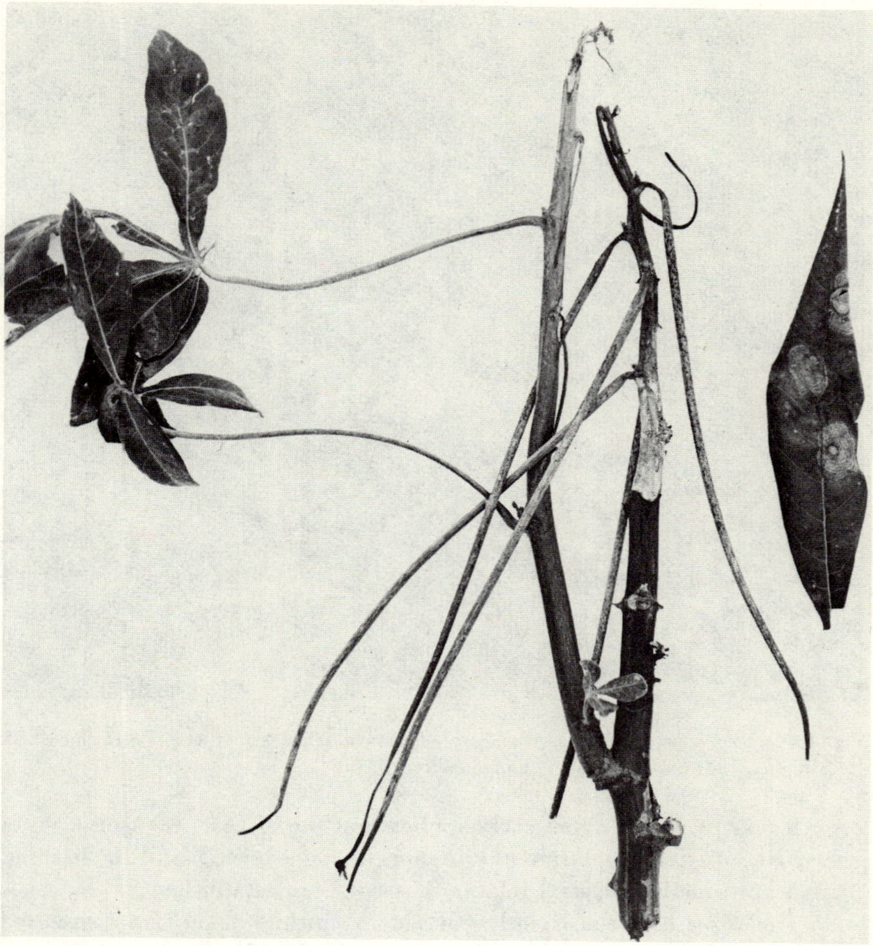

Fig. 39. Leaf lesions and stem necrosis of cassava caused by the *Phyllosticta* leaf spot fungus. (Courtesy of J. C. Lozano, Centro Internacional de Agricultura Tropical, Palmira.)

Superbrotamento (Envassouramento)

A disease of cassava observed first in 1939 in Brazil has been called "superbrotamento" (excessive budding) or "envassouramento" (witches' broom). The disease now is known to occur also in Venezuela and Mexico and can cause severe reduction in yield.

Affected plants are noticeably stunted because of shortened internodes, and excessive proliferation of branches from axillary buds imparts a witches' broom appearance (Fig. 40). Leaves on diseased plants often are somewhat chlorotic and smaller than on healthy plants.

Fig. 40. Superbrotamento (envassouramento) of cassava. (Courtesy of A. S. Costa and E. W. Kitajima, Instituto Agronomico, Campinas.)

It was presumed from early studies that the disease was caused by a virus. Recently, mycoplasma-like organisms have been found in diseased plants, but a definite causal relation has not been established.

The disease is consistently carried in cuttings taken from diseased plants and has been transmitted by grafting but not by sap inoculation.

Careful attention should be given to selection of healthy plants for cuttings or propagating materials.

1. Costa, A. S., E. W. Kitajima, A. S. Pereira, J. R. Silva and C. A. Carvalho Diaz. 1970. Molestias de virus de micoplasma da mandioca no estado, de São Paulo. *Bol. Secretar. Agr. Ind. Com., São Paulo.* 18 p.

2. Kitajima, E. W. and A. S. Costa. 1971. Corpusculos do tipo micoplasma associados a diveisas molestias de plantas, do grupo amarelo, no Estado de São Paulo. *Cienc. Cult. 23:* 285–291.

3. Kitajima, E. W., E. S. Normanha and A. S. Costa. 1972. Corpusculos do tipo micoplasma associados a una forma de superbrotamento da mandioca, na regiao de Tapachula, Chiapas, Mexico. *Cienc. Cult. 24:* 852–854.

4. Normanha, E. S., O. J. Boock and J. B. De Castro. 1946. Observaçoes de campo como contribuiçao ao estudo deo superbrotamento ou envassouramento da mandioca. *Rev. Agr. (Piracicaba) 21:* 271–302.

5. Silberschmidt, K. and A. R. Campos. 1944. Estudos relativos a doença superbrotamento ou envassouramento da mandioca. *Arq. Inst. Biol. 15:* 1–26.

African cassava mosaic (Kräuselkrankheit)

A cassava mosaic disease was reported from Africa in 1894 as "kräuselkrankheit," and is now found throughout tropical countries on that continent, including Madagascar and other nearby islands. The disease is confined to Africa, where yield losses ranging to 90 percent have been reported. A disease similar in many respects has been reported from India, but identity with African cassava mosaic remains to be established.

Symptoms of African cassava mosaic appear on leaflets early in the development of laminae as light colored or chlorotic areas, variable in size, that often contain small islands of green tissue. Chlorotic areas are well defined and distinct from unaffected leaf tissues and vary in size from small spots to entire leaflets. Affected leaf areas do not enlarge at the same rate as apparently healthy portions of laminae and result in twisted and misshapen laminae with the degree of distortion often related to the size of chlorotic areas (Fig. 41). Reduced size of leaflets is common, and affected plants are generally stunted.

African cassava mosaic is presumed to be caused by a virus, because transmission of the disease has been accomplished by grafting and by whitefly (*Bemisia* sp.). Attempts to demonstrate the presence of virus particles in diseased plants have been unsuccessful, however. The disease in Africa is restricted to species of *Manihot*, but the disease reported from India is transmissible to cucumber.

Whiteflies require an acquisition feeding time of at least 4 hours and a further 4 hours incubation to become infective, after which time they can infect in a 15 minute feeding. Only immature leaves are susceptible to successful infection by feeding of whiteflies, and plants may show first symptoms 12–20 days after inoculation. Approximately 8 days are required for infection to progress from a leaf into the stem, but systemic infection develops more rapidly thereafter. Collections of diseased plants with symptoms of different severity have prompted identification of severe and mild forms of the disease (Fig. 41). Such distinctions have insufficient validity, except for failure of the mild form to protect against infection by the severe form.

Selection of resistant varieties in breeding programs has been in progress for several years, with the most promising hybrids derived from *M. glaziovii*. Such varieties presently offer the only effective method for control.

1. Beck, B. D. A. 1971. *The breeding goals in a cassava breeding program in West Africa.* Ford Foundation, Lagos, Nigeria. 5 p.

2. Beck, B. D. A. and S. R. Chant. 1958. A preliminary investigation on the effect of

Fig. 41. Foliar symptoms of African cassava mosaic (above) and four degrees of severity of leaflet malformation (below). (Courtesy of International Institute of Tropical Agriculture, Ibadan.)

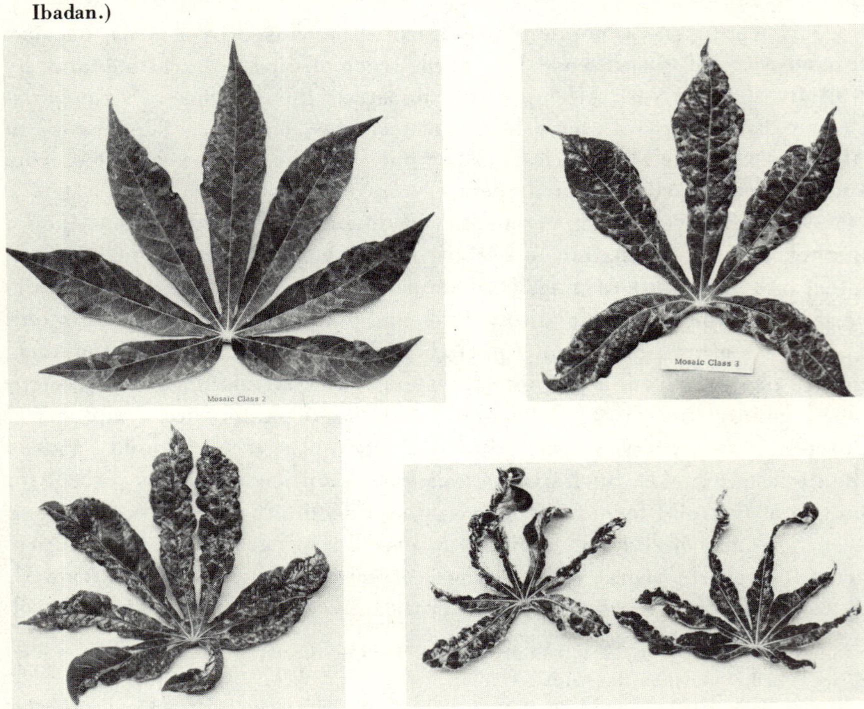

mosaic virus on *Manihot utilissima* Pohl. in Nigeria. *Trop. Agr. (Trinidad) 35:* 59–64.

3. Chant, S. R. 1958. Studies on the transmission of cassava mosaic virus by *Bemisia* sp. (*Aleyrodidae*). *Ann. Appl. Biol. 46:* 210–215.

4. Chant, S. R. 1959. A note on the inactivation of mosaic virus in cassava (*Manihot utilissima* Pohl.) by heat treatment. *Empire J. Expt. Agr. 27:* 55–58.

5. Doku, E. V. 1965. Breeding for yield in cassava. I. Indices of yield. *Ghana J. Sci. 5:* 42–59.

6. Golato, C. 1971. Virosi dell manioca in Ghana. *Riv. Agr. Subtrop. Trop. 65:* 281–287.

7. Jennings, D. L. 1960. Observations on virus diseases of cassava in resistant and susceptible varieties. I. Mosaic disease. *Empire J. Expt. Agr. 28:* 23–34.

8. Lefevre, P. 1935. Quelques considerations sur la mosaique du manioc. *Bull. Agr. Congo Belge 26:* 442–447.

9. Menon, M. R. and S. P. Raychaudhuri. 1970. Cucumber—a herbaceous host of cassava mosaic virus. *Plant Dis. Reptr. 54:* 34–35.

10. Pascalet, M. 1932. La mosaïque cu lepre du manioc. *Agron. Colon. 21:* 117–131.

11. Storey, H. H. and R. F. W. Nichols. 1938. Studies of the mosaic disease of cassava. *Ann. Appl. Biol. 25:* 790–806.

Brown streak

Brown streak disease of cassava was described from Tanganyika in 1936 and now has been reported from all of the important cassava growing areas of east Africa below 1050 m elevation. Present range and occurrence of the disease are not well defined because symptoms of African mosaic, which occurs wherever brown streak has been found, usually predominate, making visual diagnosis of the later sometimes difficult. However, the disease is confined to Africa, where losses from brown streak have caused serious reduction in quality and quantity of harvested roots.

Symptoms of brown streak appear on leaves, fruits and roots, in addition to lesions on young, green stems, which were the first symptoms recognized and resulted in the common name of the disease. Narrow, black or brown streaks appear on young, green stems and elongate as the disease progresses. Streaks may coalesce into dark blotches, which may not be completely obvious once the bark forms. Necrotic lesions, which are diagnostic, are also found in leaf scars, from which they eventually extend to kill nearby axillary buds. Leaves on infected plants become chlorotic and reminscent of virus infection, but only mature leaves exhibit symptoms and with no distortion of the laminae. Also, chlorosis of leaves caused by brown streak is distinguishable from normal yellowing of senescence because one or more lobes retain some green areas. During periods of drought, lower leaves fall progressively, and diseased plants with normal appearing upper leaves may look quite healthy unless there are stem symptoms. Fruits develop numerous necrotic spots that vary in size and extend into the flesh and are scattered over the pericarp.

Root symptoms include external cracks as deep as they are long, always surrounded by dark, discolored tissues. Internal lesions of various size and coloration develop without accompanying rot but serve to make the roots unfit for consumption.

Brown streak disease is caused by a virus transmissible mechanically (from older leaves with symptoms) and by grafting and is infectious to a number of plants other than cassava, including petunia, jimson weed and tobacco, *Nicotiana rustica* and *N. glutinosa*. The virus in sap from infected petunia is inactivated by storage for 24 hours at 20 C or dilution to 10^{-3} with water, but infectivity is not completely lost by exposure to 50 C for 10 minutes. Rod-shaped particles approximately 60 nm in length have been found in dried, diseased material.

Incidence of brown streak is limited primarily to elevations below 1050 m, but temperature does not influence severity of disease more than inherent (genetic) resistance or susceptibility.

Control can be achieved by use of healthy propagating materials and roguing of infected plants. Differences in varietal susceptibilities have been noted, but fully resistant varieties remain to be developed.

1. Jennings, D. L. 1960. Observations on virus diseases of cassava in resistant and susceptible varieties. II. Brown streak disease. *Empire J. Expt. Agr. 28:* 261–270.

2. Kitajima, E. W. and A. S. Costa. 1964. Elongated particles found associated with cassava brown streak. *East Afr. Agr. J. 30:* 28–30.

3. Lister, R. M. 1959. Mechanical transmission of cassava brown streak virus. *Nature 183:* 1588–1589.

4. Lozano, J. C. and R. H. Booth. 1974. Diseases of cassava (*Manihot esculenta* Crantz). *PANS 20:* 30–54.

5. Nichols, R. F. W. 1950. The brown streak disease of cassava—distribution, climatic effects and diagnostic symptoms. *East Afr. Agr. J. 15:* 154–160.

6. Storey, H. H. 1936. Virus diseases of East African plants—VI. *East Afr. Agr. J. 2:* 34.

South American (common) cassava mosaic

Although it had been observed there earlier, a mosaic disease of cassava was described from Brazil in 1940. It has been found since in cuttings imported from Peru and in Argentina, probably as a result of importing infected cuttings from Brazil. The disease can cause yield losses to 30 percent, but, because effective control measures are available, the disease is not considered of major importance.

Symptoms are confined to leaves and consist of mosaic of the laminae, sometimes with distortion (Fig. 42).

The disease is caused by a virus of the potato X group and is characterized by flexuous rods 15 × 495 nm inactivated by exposure for 10 minutes to 75 C, dilution above 10^{-6}, but not by storage for 128 days at room

Fig. 42. Three degrees of severity of foliar symptoms of South American (common) cassava mosaic. (Courtesy of J. C. Lozano, Centro Internacional de Agricultura Tropical, Palmira.)

temperature. The virus is readily transmissible mechanically and by grafting, but no insect vector is known, and the disease is not transmitted through seeds of cassava. Plants in 6 dicotyledonous families are susceptible to infection, including *Chenopodium* sp., castor bean (*Ricinus communis* L.), globe amaranth and cotton (*Gossypium hirsutum* L.)

Dissemination results from use of infected propagating materials, and effective control can be achieved through selection of healthy seed materials and roguing of diseased plants.

1. Costa, A. S. 1940. Observacoes sobre o mosaico comun e mosaico das nervuras de mandioca (*Manihot utilissima* Pohl.). *J. Agron. (Piracicaba) 3:* 239–248.

2. Costa, A. S. and E. W. Kitajima. 1972. Cassava common mosaic virus. *Commonwealth Mycol. Inst. Descriptions Plant Viruses 90.* 4 p.

3. Costa, A. S. and E. Normanha. 1939. Nota sobre o tratamento de manivas de mandioca (*Manihot utilissima* Pohl.). *Rev. Agr. (Piracicaba) 14:* 227–230.

4. Kitajima, E. W. and A. S. Costa. 1966. Microscopia electronica de tecidos foliares de mandioca infectados pelo virus do mosaico comun da mandioca. *Bragantia 25:* 23–28.

5. Kitajima, E. W., C. Wetter, A. R. Oliveira, D. W. Silva and A. S. Costa. 1965. Morfologia do virus do mosaico comun da mandioca. *Bragantia 24:* 247–260.

6. Silberschmidt, K. 1938. O mosaico da mandioca, *Biológico 4:* 177–181.

7. Silva, D. M. 1962. Obtençao de antissoro contra o virus do mosaico da mandioca. *Bragantia 21:* 49–52.

8. Silva, D. M., E. W. Kitajima and A. R. Oliveira. 1963. Obtençao do virus do mosaico comun da mandioca purificado. *Cienc. Cult. 15:* 304.

9. Tascon, G. A., E. W. Kitajima and A. S. Costa. 1974. Intracellular behavior of the cassava common mosaic virus. *Fitopatologia 9:* 73.

Vein mosaic

The disease of cassava known as vein mosaic was reported in 1940 in Brazil, where it was found widely dispersed but not of major economic importance.

Systemic symptoms consist of clearing in leaflet veins accompanied by downward curling or cupping of the laminae (Fig. 43). Leaflets are not noticeably reduced in size.

Spherical particles 50–60 nm in diameter indicative of virus infection have been found in infected leaves. Transmission of the disease by mechanical inoculation and grafting further support the probability that vein mosaic is caused by a virus. Only jimson weed, in addition to *Manihot* sp., is known to be susceptible to infection.

Selection of healthy stocks for propagating materials and roguing of infected plants will reduce incidence of the disease.

Fig. 43. Vein mosaic of cassava. (Courtesy of A. S. Costa, Instituto Agronomico, Campinas.)

1. Costa, A. S. 1940. Observacoes sobre o mosaico comun e mosaico das nervuras de mandioca (*Manihot utilissima* Pohl.). *J. Agron. (Piracicaba) 3:* 239–248.

2. Costa, A. S., E. W. Kitajima, A. S. Pereira, J. R. Silva and C. A. Carvalho Diaz. 1970. Molestias de virus de micoplasma da mandioca no estado de São Paulo. *Bol. Secretar. Agr. Ind. Com., São Paulo.* 18 p.

3. Kitajima, E. W. and A. S. Costa. 1966. Particulas esferoidais associadas ao virus do mosaico das nervuras da mandioca. *Bragantia 25:* 211–221.

OTHER DISEASES

Bacterial wilt (*Pseudomonas solanacearum* E. F. Smith) (See Potato—Bacterial wilt)

1. Bondar, G. 1912. Uma nova molestia bacteriana das hastes da mandioca. *Chacaras Quintaes 5:* 15–18.

2. Bondar, G. 1915. Molestia bacteriana da mandioca. *Bol. Agr. 16:* 513–524.

3. Franco do Amaral, J. 1942. Estudo do organismo causador da bacteriose da mandioca. *Arq. Inst. Biol. 13:* 119–126.

4. Goncalves, R. D. 1939. A bacteriose da mandioca no Vale do Paraiba. *Biológico 5:* 117–118.

5. Hayward, A. C. and J. M. Waterston. 1964. *Pseudomonas solanacearum. Commonwealth Mycol. Inst. Descriptions Pathogenic Fungi Bacteria 15.* 2 p.

6. Kelman, A. 1953. The bacterial wilt caused by *Pseudomonas solanacearum.* A literature review and bibliography. *N. C. Agr. Expt. Sta. Tech. Bull. 99.* 194 p.

Anthracnose

Leaf spots to 10 mm in diameter at the bases of leaves that are killed later, a wilt of young stems and cankers on older stems (Fig. 44) have been attributed

Fig. 44. Anthracnose of cassava. (Courtesy of J. C. Lozano, Centro Internacional de Agricultura Tropical, Palmira.)

to *Colletotrichum manihotis* Henn. (*Glomerella manihotis* Chev.). The disease is more severe on young leaves and during rainy seasons.

1. Affran, D. K. 1968. Cassava and its economic importance. *Ghana Farmer 12:* 172–178.

2. Bouriquet, G. 1946. Les maladies du manioc a Madagascar. *Bull. Econ. Madagascar, Tananarive 65:* 198–237.

3. Doku, E. V. 1969. *Cassava in Ghana.* Ghana Univ. Press, Accra. 44 p.

4. Drummond, O. 1946. Doenças da mandioca. *Rev. Ceres 7:* 24–33.

Latent virus

1. Bock, K. R. 1975. Maize streak, cassava latent and similar viruses. *Abstracts, III International Congress for Virology (Madrid):* 93.

2. Costa, A. S., E. W. Kitajima, A. S. Pereira, J. R. Silva and C. A. Carvalho Diaz. 1970. Molestias de virus de micoplasma da mandioca no estado de São Paulo. *Bol. Secretar. Agr. Ind. Com., São Paulo.* 18 p.

Root rots

Phytophthora

1. Oliveros, B., J. C. Lozano and R. H. Booth. 1974. A *Phytophthora* root rot of cassava in Colombia. *Plant Dis. Reptr. 58:* 703–705.

Rhizoctonia

1. Drummond-Goncalves, R. and J. Franco do Amaral. 1941. Rizotoniose em mandioca e podridão das raizes (*Diplodia*) em tungue. *Biológico 7:* 360–361.

Rosellinia (Fig. 45)

1. Drummond-Goncalves, R. 1950. Podridão das raizes. *Biológico 16:* 17–18.

Rust [*Uromyces janiphae* (Wint.) Arth.] (Fig. 46)

1. Arthur, J. C. 1915. *Uredinales* of Porto Rico based on collections by F. L. Stevens. *Mycologia 7:* 168–196.

2. Arthur, J. C. and J. R. Johnston. 1918. *Uredinales* of Cuba. *Mem. Torrey Bot. Club 17:* 97–175.

3. Baker, R. E. D. and W. T. Dale. 1951. Fungi of Trinidad and Tobago. *Commonwealth Mycol. Inst. Mycol. Paper 33.* 123 p.

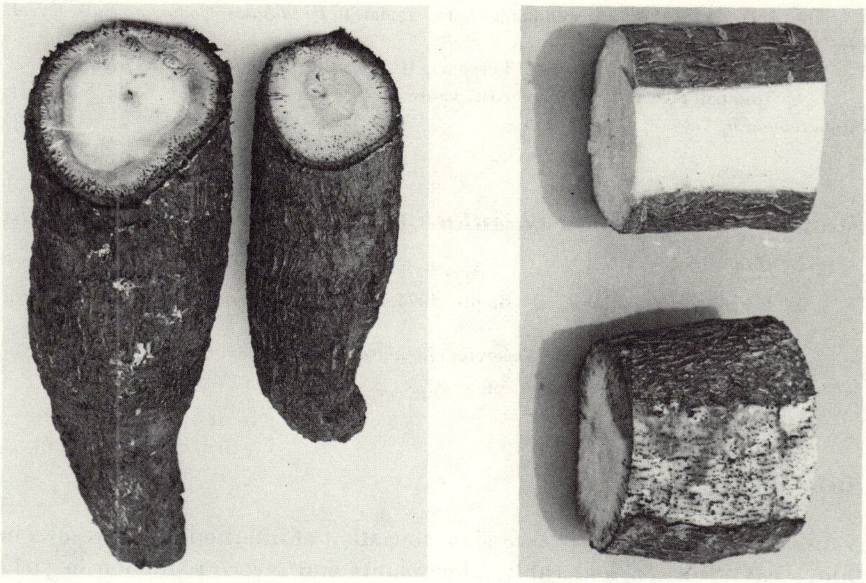

Fig. 45. *Rosellinia* root rot of cassava. (Courtesy of J. C. Lozano, Centro Internacional de Agricultura Tropical, Palmira.)

Fig. 46. Rust pustules on midveins, petiole and stems of cassava. (Courtesy of J. C. Lozano, Centro Internacional de Agricultura Tropical, Palmira.)

4. Dale, W. T. 1955. A preliminary list of Jamaican *Uredinales*. *Commonwealth Mycol. Inst. Mycol. Paper 60*. 21 p.

5. Franco do Amaral, J. 1942. Ferrugem (*Uromyces*) da mandioca. *Biológico 8:* 148.

6. Takatsu, A. 1974. Manihot rust, caused by *Uromyces* sp. in the Federal District. *Fitopatologia 9:* 73.

Stem rots [*Glomerella cingulata* (Fig. 47) and *Botryodiplodia theobromae* Pat.]

1. Lozano, J. C. and R. H. Booth. 1974. Diseases of cassava (*Manihot esculenta* Crantz). *PANS 20:* 30–54.

2. Mordue, J. E. M. 1971. *Glomerella cingulata. Commonwealth Mycol. Inst. Descriptions Pathogenic Fungi Bacteria 315*. 2 p.

Superelongation

A disease characterized by excessive elongation of internodes that results in thin, weak stems on noticeably taller plants and severe reduction in yield

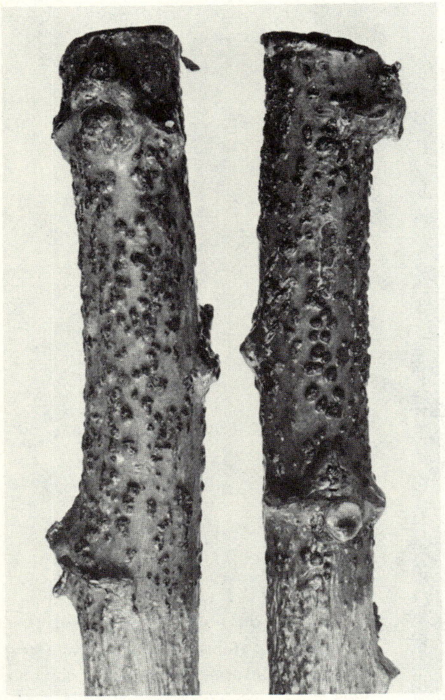

Fig. 47. Stem rot (*Glomerella cingulata*) of cassava. (Courtesy of J. C. Lozano, Centro Internacional de Agricultura Tropical, Palmira.)

Fig. 48. Foliar and root symptoms (infected immediately after planting, right; infected 2 months after planting in middle; healthy on left) of superelongation disease of cassava. (Courtesy of J. C. Lozano, Centro Internacional de Agricultura Tropical, Palmira.)

(Fig. 48) has been observed in serious proportions in Colombia. A similar disease has been observed in Mexico.

Leaves, petioles and shoots exhibit distortion, often associated with elongated cankers. Irregular, white spots develop on young leaves, causing necrosis of the laminae and defoliation. The disease is caused by *Sphaceloma manihoticola* Bitancourt & Jenkins and is most prevalent and severe during periods of high humidity and rainfall. Foliar application of fungicides may reduce disease severity. Varietal resistance has been noted.

1. Bitancourt, A. A. and A. E. Jenkins. 1950. Estudos sobre as *Miriangiales*. II. Vinte novas especies de Elsinoaceas neotropicais. *Arq. Inst. Biol. 20:* 1–28.

2. Lozano, J. C. and R. H. Booth. 1974. Diseases of cassava (*Manihot esculenta* Crantz). *PANS 20:* 30–54.

CASSIA: *Cassia tora* L.

Bacterial leaf spot

Small, round water-soaked lesions with halos were noted on leaves of *Cassia tora* in India in 1948. Lesions, also noted on pods and seeds, frequently enlarged to 5 mm and coalesced to form irregular spots that caused defoliation when numerous. Infection sometimes extended via leaf petioles into stems where dark, vertical streaks to 5 cm developed.

The bacterium was identified as a gram negative, short rod 0.8–1.0 × 1.2–2.1 μm, motile by a single polar flagellum and named *Xanthomonas cassiae* Kulkarni, Patel & Dhande. Optimum temperature for growth *in vitro* was 26–31 C.

Infection of uninjured leaves occurred consistently, and the bacterium survived in lesions on seed from infected pods for 4 months and in infected plant debris for a year. The bacterium also infected pea but not other species of *Cassia* nor species in 20 other genera.

Specific control methods have not been devised, but care should be taken to select seed from plants free of the disease. New plantings should be isolated from earlier planting sites where the disease has occurred, and debris from infected plants should be destroyed.

1. Kulkarni, Y. S., M. K. Patel and G. W. Dhande. 1951. *Xanthomonas cassiae*, a new bacterial disease of *Cassia tora* L. *Current Sci. 20:* 47.

2. Kulkarni, Y. S., M. K. Patel and G. W. Dhande. 1952. Bacterial leaf-spot of *Cassia tora* L. *Indian Phytopathol. 5:* 76–88.

Fusarium wilt

A *Fusarium* sp. isolated from infected *Cassia tora* in South Carolina and Georgia (United States) in 1946–1948 was pathogenic to alfalfa (*Medicago sativa* L.) and *Physalis alkekengi* L.

1. Armstrong, G. M. and J. K. Armstrong. 1954. Alfalfa—a common host for the wilt fusaria from alfalfa, cotton, and *Cassia*. *Plant Dis. Reptr. 38:* 221–222.

2. Armstrong, J. K. and G. M. Armstrong. 1959. *Physalis alkekengi* L., a new host for the U. S. cotton, *Cassia*-, and sweet potato-wilt fusaria. *Plant Dis. Reptr. 43:* 509–510.

Leptosphaerulina leaf spot

Small, dark spots that enlarged and coalesced to form large, irregular lesions with prominent spherical ascocarps were found on *Cassia tora* and other *Cassia* sp. in India in 1963. The causal fungus, identified as *Leptosphaerulina trifolii* (Rost.) Petr., produced membranous, brown ascocarps 77–206 μm diameter, each with 5–15 asci that measured 27–48 × 55–115 μm. The 8 muriform ascospores in each ascus were 10–17 × 20–45 μm.

1. Booth, C. and K. A. Pirozynski. 1967. *Leptosphaerulina trifolii. Commonwealth Mycol. Inst. Descriptions Pathogenic Fungi Bacteria 146.* 2 p.

2. Graham, J. H. and E. S. Luttrell. 1961. Species of *Leptosphaerulina* on forage plants. *Phytopathology 51:* 680–693.

3. Satya, H. N. and V. K. Rajalakshmy. 1964. *Leptosphaerulina trifolii* (Rost.) Petr., a new record for India. *Current Sci. 33:* 409.

Mosaic

Found in 1956 and again 3 years later, *Cassia* mosaic was reported from Papua New Guinea in 1961. The causal virus, infectious only to *C. occidentalis* L., *C. tora* and *Crotalaria anagyroides* H. B. & K. of more than 30 species tested, was inactivated by dilution of sap from infected plants 2 × 10^{-3}, exposure for 10 minutes to 62 C or storage *in vitro* 168 hours. No insect vector was found nor was seed transmission demonstrated, but the virus was readily sap transmissible.

1. Van Velsen, R. J. 1961. *Cassia* mosaic, a plant virus of *Cassia* spp. in New Guinea. *Papua New Guinea Agr. J. 14:* 124–127.

"Negro coffee" mosaic

A mottle disease of *Cassia occidentalis* leaves (no symptoms on flowers) described from India in 1974 was transmitted only by *Aphis solanella* Theob. of 7 aphid species tested and was infectious only to that host. Virus particles measured 21 × 550–580 nm and were flexuous.

1. Verma, V. S. and F. R. Niazi. 1974. Negro coffee mosaic-virus—eine neue viruskrankheit an *Cassia occidentalis* L. *Z. Pflanzenkrankh. Pflanzenschutz 81:* 608–610.

OTHER DISEASES

Cercospora leaf spot [*Cercospora cassiocarpa* (Sacc.) Chupp]

Spots on leaves of *Cassia occidentalis* were reported from India.

1. Gupta, P. C. 1973. Parasitic fungi on medicinal plants from India. I. Some *Cercosporae*. *Z. Pflanzenkrankh. Pflanzenschutz 80:* 328–330.

Ringspot

Mosaic symptoms interspersed with ringspots on leaves of stunted plants were caused by a virus infectious to *Cassia tora*, *Phaseolus mungo*, mung bean, cowpea, peanut (*Arachis hypogaea* L.) and pea but not infectious to tobacco, tomato, pepper, pigeon pea, soybean, maize or zinnia. The virus was inactivated by exposure to temperature between 70–80 C, storage at room temperature *in vitro* for 18 hours (4–6 days at 9–10 C) or dilution beyond 10^{-5} – 10^{-6}.

1. Mathur, R. S. and B. R. Singh. 1972. Ringspot disease of coffee senna (*Cassia occidentalis*) in India. *Indian Phytopathol. 25:* 314–315.

Tobacco etch

1. Anderson, C. W. 1954. *Cassia tora*, a leguminous host of tobacco etch virus. *Plant Dis. Reptr. 38:* 736–738.
2. Shepherd, R. J. and D. E. Purcifull. 1971. Tobacco etch virus. *Commonwealth Mycol. Inst. Descriptions Plant Viruses 55.* 4 p.

Tobacco mosaic

1. Capoor, S. P. 1961. *Cassia tora:* local lesion host of tobacco mosaic virus. *Current Sci. 30:* 353.
2. Zaitlin, M. and H. W. Israel. 1975. Tobacco mosaic virus (type strain). *Commonwealth Mycol. Inst. Descriptions Plant Viruses 151.* 5 p.

CHICKPEA (GRAM, GARBANZO BEAN): *Cicer arietinum* L.

Rust

Rust occurs in Africa, Asia, particularly India, is widespread in southern Europe and has been found in Mexico in recent years. Losses may be sizable but have not been accurately determined.

Pustules appear first as small, round, brown spots which may coalesce and turn dark later. Defoliation results from severe infection.

Urediospores of *Uromyces ciceris-arietini* Jacz. are 20–28 μm in diameter, globose, light brown in color, echinulate with 4–8 germpores and walls 3–4 μm in thickness. Teliospores vary from 18–24 × 18–30 μm, are round to angular, verrucose and dark brown. Neither pycnia nor aecia have been found.

Urediospore dissemination, largely by wind, determines disease incidence since teliospores have not been shown to cause infection. Optimum temperature for urediospore germination is 15–20 C and is prevented above 30 C, which accounts for low disease incidence in summer when high temperatures prevail. *Trigonella polycerata* L. is also susceptible in controlled inoculations.

Planting of resistant varieties is recommended (cv. IP 82 is susceptible as a seedling but only mildly affected later; cv. Karnal 29 had no pustules in 1939), since no fungicidal treatment has proven practicable or given satisfactory control.

1. León G., H. M. 1963. El chahuixtle del garbanzo: una nueva enfermedad en México. *Agr. Tec. Mex. 2:* 71.
2. Mehta, P. R. and B. B. Mundkur. 1946. Some observations on the rust of gram (*Cicer arietinum* L.). *Indian J. Agr. Sci. 16:* 186–192.
3. Payak, M. M. 1962. Natural occurrence of gram rust in uredial stage on *Trigonella polycerata* L., in Simla hills. *Current Sci. 31:* 433–434.
4. Punithalingam, E. 1968. *Uromyces ciceris-arietini. Commonwealth Mycol. Inst. Descriptions Pathogenic Fungi Bacteria 178.* 2 p.

Ascochyta blight (Anthracnose)

Occurrence of blight or anthracnose of chickpea in India has been recognized since 1920–30, when crop losses were estimated at 50 percent annually. Blight now has been reported from Iran, Israel and Greece and is considered a major disease that exacts heavy losses under environmental conditions conducive to disease development.

Spots that are circular on leaves and pods and elongated on stems vary in size but are consistently dark in color. Pycnidia, which appear as small black dots, develop in the lesions and often are arranged in concentric rings in pod spots. Infection from pod lesions frequently extends inward to completely prevent formation of seeds, or seeds may continue to full development with dark brown lesions on the outer surfaces. Lesions may enlarge to girdle either branches or the main stem, and plant parts above or distal to such lesions are killed. During early development of the disease, affected plants may be scattered throughout the field and not be obvious from a distance. The disease can spread quite rapidly, often in a circular pattern, from infection centers to destroy an entire crop in a few days.

The causal fungus is *Ascochyta rabiei* (Pass.) Lab. that produces spherical to pyriform pycnidia from which emerge, upon wetting, oval, unicellular spores 4×10 μm in mucilaginous cirri. Cardinal temperatures for growth of the fungus *in vitro* are 33, 20 and less than 10 C. The sexual stage of the fungus (*Mycosphaerella rabiei*) has been reported on overwintered debris from Bulgaria and Greece.

Primary inoculum for blight may come either from planting infected seed, which often results in infected seedlings, or from debris remaining from diseased plants in the previous crop. Spread of the disease within the field is largely dependent on rain to induce extrusion of spores by wetting of pycnidia. The mucilaginous material in cirri is soluble in water, which permits liberation of individual spores in free water so they can be further dispersed by splasing rain and wind. Dissemination of the disease over considerable distances can be accomplished by wind carrying dead leaves and other debris with pycnidia. Disease development is retarded at temperatures near the minimum for growth of the fungus *in vitro* because of reduced spore germination. However, spores remain viable, having been reported to withstand freezing, and germinate later when conditions become more favorable.

It is preferable to make new plantings some distance from existing crops and on land not planted recently in chickpea. If successive plantings of chickpea are to be made on the same land, care should be taken to completely turn under or burn all debris from the previous crop. Only seed from disease-free plants should be planted, and treatment of all seeds with benzimidazole (either thiabendazole or benomyl at 3 g/kg seed) is recommended. Foliar applications of these materials or other fungicides, such as zineb or maneb, where water is available may prove beneficial. Differences in varietal susceptibility have been reported and offer some promise for development of resistant varieties.

1. Aujla, S. S. 1964. Study on eleven isolates of *Phyllosticta rabiei* (Pass.) Trot., the causal agent of gram blight in the Punjab. *Indian Phytopathol. 17:* 83–87.

2. Bedi, K. S. 1949. Infectivity and longevity of dried cultures of *Ascochyta rabiei* (Pass.) Lab., the causal fungus of gram blight. *Indian Phytopathol. 2:* 6–10.

3. Hafiz, A. 1952. Basis of resistance in gram to *Mycosphaerella* blight. *Phytopathology 42:* 422–424.

4. Halfon-Meiri, A. 1970. Infection of chickpea seeds by *Ascochyta rabiei* in Israel. *Plant Dis. Reptr. 54:* 442–445.

5. Kaiser, W. J. 1973. Factors affecting growth, sporulation, pathogenicity, and survival of *Ascochyta rabiei. Mycologia 65:* 444–457.

6. Kaiser, W. J., M. Okhovat and G. H. Mossahebi. 1973. Effect of seed-treatment fungicides on control of *Ascochyta rabiei* in chickpea seed infected with the pathogen. *Plant Dis. Reptr. 57:* 742–746.

7. Kovačhevski, I. 1936. *The blight of chick pea,* Mycosphaerella rabiei *n. sp.* Min. Agr. Nat. Domains, Sofia. 80 p. (See also *Rev. Appl. Mycol. 15:* 700.)

8. Labrousse, F. 1931. L'anthracnose du pois-chiche. *Rev. Pathol. Veg. Entomol. Agr. 18:* 226–231.

9. Luthra, J. C. and K. S. Bedi. 1932. Some preliminary studies on gram blight with reference to its cause and mode of perennation. *Indian J. Agr. Sci. 2:* 499–515.

10. Luthra, J. C., A. Sattar and K. S. Bedi. 1935. Life-history of gram blight [*Ascochyta rabiei* (Pass.) Lab.=*Phyllosticta rabiei* (Pass.) Trot.] on gram (*Cicer arietinum* L.) and its control in the Punjab. *Agr. Livestock India 5:* 489–498.

11. Luthra, J. C., A. Sattar and K. S. Bedi. 1939. Variation in *Ascochyta rabiei* (Pass.) Lab., the causal fungus of blight of gram (*Cicer arietinum* L.). *Indian J. Agr. Sci. 9:* 791–805.

12. Luthra, J. C., A. Sattar and K. S. Bedi. 1941. Determination of resistance to blight disease [*Mycosphaerella rabiei* Kovačhevski-*Ascochyta rabiei* (Pass.) Lab.] in gram types. *Indian J. Agr. Sci. 11:* 249–264.

13. Punithalingam, E. and P. Holliday. 1972. *Ascochyta rabiei. Commonwealth Mycol. Inst. Descriptions Pathogenic Fungi Bacteria 337.* 2 p.

14. Sattar, A. 1933. On the occurrence, perpetuation and control of gram (*Cicer arietinum* L.) blight caused by *Ascochyta rabiei* (Pass.) Labrousse, with special reference to Indian conditions. *Ann. Appl. Biol. 20:* 612–632.

15. Solel, Z. and J. Kostrinski. 1964. The control of *Ascochyta* anthracnose of chickpea. *Phytopathol. Mediter. 3:* 119–120.

16. Sprague, R. 1930. Notes on *Phyllosticta rabiei* on chick-pea. *Phytopathology 20:* 591–593.

17. Zachos, D. G. 1951. Recherches sur la désinfection des graines du pois-chiche (*Cicer arietinum* L.) infectées par l'*Ascochyta rabiei* (Pass.) Labr. *Ann. Inst. Phytopathol. Benaki 5:* 76–87.

18. Zachos, D. G., C. G. Panagopoulos and S. A. Makris. 1963. Recherches sur la biologie, l'épidémiologie et la lutte contre l'anthracnose du pois-chiche. *Ann. Inst. Phytopathol. Benaki 5:* 167–192.

Fusarium wilt

The disease of chickpea now known as *Fusarium* wilt was reported from India in 1910, but confusion remained for the next 20 years about the cause. Wilt has been identified more recently in Burma and California. Incidence and economic importance of wilt, estimated at 10 percent annual loss in Uttar Pradesh, have been difficult to assess accurately because symptoms are similar to other root diseases.

Leaves on infected plants turn dull green and later chlorotic before wilting develops. Often wilt symptoms are confined to one side of the plant before it is systemically affected and killed. Dark discoloration of xylem tissue extends from the roots into the stems, where some parenchymatous tissue may be involved.

The fungus that causes wilt has been identified as *Fusarium lateritium* Nees emend. Snyd. & Hans. f. *ciceri* (Padwick) Erwin (syn. *F. orthoceras* var. *ciceri* Appel & Wr.), which is not infectious on pigeon pea, cowpea, alfalfa or sunn hemp (*Crotalaria juncea* L.).

Annual recurrence of the disease in the same areas of fields is characteristic, and the pathogen may remain viable in the soil for a number of years. Concentration of diseased plants in specific locations distinguishes the pattern of incidence from other diseases of chickpea with similar symptoms. Another fungus, *F. solani* (Mart.) Appel & Wr. f. sp. *pisi* (F. R. Jones) Snyd. & Hans., has been reported more recently attacking chickpea in Washington (United States).

Differences in varietal susceptibility have been noted and offer the most promising means for control.

1. Ayyar, V. R. and R. B. Iyer. 1936. A preliminary note on the mode of inheritance of reaction to wilt in *Cicer arietinum*. *Proc. Indian Acad. Sci., Ser. B. 3:* 438–443.

2. Erwin, D. C. 1958. *Fusarium lateritium* f. *ciceri*, incitant of *Fusarium* wilt of *Cicer arietinum*. *Phytopathology 48:* 498–501.

3. Kraft, J. M. 1969. Chickpea, a new host of *Fusarium solani* f. sp. *pisi*. *Plant Dis. Reptr. 53:* 110–111.

4. Luthra, J. C. and K. S. Bedi. 1932. Some preliminary studies on gram-blight with reference to its cause and mode of perennation. *Indian J. Agr. Sci. 2:* 499–515.

5. Mathur, R. S., J. S. Jain and S. C. Atheya. 1960. Resistance of gram varieties to *Fusarium* wilt in Uttar Pradesh, 1949–1958. *Current Sci. 29:* 403.

6. McKerral, A. 1923. A note on *Fusarium* wilt of gram in Burma and measures taken to combat it. *Agr. J. India 18:* 608–613.

7. Padwick, G. W. 1940. The genus *Fusarium*. III. A critical study of the fungus causing wilt of gram (*Cicer arietinum* L.) and of the related species in the subsection *Orthocera*, with special relation to the variability of key characteristics. *Indian J. Agr. Sci. 10:* 241–284.

8. Prasad, N. and G. W. Padwick. 1939. The genus *Fusarium*. II. A species of *Fusarium* as a cause of wilt of gram (*Cicer arietinum* L.). *Indian J. Agr. Sci. 9:* 371–380.

Stemphylium leaf spot

Small, irregular, dark lesions on leaves, pods and stems have been described twice from India (1961 and 1969) and once later from Iran. Leaves with several lesions become chlorotic and fall.

Conidia of *Stemphylium sarcinaeforme* (Cav.) Wilts. are muriform, dark, variable, with 3–4 septa and measure 14–27 × 20–38 μm.

The fungus can be borne without symptoms in infected seed. Differences in varietal susceptibilities have been noted, but only seed from healthy plants should be selected for planting.

1. Das, G. N. and P. K. Sen Gupta. 1961. A *Stemphylium* leaf spot disease of gram. *Plant Dis. Reptr. 45:* 979.

2. Kaiser, W. J. 1972. Occurrence of three fungal diseases of chickpea in Iran. *FAO Plant Protect. Bull. 20:* 73–78.

3. Prasad, H., M. G. Haider and K. D. Prasad. 1969. Blight disease of gram. *Indian Phytopathol. 22:* 405–406.

OTHER DISEASES

Crown blight (*Sclerotinia sclerotiorum*)

1. Kaiser, W. J. 1972. Occurrence of three fungal diseases of chickpea in Iran. *FAO Plant Protect. Bull. 20:* 73–78.

Floral blight [*Alternaria alternata* (Fr.) Keissler]

Observed in India in 1971, floral blight was characterized by shrivelled and brown necrotic buds and flowers that fell prematurely.

1. Vishwakarma, S. N. and K. C. B. Chaudhary. 1974. Floral blight of gram incited by *Alternaria alternata*. *Neth. J. Plant Pathol. 80:* 110–112.

Foot blight (*Phytophthora megasperma* Drechsler)

1. Suryanarayana, D. and S. R. Pathak. 1968. Foot blight—a new disease of gram. *FAO Plant Protect. Bull. 16:* 71–73.

2. Waterhouse, G. M. and J. M. Waterston. 1966. *Phytophthora megasperma*. *Commonwealth Mycol. Inst. Descriptions Pathogenic Fungi Bacteria 115.* 2 p.

Foot rot (*Operculella padwickii* Kheswalla)

1. Kheswalla, K. F. 1941. Foot-rot of gram (*Cicer arietinum* L.) caused by *Operculella padwickii* nov. gen. nov. sp. *Indian J. Agr. Sci. 11:* 316–318.

2. Williams, F. J., J. S. Grewal and K. S. Amin. 1968. Serious and new diseases of pulse crops in India in 1966. *Plant Dis. Reptr. 52:* 300–304.

Ozonium wilt (*Ozonium texanum* Neal and West)

1. Mishra, J. N. 1955. Wilt of gram (*Cicer arietinum* L.) in Bihar. *Current Sci. 24:* 210.

Powdery mildew (*Erysiphe* sp.)

1. Kaiser, W. J., D. Danesh, M. Okhovat and H. Mossahebi. 1968. Diseases of pulse crops (edible legumes) in Iran. *Plant Dis. Reptr. 52:* 687–691.

Root rots

Pathogenicity on chickpea of *Rhizoctonia solani, Macrophomina phaseolina, Pythium* sp. and *Fusarium* sp. was verified in greenhouse inoculations in Iran.

1. Dastur, J. F. 1935. Gram-wilts in the central provinces. *Agr. Livestock India 5:* 615–627.

2. Holliday, P. and E. Punithalingam. 1970. *Macrophomina phaseolina. Commonwealth Mycol. Inst. Descriptions Pathogenic Fungi Bacteria 275.* 2 p.

3. Kaiser, W. J., D. Danesh, M. Okhovat and H. Mossahebi. 1968. Diseases of pulse crops (edible legumes) in Iran. *Plant Dis. Reptr. 52:* 687–691.

Verticillium wilt (*Verticillium albo-atrum* Reinke & Berth.)

Foliage of affected plants becomes yellow (xylem discolored light brown), and the plants eventually wilt and die.

1. Erwin, D. C. 1957. *Fusarium* and *Verticillium* wilt diseases of *Cicer arietinum. Phytopathology 47:* 10.

2. Erwin, D. C. 1958. *Verticillium* wilt of *Cicer arietinum* in southern California. *Plant Dis. Reptr. 42:* 1111.

3. Hawksworth, D. L. and P. W. Talboys. 1970. *Verticillium albo-atrum. Commonwealth Mycol. Inst. Descriptions Pathogenic Fungi Bacteria 255.* 2 p.

Virus diseases

Four virus diseases (alfalfa mosaic, bean yellow mosaic, cucumber mosaic and pea leaf roll) affect chickpea in Iran and cause wilting, yellowing, foliage mottling and stunting, in addition to 50–90 percent yield reduction from infected plants. Bean yellow mosaic also has been found to affect chickpeas in California. All 4 viruses are transmissible by *Aphis craccivora,* and all but pea leaf roll virus can be transmitted mechanically. Alfalfa mosaic and cucumber mosaic cause local lesions on globe amaranth and systemic infection in cucumber, and all but cucumber mosaic, which causes local lesions only, systemically infect broad bean. Lentil is infected systemically by each of the 4 viruses.

1. Kaiser, W. J. and D. Danesh. 1971. Etiology of virus-induced wilt of *Cicer arietinum*. *Phytopathology 61:* 453–457.

2. Kaiser, W. J., D. Danesh, M. Okhovat and H. Mossahebi. 1968. Diseases of pulse crops (edible legumes) in Iran. *Plant Dis. Reptr. 52:* 687–691.

3. Snyder, W. C., A. O. Paulus and A. H. Gold. 1956. Virus yellows of garbanzo. *Phytopathology 46:* 27.

CINCHONA: *Cinchona officinalis* L. (*C. ledgeriana* Moens.)

Mopog (Mopo)

Recognized in 1882 in Java and later in Sumatra and India, this disease has caused 10–100 percent losses in seedbeds. It is the most serious disease of cinchona seedlings 2 inches or less in height. The same fungus causes a root rot disease in Puerto Rico.

Seeds may be attacked before germination and rot, while seedlings are infected any time after germination. The stems of seedlings become rotted and blackened near the soil line, and the plant falls easily but may remain green foı several days. Occasionally, cotyledons that touch the soil are infected, rot, and drop, but the remainder of the plant is not affected. Areas affected by mopog are easily detected as localized clusters of dead plants in the center of the seedbed. These areas enlarge rapidly under moist conditions, and many plants at the margins of such areas exhibit stem lesions near the soil line. Mycelium of the causal fungus often covers the dead plants and soil underneath them.

Mycelium of *Rhizoctonia solani* is yellow to brown in color, septate, with cells 100–200 μm in length, 8–12 μm in width, and forms compact masses that produce sclerotia. Sclerotia are smooth, brown, somewhat flattened and irregular in shape and size up to 1 cm or more in diameter.

The causal fungus is worldwide in distribution and remains viable in soil for extended periods. It is generally considered to be pathogenic to a large number of plants, although isolates from cinchona are reported to have a rather narrow host range. Incidence and progress of the disease on cinchona are enhanced by excessively moist soil.

Application of 1.25 ℓ formaldehyde (0.008 percent)/0.1 m^2 or steam sterilization of soil for 2 hours at 5.6 kg pressure prior to seeding has been used to control mopog disease. Also, drench application of affected seedbed areas with wettable Spergon (48 percent) at 1 g/0.5 ℓ water/0.1 m^2 or Yellow Cuprocide (21 g/ 3.5 ℓ water/m^2) or Perenox (28 g/ 3.5 ℓ water m^2) has been recommended.

Similar (damping-off) symptoms may be caused by *Cladosporium her-*

barum (Pers.) Lk., *Clitocybe* sp., *Fusarium* sp., *Pestalozzia* sp. and *Pythium* sp.

1. Crandall, B. S. 1950. Coffee and cinchona damping-off controlled by post-germination seedbed treatment. *Plant Dis. Reptr. 34:* 263.

2. Lombard, F. F. 1947. Review of literature on cinchona diseases, injuries, and fungi. *U. S. Dept. Agr. Bibliogr. Bull. 9.* 70 p.

3. Madarang, S. A. 1941. *Rhizoctonia* damping-off of cinchona seedlings. *Philippine J. Forest. 4:* 105–121.

4. Winters, H. F. 1950. Preventing damping-off in cinchona seedbeds. *Trop. Agr. (Trinidad) 27:* 123–126.

Leaf spot diseases

Algal leaf spot was reported on cinchona in 1939 and has been observed in the Belgian Congo (Republic of Congo), Colombia, Guatemala and Peru. A similar disease has been reported from Java. Incidence of this disease usually does not warrant control measures.

Lesions are mostly on the upper leaf surfaces and are circular, raised, light red-brown with some green coloring. The causal alga, *Cephaleuros virescens* O. Kunze, produces a round, reddish, yellow to brown thallus up to 1 cm in diameter attached to the leaf by 1-3-celled rhizoids. Young thalli are devoid of hairs, but older thalli are covered with stiff, simple, sterile hairs and branched, fertile filaments that bear lateral sporangia 38–40 × 25–30 μm. Biciliate zoospores measure 5–8 μm in length.

Leaf spots 2–3 mm in diameter that appeared black on the upper leaf surface but brown to black from below were described from the United States in 1887 and later found in Uganda, and attributed to *Cercospora cinchonae* Ell. & Ev. The fungus produced cylindrical conidia that were nearly hyaline, straight, 3-septate and measured 2.5 × 25–35 μm.

Red-brown leaf lesions caused by *Colletotrichum cinchonae* Koord. have been reported from Java and Russia. Acervuli of the fungus are sparse, 0.1 mm in diameter and spores oval to round, hyaline, and 3–5 × 9–18 μm. Setae are 1-many septate, erect to flexuous, 4–7 × 60–75 μm.

Circular, brown leaf spots that later turn yellow reported from Laos and India are caused by *Gloeosporium* sp. that produces round to irregular sori and unicellular, cylindrical, hyaline conidia 3.3–4.5 × 8–15 μm.

Leaf spots 0.5 cm in diameter and yellow to brown, observed in Ecuador, Java and Russia since 1892, are caused by *Phyllosticta cinchonae* Pat. Bordeaux mixture spray applications have provided satisfactory control.

Leaf lesions (1 cm), caused by *Prillieuxina cinchonae* Stevenson, were observed in Costa Rica and the causal fungus described in 1943. The fruiting structure of the fungus is 150–300 μm wide and asci 25–35 × 30–45 μm, each with 8 uniseptate ascospores 12–15 × 24–27 μm.

Small, nearly translucent lesions that turned brown in 1–2 days were reported on greenhouse cinchona plants from France in 1904. The infections enlarged to completely involve affected laminae in 5–6 days, after which the leaves shrivelled and dropped. Apothecia of the causal fungus *Sclerotinia fuckeliana* (de Bary) Fuckel are light brown in color and 5–10 mm in height. Asci measure 12–13 × 130 μm and the oval, hyaline spores 6–7 × 10–11 μm.

Lapp disease in Java and Sumatra, and possibly Vietnam, is characterized by brittle, distorted young leaves with red margins and interveinal brown lesions with colored margins from which the centers fall. Shoots also become malformed, and affected plants eventually die, but the cause of the disease has not been determined.

Chlorosis of leaves in Puerto Rico was attributed to iron deficiency in 1943, but no cause was determined for unilateral development of laminae on leaves irregular in size and shape and colored light pink.

Various types of leaf mottling with or without sterile flowers and reduced corollas have been noted in Eritrea (Ethiopia), Guatemala, Java and Puerto Rico but no cause determined.

1. Crandall, B. S. 1954. The diseases of the cinchona tree. *FAO Plant Protect. Bull. 3:* 33–37.

2. Crandall, B. S. and W. C. Davis. 1944. *Cephaleuros virescens* on cinchona in Central and South America. *Plant Dis. Reptr. 28:* 926.

3. Lombard, F. F. 1947. Review of literature on cinchona diseases, injuries, and fungi. *U. S. Dept. Agr. Bibliogr. Bull. 9.* 70 p.

4. Reinking, O. A. 1945. Report on cinchona diseases in Guatemala. *Plant Dis. Reptr. 29:* 432–439.

Thread blight

Thin mycelial mats of *Ceratobasidium koleroga* on the trunk, branches and leaves of cinchona plants in Guatemala were described in 1944. Leaves killed by the infection remained hanging by threads of the fungus that sometimes traversed entire plants. Similar symptoms were described from Colombia and attributed to *Marasmius* sp. infection.

1. Davis, W. C. and B. S. Crandall. 1944. Some cinchona diseases in the western hemisphere. *Plant Dis. Reptr. 28:* 996–997.

Scab

Leaf spots and erumpent cankers on stems and inflorescent branches caused by *Elsinoë cinchonae* Jenkins were reported from Colombia and Peru in 1945. Infected fruit capsules were often quite misshapen.

Stem lesions were commonly elliptical, 4 × 5 mm, with dark, elliptical ascomata 300 μm in diameter and globose asci 18–20 μm in diameter. Ascospores were hyaline, 1–3 septate, 15 × 15 μm while conidia, which were somewhat infrequent and borne on conidiophores 3.5–5 × 5–15 μm, were brown, elliptical and 4–5 × 8–10 μm.

1. Jenkins, A. E. 1945. Scab of *Cinchona* in South America caused by *Elsinoë*. *J. Washington Acad. Sci. 35:* 344–352.

Stem blights

Diplodia theobromae infected plants in Java through pruning wounds and subsequently invaded systemically and killed affected plants. Elliptical, uniseptate, hyaline spores 12–15 × 25–35 μm were produced in perithecia 200 μm in diameter.

Fusarium sp. infection of seedlings 2–3 years old causes leaves to turn dull yellow and later red before they wilt. Bark depressions about 5 mm wide extend from the soil line to the lowest branch on diseased seedlings in India where the disease has been observed. Discoloration of the bark extends into the wood under such lesions. Incubation is longer than 3 months, which precludes selection of healthy plants for transplanting. Continual development of new plant bed sites, planting only after sterilization of soil, and transplanting at an early age have reduced losses but are less than satisfactory control measures.

Species of *Phytophthora*, including *P. cinchonae* Sawada, *P. cinnamomi* Rands and *P. palmivora* (*P. faberi* Maub.) have been reported since 1928 to attack seedling cinchona plants in Formosa, Malaya, Philippines, India, Peru, Puerto Rico and Guatemala. Losses have been extensive in some instances. Symptoms vary from darkening of the stem with subsequent loss of foliage and death of plant (root infection) to infection developing first in the tips of shoots and progressing downward to the roots before plants are killed. Also, stem cankers and bark rot at the soil line followed by root rot and dieback symptoms have been associated with *Phytophthora* infections. Control can be accomplished by providing adequate drainage and aeration in seedbed areas.

1. Crandall, B. S. 1947. A new *Phytophthora* causing root and collar rot of *Cinchona* in Peru. *Mycologia 39:* 218–223.

2. Crandall, B. S. 1954. The diseases of the cinchona tree. *FAO Plant Protect. Bull. 3:* 33–37.

3. Crandall, B. S. and W. C. Davis. 1945. *Phytophthora* wilt and stem canker of cinchona. *Phytopathology 35:* 138–140.

4. Kheswalla, K. F. 1935. Seedling blight of *Cinchona ledgeriana* caused by *Phytophthora palmivora* Butl. in the Darjeeling district. *Indian J. Agr. 5:* 485–495.

5. Sawada, K. 1936. *Phytophthora* blight of cinchona seedlings occurring in Formosa, caused by *Phytophthora cinchonae* Saw. n. sp. *Formosan Agr. Rev. 32:* 326–346.

6. Waterhouse, G. M. and J. M. Waterston. 1966. *Phytophthora cinnamomi. Commonwealth Mycol. Inst. Descriptions Pathogenic Fungi Bacteria 113.* 2 p.

Pink disease

One of the more important and frequent diseases of cinchona, pink disease has been found in Burma, Ceylon, India, Africa, South and Central America in addition to Java. It has been recognized as a serious disease of cinchona since 1905.

Pink or white crusts on bark, ringed or cracked bark, and callouses on affected branches or trunks are characteristic symptoms of pink disease. Leaves on infected trees become reddened and dry but remain attached when young twigs are killed quickly. Branches sometimes exhibit rust-colored spots which result from bleeding and drying of sap that exudes from infected areas. Woody portions of trees are often girdled and young plants frequently killed, while older trees may not be killed but severely stunted instead.

Mycelial mats of *Botryobasidium salmonicolor* found on the surface of plant tissues are orange to pink and 1–2 mm thick, consisting of loosely interwoven hyphae 4–7 μm in width, oriented longitudinally. Basidiospores are hyaline, 6–8 × 9–12 μm. Catenulate conidia of *Necator decretus* Massee are 7–8 × 14–18 μm.

Infection occurs through the bark of young, tender woody growth under conditions of high humidity, and the fungus penetrates to the pith, where the infection may progress in advance of external symptoms. While the mycelium of the fungus may be found on nearly any external plant surface, the orange-red fruiting bodies occur most frequently on the upper surfaces of branches and twigs. Excessive moisture, close planting or dense, compact foliage enhances disease development, which commonly is retarded in dry seasons.

Pruning of diseased tissue during periods of dry weather has proven beneficial for reducing the severity of the disease.

1. Davis, W. C. and B. S. Crandall. 1944. Some cinchona diseases in the western hemisphere. *Plant Dis. Reptr. 28:* 996–997.

2. Petch, T. 1912. Pink disease in Java. *Trop. Agr. (Ceylon) 39:* 44–45.

Dasyscypha canker

A serious canker disease of cinchona in Java, known since 1887, has been attributed to *Dasyscypha warburgiana* P. Hennings. Sparse ascomata are

0.5–1 mm in diameter, asci hyaline and cylindrical to subclavate, 7–9 × 50–70 μm, each with 8 ascospores 3–4 × 5–10 μm.

1. Lombard, F. F. 1947. Review of literature on cinchona diseases, injuries and fungi. *U. S. Dept. Agr. Bibliogr. Bull. 9.* 70 p.

Stripe canker (Stem rust)

Symptoms probably caused by stripe canker have been observed in Java and Sumatra since 1886, but the association of pathogenic fungi was not established until about 1940. Of common occurrence under favorable conditions, similar symptoms caused by the same or related fungal species have been reported from Central and South America and Puerto Rico.

Characteristic symptoms consist of parallel depressions that extend along the (nearly vertical) longitudinal axis of the trunk bark. These areas often exhibit rust-colored spots (stem rust) that result from drying of the exudate from these bleeding lesions. Eventually, the lesions become cankerous, and the infection progresses into, and kills, the tissues under affected bark areas and the roots. Variations in symptom expression may include stem girdling cankers, dieback and leaf infections.

Two species of *Phytophthora*, *P. palmivora* and *P. parasitica* have been established as the fungi that cause stripe canker disease. Disease incidence is enhanced by repeated use of old seedbed sites, improper terracing that does not permit adequate drainage, and dense foliage that prevents aeration. Excessive soil moisture and high humidity favor development of the disease.

Pruning diseased tissues thoroughly, followed by application of 5 percent carbolineum in grease to the tissues exposed by excision, can be used as a remedial measure for minor infections. Selection of new seedbed sites and provision for adequate drainage and pruning dense foliage to permit aeration also are strongly recommended.

1. Szkolnik, M. 1951. *Phytophthora parasitica* disease of cinchona in Central American field plantings. *Plant Dis. Reptr. 35:* 16–24.

Armillaria root rot

Root rot of seedling and plantation trees caused by *Armillariella mellea* (Vahl ex Fr.) P. Karst has been reported from Java and Sumatra. Aboveground symptoms are absent until the fungus is well established in the roots, and, when the roots are killed, leaves turn color and drop. Tough, white mycelium can be found between the bark and wood, and black rhizomorphs sometimes occur on the outer surfaces of roots.

Infected plants, including the roots, should be removed and destroyed, and application of slaked lime in the planting site may be beneficial before replanting. Use of *Cinchona succirubra* Pavon as rootstock material may serve to alleviate losses from *Armillaria* root rot.

1. Crandall, B. S. 1954. The diseases of the cinchona tree. *FAO Plant Protect. Bull. 3:* 33–37.

2. Pegler, D. N. and I. A. S. Gibson. 1972. *Armillariella mellea. Commonwealth Mycol. Inst. Descriptions Pathogenic Fungi Bacteria 321.* 2 p.

Gray root rot

Reported from India, Republic of Congo, Indochina, Java and Sumatra, gray root rot caused by *Rosellinia arcuata* Petch and *R. bunodes* is the most important root disease of cinchona. Similar symptoms have been noted in Puerto Rico and Guatemala. Plants of all ages are susceptible, although seedling trees are more often attacked.

Tips of primary and secondary roots are affected first, from which points the fungi spread over the surface of roots as a gray mycelium that later turns black. Fan-shaped, white mycelial mats 1.5 cm in diameter may be found between the bark and wood, and external mycelium sometimes covers the base of the trunk. Above-ground symptoms do not appear until the root infection is well established, when plants suddenly wilt and leaves dry out but remain hanging for some time before falling.

Ostiolate perithecia of *R. arcuata*, 1.5–2.4 mm in diameter (100 μm in height) are scattered in the purple-brown mycelium and bear cylindrical asci 8 × 300 μm. The dark ascospores measure 5–7 × 30–47 μm. *R. bunodes*, which causes root rot of cinchona in Java and Sumatra, produces perithecia to 2 mm in diameter.

Diseased trees should be removed and destroyed and the planting site abandoned or the soil sterilized before replanting. *Cinchona succirubra* is less susceptible to infection and should be used for rootstock.

1. Crandall, B. S. 1954. The diseases of the cinchona tree. *FAO Plant Protect. Bull. 3:* 33–37.

2. Reinking, O. A. 1945. Report on cinchona diseases in Guatemala. *Plant Dis. Reptr. 29:* 432–439.

3. Sivanesan, A. and P. Holliday. 1972. *Rosellinia arcuata. Commonwealth Mycol. Inst. Descriptions Pathogenic Fungi Bacteria 353.* 2 p.

4. Sivanesan, A. and P. Holliday. 1972. *Rosellinia bunodes. Commonwealth Mycol. Inst. Descriptions Pathogenic Fungi Bacteria 351.* 2 p.

Red root rot

Ganoderma pseudoferreum remains localized and damages roots of plants 4 years old or less. It has been reported from Java, Sumatra and Republic of

Congo. The fungus produces red to violet-black, flattened rhizomorphs that are closely appressed to root surfaces, and subsequent infection by white mycelia induces roots to rot. The stalked sporophores, up to 30 cm in diameter, are brown-black or brown-red with interior cap light brown in color. Pores in the cap are 90–140 μm in diameter and hyaline basidia 7 × 20 μm. Basidiospores are hyaline to yellow-brown, ellipsoid, 2.8–4 × 6–9.5 μm.

1. Crandall, B. S. 1954. The diseases of the cinchona tree. *FAO Plant Protect. Bull. 3:* 33–37.

Other root rots

A root and collar rot of seedling cinchona plants, caused by *Sclerotium rolfsii,* has been reported from Peru and Burma.

Sporodesmium cinchonae Koorders has been recognized as the incitant of collar rot and subsequent death of young plants in Java since 1891. Diseased plants continue to appear healthy for a time after root infection can be noted, but leaves later turn red, and death of the plants soon follows. Planting of seedbeds in well-drained, new soil is recommended.

1. Crandall, B. S. 1945. Nursery root and collar rot of cinchona and root rot of canavalia caused by *Sclerotium rolfsii. Plant Dis. Reptr. 29:* 536–537.

2. Crandall, B. S. and W. C. Davis. 1944. Occurrence of cinchona root rots in the Americas. *Plant Dis. Reptr. 28:* 926–929.

CINNAMON: *Cinnamomum zeylanicum* L.

Exosporium leaf spot (*Exosporium* sp.)

Obclavate conidia 8.5–11 × 26–61 μm, each with 3 pseudoseptae, are born on fasciculate, multiseptate, simple conidiophores 4.5–6.5 × 200–720 μm.

1. Muthappa, B. N. 1966. A new species of *Exosporium* on *Cinnamomum zeylanicum* L. from India. Sydowia 19: 146–147.

Rust (*Uraecium nothopegiae* T. S. Ramakrishnan & K. Ramakrishnan) (See also Tejpat—Rust)

Malformations like witches' brooms were induced on red-colored, young shoots in India. Only uredia were observed with urediospores unicellular, pedicellate, light brown and 5–10 × 12–20 μm.

1. Ramakrishnan, T. S. 1965. Notes on some fungi from South India. IX. *Proc. Indian Acad. Sci., Sect. B., 62:* 32–35.

CLOVE: *Eugenia aromatica* Baill.

Dieback

Similarity of described symptoms that have been designated dieback and that may have resulted from any of several causes has created considerable confusion about the cause and distinguishing characters of the disease(s). Dieback of clove has been widespread and probably the most important disease of this crop in Zanzibar since early in this century, but descriptions of the symptoms in the various accounts were not sufficient to attribute losses to a single disease with accuracy.

Development of the disease is essentially the same on trees of all ages but varies largely from location of primary infection site. Young trees infected on the lower portion of the trunk are quickly killed. Many leaves dry to a bright brown color and remain hanging. Similar symptoms, including death of affected trees, result from infection of roots damaged during cultivation. The terminal portions of branches on mature trees die suddenly after becoming infected following injury during harvesting, and infection then progresses toward the main trunk. Progress of the disease, until girdling of the trunk is completed, may leave a tree with only one or a few branches still alive before the entire tree is eventually killed.

The internal appearance of the disease is quite noticeable, with affected wood stained a deep red-brown and sharply delimited from the light colored, healthy wood. Infection is not confined to any specific tissues, but the discoloration of the wood is a diagnostic symptom of infection. Inconspicuous fruiting bodies of the causal fungus, the necks of which may protrude from the bark, develop first around the site of original infection but may not be readily noticed until yellow cirri of spores are extruded under moist conditions.

Pycnidia (500–750 μm in diameter) of the causal fungus *Endothia eugeniae* (Nutman & Roberts) Reid & Booth (syn. *Cryptosporella eugeniae*) are produced in stromata. Hyaline, ovoid pycnidiospores 1.5 × 3–5 μm are extruded in yellowish cirri 1 cm or more in length. Ascospores 3–4 × 7–8 μm are also hyaline and oblong and are borne in obclavate to fusiform asci 4.5–6 × 24–30 μm. Clustered (2–10) perithecia 500–900 μm with necks 120 × 470–680 μm that converge and which are also produced in stromata develop first in elongated bark fissures.

The causal fungus readily invades through wounds resulting from harvest or cultivation injury and sporulates within a few weeks after infection, pycnidia first and perithecia later. Sporulation, particularly perithecial, may be abundant, and dissemination is achieved by rain dispersal of spores and by contamination of workers and their clothing during harvest. Development of disease is much more rapid in twigs, where symptoms may be evident in a few weeks, while dieback may not appear in infected, mature trees for a year or more.

Satisfactory control has been achieved by sanitation that includes removing and destroying dead and moribund trees and by excising infected wood, followed by immediate painting of the resulting exposed surfaces. Wounding incurred during harvesting and cultivation should be kept to a minimum, and treatment of wounds that do result accomplished as soon as possible.

1. Booth, C. and I. A. S. Gibson. 1973. *Endothia eugeniae. Commonwealth Mycol. Inst. Descriptions Pathogenic Fungi Bacteria 363.* 2 p.

2. Nutman, F. J. and F. M. Roberts. 1952. Acute die-back of clove trees in the Zanzibar Protectorate. *Ann. Appl. Biol. 39:* 599–608.

3. Nutman, F. J. and F. M. Roberts. 1952. Two new species of fungi on clove trees in the Zanzibar Protectorate. *Trans. Brit. Mycol. Soc. 36:* 229–234.

4. Nutman, F. J. and F. M. Roberts. 1953. Investigations into diseases of the clove tree in Zanzibar. *East Afr. Agr. J. 18:* 146–154.

5. Reid, J. and C. Booth. 1969. Some species segregated from the genera *Cryptospora, Cryptosporella,* and *Sillia. Can. J. Bot. 47:* 1055–1060.

Sudden death

Loss of clove trees was a recognized problem in Zanzibar and Pemba when a disease of unknown cause was first reported in 1892. A similar condition was described from Madagascar in 1907, and only sporadic attention was given the disease for the next 40 years until it assumed serious proportions in Zanzibar. During this period, similarity of symptoms prompted frequent confusion with dieback disease. The disease is not known to occur other than in Sri Lanka, India, the Moluccas or the Seychelles.

Early symptoms consist only of a general chlorosis of leaves, but subsequent disease development is quite rapid, as the name implies. Leaf fall develops quickly and may be nearly complete in a few days after which a slight wilt develops, and the remaining leaves do not fall but turn a bright red-brown. Typically, the feeder roots of affected trees are completely deteriorated by the time the first symptoms are evident in the foliage. Dead trees left standing assume a silvery appearance after all the dead leaves have fallen.

Internal, yellow discoloration of the cambium region, characteristic of sudden death infection, can be found in main lateral roots immediately after the tree is killed, and some smaller roots may be decayed but bright yellow in color. Discoloration of the cambial region extends to the crown in a few days and continues to spread upward in a widening band in the dead tree until it is generally distributed throughout the tree by about 9 months after the tree was killed. Fructifications of the causal fungus occur only on the discolored wood, with pycnidia developing first and perithecia 3–4 months after the tree is killed.

Valsa eugeniae Nutman & Roberts causes sudden death disease and produces in stromata groups of black, globular perithecia 200–280 μm in diameter with elongated necks 50–60 × 600–900 μm. Numerous, subclavate asci 4–7 × 17–26 μm each contain 8 hyaline, unicellular ascospores 1–1.5 × 4–5 μm. Black, flattened unilocular pycnidia 200–300 μm in diameter and 280–500 μm in height contain hyaline conidiophores 1–1.5 × 10–42 μm, on which are borne unicellular pycnidiospores 0.5–1 × 2–4 μm that are extruded in yellow globules.

Infection, typically in the root system of mature trees, has been shown experimentally to result from spores washed from the soil surface. The causal fungus is quite capable of invading uninjured plants and actually does not develop well from inoculated wounds. Further, the fungus is not important as a cause of dieback symptoms. Seedlings and immature trees apparently are not susceptible to infection, although a condition known as "slow decline" of trees up to 20 years of age strongly resembles sudden death in its later stages and occurs in areas where mature trees have been killed by sudden death disease. The incubation period in mature trees may extend over several months, since the clove tree is unique in that foliar evidence of root damage often does not appear until a large portion of the root system is no longer functional. Dissemination of spores presumably is by rain and surface water.

Control procedures must include removal of all dead and moribund trees, including stumps, which serve only as sources of inoculum. Removal of all roots from sites where trees have been killed before replanting with young healthy trees is recommended as a precaution against slow decline disease.

1. Nutman, F. J. 1950. Studies of the clove tree. III. The effect of the sudden-death disease on water relations. *Ann. Appl. Biol. 37:* 584–590.

2. Nutman, F. J. and F. M. Roberts. 1952. Two new species of fungi on clove trees in the Zanzibar Protectorate. *Trans. Brit. Mycol. Soc. 36:* 229–234.

3. Nutman, F. J. and F. M. Roberts. 1953. Investigations into the diseases of the clove tree in Zanzibar. *East Afr. Agr. J. 18:* 146–154.

4. Nutman, F. J. and F. M. Roberts. 1953. Sudden death disease of the clove tree, *Eugenia aromatica. Nature 171:* 128.

5. Nutman, F. J. and F. M. L. Sheffield. 1949. Studies of the clove tree. I. Sudden-death disease and its epidemiology. *Ann. Appl. Biol. 36:* 419–439.

6. Nutman, F. J., F. M. L. Sheffield, O. S. Swainson and D. W. Winter. 1951. The sudden-death disease of cloves and its economic and agricultural significance. *Empire J. Expt. Agr. 19:* 145–159.

7. Sheffield, F. M. L. 1950. Studies of the clove tree. II. Histology, with special reference to sudden death disease. *Ann. Appl. Biol. 37:* 23–45.

8. Sheffield, F. M. L. 1952. Studies of the clove tree. IV. Natural grafting and its bearing on sudden-death disease. *Ann. Appl. Biol. 39:* 103–110.

9. Sivanesan, A. and P. Holliday. 1970. *Valsa eugeniae. Commonwealth Mycol. Inst. Descriptions Pathogenic Fungi Bacteria 230.* 2 p.

OTHER DISEASES

Gloeosporium leaf spot

1. Jose, P. C. and P. V. Paily. 1966. A *Gloeosporium* leaf spot of clove (*Eugenia caryophyllata*) in Kerala. *Kerala J. Agr. Res. 4:* 115.

Matibudjang, Apoplexy, Slow decline

1. Hadiwidjaja, T. 1956. Matibudjang disease of the clove trees. *Contrib. Gen. Agr. Res. Sta. Bogor (Indonesia) 143.* 73 p.

2. Hubert, F. P. 1957. Diseases of some export crops in Indonesia. *Plant Dis. Reptr. 41:* 55–63.

Sumatra disease

1. Waller, J. M. and D. Sitepu. 1975. Sumatra disease of cloves in Indonesia. *PANS 21:* 141–147.

COFFEE: *Coffea arabica* L.

Halo blight (Mancha aureolada) (*Pseudomonas garcae* Amaral, Teixeira and Pinheiro)

A leaf disease of coffee characterized by necrotic spots to 1.5 cm surrounded by large, diffuse, yellow halos (Fig. 49) appeared in serious proportions in São Paulo, Brazil, in late 1955 and the following year. The disease could be found on seedlings in seedbeds, but occurred more often in new field plantings, where plants 2–5 years old were most susceptible. The disease was caused by a bacterium identified as a species of *Pseudomonas* pathogenic also on citrus, potato (*Solanum tuberosum* L.), bean and olive (*Olea europaea* L.). Two pathotypes were distinguished. Preliminary attempts at control gave inconclusive results, but resistant (Harar, Dilla, Alghe and Geisha) and susceptible (Mundo, Novo, KP 532 and Matari) cultivars were identified.

1. Amaral, J. F., G. C. Teixeira and E. D. Pinheiro. 1956. A bacteria causadora de "mancha aureolada" do cafeeiro. *Arq. Inst. Biol. 23:* 151–155.

Fig. 49. Halo blight symptoms on coffee leaves. (Courtesy of S. A. de Moraes, Instituto Agronomico, Campinas.)

2. Anonymous. 1956. Mancha aureolada, nova moléstia bacteriana do cafeeiro. *Biológico 22:* 93–98.

3. Costa, A. S., J. F. D. Amaral, A. P. Viegas, D. M. Silva, C. G. Teixeira and E. D. Pinheiro. 1957. Bacterial halo blight of coffee in Brazil. *Phytopathol. Z. 28:* 427–444.

4. Costa, A. S. and D. M. Da Silva. 1960. Estudos sobre a mancha aureolada do cafeeiro. *Bragantia 19:* 63–68.

5. Moraes, S. A., M. H. Sugimori, M. T. Filho and P. C. T. Carvalho. 1975. Resistencia de cafeeiros a *Pseudomonas garcae. Summa Phytopathol. 1:* 105–110.

6. Paradela F., O., I. J. A. Ribeiro and M. H. Sugimori. 1974. Differencial behavior of coffee plant progenies in relation to two isolates of the agent of halo blight. *Fitopatologia 9:* 64.

Canker (Llaga macana)

The disease of coffee originally called "kanker" and described from Java (1900) is now known to occur also in Colombia, Costa Rica, Ecuador, Guatemala and Venezuela. Although it is widespread, exact distribution and economic importance of the disease is indefinite because symptoms often appear in older trees and are easily confused with symptoms of other diseases. However, canker is one of the most important diseases of coffee because of progressively reduced yields from trees that survive.

Affected trees are first noticeable by loss of green color in the leaves that gradually become yellow and fall. Branches above and/or below the original infection site, which may not show through the bark on the trunk, may be killed as the infection progresses. Upon removal of the bark, one or more dark irregular cankers are found which may occur at any height on the trunk but are usually near primary wounds and often within a few cm of the soil line. Infections progress slowly and extend more rapidly downward than upward as the lesions increase in size, but penetration and discoloration of the wood usually is no more than 2 cm deep. Diseased trees are generally more than 5 years old and infection of limbs or trunks confined to bark surfaces 3 or more years old. Infected trees usually are killed but steadily decline and become unprofitable if they survive.

The causal fungus is *Ceratocystis fimbriata* (see Cacao—Mal del machete). Conidia and ascospores are disseminated by rain and are also wind-borne. Infection readily occurs through pruning wounds or other types of injuries, i.e. accidental machete cuts, and sporulation (ascospores) has been observed 11 days after inoculation. Incidence of the disease in Costa Rica was found to be much greater (60 percent) at higher altitudes (1050–1500 m) than below 750 m (5 percent). High humidity also increased disease incidence.

Resistance to infection is related to chlorogenic acid content, which has been found higher in resistant *Coffea canephora* Pierre ex Froehner, *C. liberica* Bull ex Hiern and in young limbs compared to old trunks of *C. arabica*. Germination of conidia is stimulated by low concentrations (39–78 ppm) of chlorogenic acid but completely inhibited by 625 ppm.

Seriously diseased trees should be removed and destroyed. Injury to the remaining trees should be carefully avoided and all wounds, including pruning cuts, treated immediately with copper oxychloride. It is possible that early trunk infections could be excised successfully, and limb lesions can be removed by pruning. Both procedures should be followed by treating the exposed areas as indicated above. Attention should be given to reduce humidity by pruning, appropriate spacing of plants and frequent weeding.

1. Echandi, E. and C. E. Fernández. 1962. Relation between chlorogenic acid content and resistance to coffee canker incited by *Ceratocystis fimbriata*. *Phytopathology 52:* 544–546.

2. Echandi, E. and R. H. Segall. 1956. Trunk, branch, and stem canker of coffee trees. *Plant Dis. Reptr. 40:* 916–918.

3. Izquierdo C., E. 1963. El cancer del cafeto. *Cafe Peruano 1:* 10, 16.

4. Morgan-Jones, G. 1967. *Ceratocystis fimbriata. Commonwealth Mycol. Inst. Descriptions Pathogenic Fungi Bacteria 141.* 2 p.

5. Pontis, R. E. 1951. A canker disease of the coffee tree in Colombia and Venezuela. *Phytopathology 41:* 178–184.

6. Schieber, E. 1965. Enfermedades del tallo y raices del cafe en America Central. *Rev. Cafetalera 44:* 18–21.

7. Schieber, E. and E. Echandi. 1960. The canker of coffee trees in Guatemala caused by *Ceratocystis fimbriata* (Ell. & Halst.) Hunt. *Coffee 2:* 91–93.

8. Snyder, W. C., E. E. Trujillo, K. C. McOnie and C. A. Berger. 1960. Trunk and branch canker of coffee trees in Guatemala. *Plant Dis. Reptr. 44:* 566–567.

9. Szkolnik, M. 1951. Coffee trunk and stem canker in Guatemala. *Plant Dis. Reptr. 40:* 500–501.

American leaf spot (Ojo de gallo)

Early historical records of the disease now known as American leaf spot are less than complete, but the disease was observed in Colombia shortly before 1880 and recorded soon thereafter. The disease now occurs in all South and Central American countries where coffee is grown, and there is some evidence that the causal agent does not depend on coffee as a host but rather is a pathogen normally present in tropical American forests. Until recently, the disease was considered the most important on coffee in the western hemisphere. Estimates of annual losses vary because of local environmental differences but often range from 25–75 percent.

Leaf infections first appear as dark colored spots 1 mm or more in diameter surrounding a visible, small, spherical yellow body attached to the leaf surface. Spots become sunken as they grow older and may enlarge to 2 cm or more in diameter but often remain smaller. Lesions, which all have discrete margins, become light tan in color, may retain a narrow dark border and remain mostly round, although spots on veins are sometimes slightly elongated (Fig. 50). Leaves often have several lesions before falling and, while the disease is largely confined to leaves and defoliation can be severe, young stems and berries also are susceptible to infection.

The current name of the causal fungus is *Mycena citricolor* (Berk. & Curt.) Sacc., which is characterized by production of stalked (1–2 mm), yellow, gemmae ("cabecitas") or asexual reproductive bodies to 0.3 mm diameter. Bright yellow basidiocarps with radially striate pilei 1.5–2.5 mm in diameter on stipes 1–1.5 cm in length develop *in vitro* and *in vivo* and produce hyaline, ovoid basidiospores 2.5–3 × 4–5 μm. The fungus in all its forms is luminescent, which facilitates identification without culturing.

Fig. 50. American leaf spot (ojo de gallo) lesions on coffee leaves and fruits.

Disease development is directly affected by relative humidity and rainfall. Production of cabecitas from leaf lesions occurs readily but only under conditions of high moisture. Dispersal of these vegetative reproductive bodies is almost entirely by water droplets or by splashing water, and conditions of high humidity must persist for a few hours if infection is to occur. Optimum temperature for mycelial growth *in vitro* is about 24 C while 18–21 C favors infection by gemmae, but 100 percent relative humidity during incubation is essential. Low light intensity, but not complete darkness, has been shown to enhance gemmae production. Leaf lesions that develop in darkness continue to enlarge, although they produce no gemmae, while lesions exposed to light during development are limited in size but produce gemmae. Lower leaves are more severely affected than those higher on the plant. The role of basidiospores in the disease cycle has not been determined, although it has been reported that the fungus is heterothallic.

Cultivation procedures that improve air drainage and reduce intervals of high humidity, such as wider spacing, regular weeding and pruning, have been recommended. Cupric hydroxide (0.45 kg/100 ℓ) and copper oxychloride (0.68 kg/100 ℓ) were found to reduce production of gemmae.

1. Buller, A. H. R. 1934. *Researches on fungi. Vol. 6*, p. 397–442. Longmans, Green & Co., London.

2. Echandi, E. 1956. Inhibition of gemmae (cabecitas) production of *Mycena citricolor* on coffee trees. *Plant Dis. Reptr. 40:* 775.

3. Echandi, E. and R. H. Segall. 1958. The effectiveness of certain eradicant fungicides on inhibition of gemmae of *Mycena citricolor. Phytopathology 48:* 11–14.

4. Flück, V. 1961. Über ein infektionsverfahren mit *Mycena citricolor. Phytopathol. Z. 40:* 248–257.

5. Maublanc, A. and E. Rangel. 1914. Le *Stilbum flavidum* Cooke, forme avortee de l'*Omphalia flavida* n. sp. *Bull. Soc. Mycol. 30:* 41–47.

6. Moreno, J. D. and V. M. Napoles. 1961. Comportamiento de varios fungicidas en el combate de *Mycena citricolor* del cafeto en Ecuador. *Turrialba 11:* 93–97.

7. Puttemans, M. A. 1904. Sur la maladie du cafeier produite par le *Stilbella flavida. Bull. Soc. Mycol. 20:* 157–165.

8. Quirós, M. 1951. Ojo de gallo (*Omphalia flavida* Maubl. & Range.) *Suelo Tico 5:* 65–69.

9. Rodrigues, C. J., Jr. and D. C. Arny. 1967. Effect of light on gemmae formation in *Mycena citricolor. Can. J. Bot. 45:* 675–679.

10. Salas, J. A. and J. G. Hancock. 1972. Production of the perfect stage of *Mycena citricolor* (Berk. and Curt.) Sacc. *Hilgardia 41:* 213–234.

11. Sequeira, L. 1954. Nuclear phenomena in the basidia and basidiospores of *Omphalia flavida. Mycologia 46:* 470–483.

12. Sequeira, L. 1958. The host range of *Mycena citricolor* (Berk. & Curt.) Sacc. *Turrialba 8:* 136–147.

13. Sequeira, L. and T. A. Steeves. 1954. Auxin inactivation and its relation to leaf drop caused by the fungus *Omphalia flavida. Plant Physiol. 29:* 11–16.

14. Wellman, F. L. 1949. Propagacion del "ojo de gallo" por la lluvia en las fincas de café. *Suelo Tico 2:* 13–16.

15. Wellman, F. L. 1950. Dissemination of *Omphalia* leaf spot of coffee. *Turrialba 1:* 12–27.

16. Wellman, F. L., J. C. Walker, A. Cook and M. Gallegly, Jr. 1951. Effects of temperature on vegetative growth of five coffee disease fungi. *Primera Asamblea Latino-Americana Fitoparasitología. Mex. Dept. Agr. Misc. Foletto 4:* 251–257.

Leaf rust

The history of plant pathology includes only a few examples of significant social impact directly attributable to a single plant disease. Rust of coffee is one of these diseases because it has been largely responsible for significant change not only on agriculture in the Far East but on the beverage preference of an empire. Leaf rust of coffee was first recognized as a plant disease of consequence in the Royal Botanical Gardens at Perideniya, Sri Lanka in 1868, although occurrence of the disease apparently was recorded some 7 years earlier near Lake Victoria (Africa). Within 10 years, yields of coffee in Sri Lanka had been reduced by more than half, and the coffee industry that had flourished there for more than a century was in serious jeopardy. Within a few years, many coffee plantations in Sri Lanka were abandoned as the disease spread throughout southeast Asia and from Madagascar to the Philippine Islands by 1888.

Coffee produced in Sri Lanka prior to the appearance of rust there provided, in large measure, the beverage needs for much of the British Empire. However, the devastation brought on by rust caused many coffee estates to be abandoned, and the owners emigrated; but those growers who chose to remain were obliged to find another plantation crop, e.g. tea. The tea industry prospered and, as a result, the beverage taste of an empire was changed from coffee to tea.

Although leaf rust was known to occur in South Africa as early as 1878, it did not spread rapidly across that continent until the early and mid-twentieth centuries. The disease had been confined to Africa, Asia and the southwest Pacific until 1970, when it was discovered in Brazil (Bahia), where it spread in a southwesterly direction until it had invaded all of the major coffee growing areas in that country. At this time, only a few coffee growing areas, including Colombia, Hawaii, and Papua New Guinea remain free of the disease, although it was necessary to eradicate the disease from the latter country in 1965 at a cost of $ Australian 70,000.

Rust continues to be the most important disease of coffee today because it is present in most of the major coffee growing areas of the world and can cause tremendous losses in yields.

Symptoms of rust first consist of chlorotic leaf spots that become yellow-orange, powdery, more or less round and pustular on the lower surfaces of coffee leaves. They enlarge slowly from 1–3 mm to 2 cm or more and may coalesce (Fig. 51). Mature lesions consist of necrotic centers surrounded by circular bands of sori. Lesions are visible from the upper leaf surfaces as

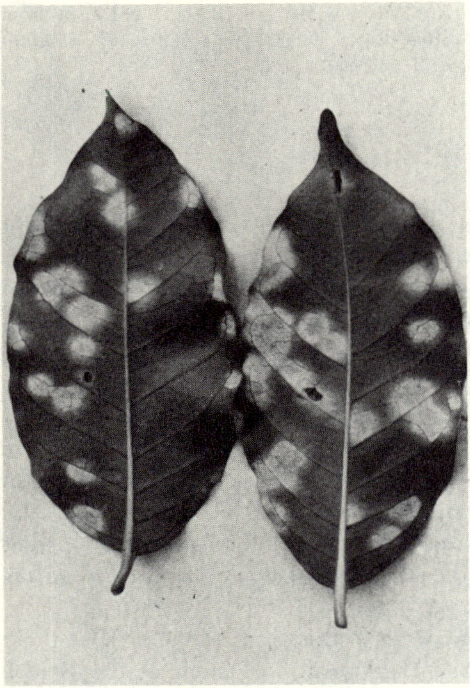

Fig. 51. Leaf rust pustules on the underside of coffee leaves. (Courtesy of D. M. Okioga, Coffee Research Foundation, Ruiru.)

yellowed areas but without the powdery appearance resulting from sporulation. Affected leaves become dark and necrotic, are killed and fall prematurely. Defoliation may be nearly complete with loss of any developing berries. Dieback of young branches and shoots may result, and growth and vigor of affected plants are seriously reduced, as is the berry crop the year following serious rust infection.

Leaf rust is caused by *Hemileia vastatrix* Berk. & Br., characterized by sori in pustular lesions consisting of clavate filaments that extend through stomata and on which spores are borne. Unicellular uredospores, 18–28 × 28–36 μm and reniform, are heavily warted on the convex surface but smooth on the straight side, while teliospores that develop in the centers of old lesions are spherical, smooth and 20–28 μm in diameter. The promycelium from a germinating teliospore is triseptate and bears 4 basidiospores each with 1-several nuclei, but most basidiospores are binucleate. (Note: A similar fungus, *H. coffeicola* Maubl. & Roger, is distinguished by sori that cover the entire lower surfaces of leaves, reniform uredospores that measure 20–28 × 34–40 μm and spherical teliospores that are 20–26 μm in diameter. This fungus is much more restricted in geographic distribution, being presently confined to São Tomé and Príncipe, Nigeria, Cameroon and Uganda.)

Uredospores, the only known infectious spore form, germinate by 1–5 germ tubes that penetrate indirectly, produce various forms of haustoria, and invade intercellularly to the pallisade tissue. Germination of uredospores occurs within 3–5 hours, appressorium formation within 6.5–8.5 hours and first symptoms within 14 days. Sporulation may begin 2–4 days after symptoms appear. Cardinal temperatures for uredospore germination, which is dependent on free water, are 15, 22 and 29 C. Basidiospores are not known to infect species of the genus *Coffea* and neither pycnia nor aecia have been described.

Spores, often dislodged in clusters, may be dispersed by wind but can be detached and distributed by splashing rain, insects and man. Debate continues concerning the importance of air-borne spores in dissemination, and many believe that dispersal over short distances, and certainly within a field, results largely from water droplets and splashing rain. However, that spores can be carried in the air to considerable heights, and that the disease appears in distant locations from known infection centers with no evident vector(s), is presumptive evidence of dissemination via air currents.

Disease development is directly influenced by environmental conditions, particularly rainfall, and mathematical formulations to predict severity of infection have been devised. The necessity of free water for spore germination and dispersal of spores by splashing rain to the undersurfaces of leaves where infection through stomata occurs are both obvious causes for rust to appear with the onset of a wet season and to increase in intensity throughout the duration of such conditions. Rust is much less important during dry seasons, when infected leaves fall and inoculum levels remain low. Length of the incubation period varies according to altitude (temperature) and is shorter in duration at lower levels. A shorter incubating period permits more sporulating cycles per season that results in more serious disease incidence at lower altitudes.

Physiologic specialization in the leaf rust fungus was first reported from India in 1932 and has been studied extensively since that time, much of the work being done at Centro de Investigacão das Ferrugens do Cafeeiro, Oeiras, Portugal. Six dominant factors (S_H1–S_H6) have been identified in *C. arabica* and hybrids and have led to differentiation of at least 30 races of the fungus based on the responses of 21 clones selected from several coffee species. The existence of additional races has been postulated on a gene-for-gene theory.

Three fungi have been found associated with uredospores of *H. vastatrix*, but only *Verticillium hemileiae* Bour. is a serious hyperparasite and then only under glasshouse conditions of relative humidity above normal and temperature below 20 C. *Cladosporium hemileiae* Stey., also probably a true hyperparasite, has been found only with *H. vastatrix* on *C. canephora* in Angola. *Penicillium waksmani* Zaleski is often found associated with the leaf rust fungus but is not known to be predacious on the rust. The first and last mentioned fungi (*V. hemileiae* and *P. waksmani*), with *Paranectria carrissiana* Cam. & Luz., have been found associated with uredospores of *H. coffeicola*.

Genetic resistance to leaf rust was recognized by 1912 in varieties of *C. canephora* ("Robusta") and *C. liberica*, but it was difficult to obtain fertile hybrids with *C. arabica*. "Arabica" types with inherent resistance, e.g. Coorg and Kents, eventually were attacked, but breeding programs to develop commercial types with broad spectrum resistance has continued. Resistant varieties have not been important in Africa and Asia yet because of the prevalence of races of rust capable of infecting plants without resistance to the complete list of races and because useful chemical control is available. Varieties such as Geisha and Blue Mountain are resistant to many, but not all, common races of rust and so can not be recommended without caution.

Application of fungicides for control of leaf rust has been recognized since 1886 but apparently was not used until 1926 in India. Fungicides containing copper, e.g. cuprous oxide or cuprous oxychloride, can be used effectively against rust as well as several other materials, particularly zineb. Timing of applications is critical, and maximum effectiveness is achieved by treatment immediately before and during the early season rains. While infection occurs only on lower leaf surfaces, application of fungicide to either leaf surface provides nearly equal control, which is directly related to concentration of fungicide available for mixing with free water on leaf surfaces during periods of spore dispersal by splashing rain. Until effective control can be expected through use of resistant varieties, copper fungicides (approximately 3.7 kg/ha each application) can provide satisfactory control where horticultural practices permit or can be adapted to efficient application. Future plantings should be made in a manner to provide for application of fungicides in anticipation of need for rust control, even though the disease may not be present at the time.

1. Anonymous. 1971. *Centro de investigacão das ferrugens do cafeeiro.* Estação Agronomica Nacional, Oeiras, Portugal. 29 p.

2. Bettencourt, A. J. and M. Noronha-Wagner. 1969. Genetic factors conditioning resistance of *Coffea arabica* L. to *Hemileia vastatrix* Berk. et Br. Agron. *Lusitana* 31: 285–292.

3. Bock, K. R. 1962. Dispersal of uredospores of *Hemileia vastatrix* under field conditions. *Trans. Brit. Mycol. Soc. 45:* 63–74.

4. Bock, K. R. 1962. Seasonal periodicity of coffee leaf rust and factors affecting the severity of outbreaks in Kenya colony. *Trans. Brit. Mycol. Soc. 45:* 289–300.

5. Bock, K. R. 1962. Control of coffee leaf rust in Kenya colony. *Trans. Brit. Mycol. Soc. 45:* 301–313.

6. Burdekin, D. A. 1964. The effect of various fungicides on leaf rust, leaf retention and yield of coffee. *East Afr. Agr. Forest. J. 30:* 101–104.

7. Chaves, G. M., J. C. Filho, M. G. de Carvalho, K. Matsuoka, D. T. Coelho and C. Shimoya. 1970. A ferrugem do cafeeiro (*Hemileia vastatrix* Berk. & Br.) Revisão de literatura com observações e comentários sobre a enfermedade no Brazil. Seiva Edicão Especial, Universidade Federal de Vicosa, Minas Gerais, Brasil. 75 p.

8. Chinappa, C. C. and M. S. Sreenivasan. 1965. Cytological studies on germinating teliospores of *Hemileia vastatrix*. *Caryologia 18:* 625–631.

9. Chinappa, C. C. and M. S. Sreenivasan. 1965. Secondary sporidia in *Hemileia vastatrix*. *Current Sci. 34:* 668–670.

10. Chinappa, C. C. and M. S. Sreenivasan. 1968. Cytology of *Hemileia vastatrix*. *Caryologia 21:* 75–82.

11. Crowe, T. J. 1963. Possible insect vectors of the uredospores of *Hemileia vastatrix* in Kenya. *Trans. Brit. Mycol. Soc. 46:* 24–26.

12. Firman, I. D. 1972. Coffee leaf rust in Brazil. *FAO Plant Protect. Bull. 20:* 122–126.

13. Firman, I. D. and J. A. N. Wallis. 1965. Low volume spraying to control coffee leaf rust in Kenya. *Ann. Appl. Biol. 55:* 123–137.

14. Hocking, D. 1968. Effects of light on germination and infection of coffee rust (*Hemileia vastatrix*). *Trans. Brit. Mycol. Soc. 51:* 89–93.

15. Laundon, G. F. and J. M. Waterston. 1964. *Hemileia vastatrix*. *Commonwealth Mycol. Inst. Descriptions Pathogenic Fungi Bacteria 1*. 2 p.

16. Laundon, G. F. and J. M. Waterston. 1964. *Hemileia coffeicola*. *Commonwealth Mycol. Inst. Descriptions Pathogenic Fungi Bacteria 2*. 2 p.

17. Maublanc, A. and L. Roger. 1934. Une nouvelle rouille du caféier au Cameroun. *Bull. Soc. Mycol. France 50:* 193–202.

18. Mayne, W. W. 1932. Physiological specialisation of *Hemileia vastatrix* B. & Br. *Nature 129:* 510.

19. Mulinge, S. K., C. J. Baker and J. L. Pereira. 1974. Effects of pyracarbolid on coffee leaf rust in Kenya. *Phytopathology 64:* 147–149.

20. Mulinge, S. K. and E. Griffiths. 1974. Effects of fungicides on leaf rust, berry disease, foliation and yield of coffee. *Trans. Brit. Mycol. Soc. 62:* 495–507.

21. Narasimhaswamy, R. L. 1961. Coffee leaf disease (*Hemileia*) in India. *Coffee 3:* 33–39.

22. Noronha-Wagner, M. and A. J. Bettencourt. 1967. Genetic study of the resistance of *Coffea* spp. to leaf rust. I. Identification and behaviour of four factors conditioning disease reaction in *Coffea arabica* to twelve physiologic races of *Hemileia vastatrix*. *Can. J. Bot. 45:* 2021–2031.

23. Nutman, F. J. and F. M. Roberts. 1962. Dispersal of coffee rust *Hemileia vastatrix* B. et B. *Nature 194:* 1296.

24. Nutman, F. J. and F. M. Roberts. 1963. Studies on the biology of *Hemileia vastatrix* Berk. & Br. *Trans. Brit. Mycol. Soc. 46:* 27–48.

25. Nutman, F. J., F. M. Roberts and K. R. Bock. 1960. Method of uredospore dispersal of the coffee leaf-rust fungus, *Hemileia vastatrix*. *Trans. Brit. Mycol. Soc. 43:* 509–515.

26. Oliveira, B. D. 1954–1957. As ferrugens do cafeeiro. *Rev. Cafe Portugues 1 (4):* 5–13; *2 (5):* 5–12, *(6):* 5–13, *(7):* 9–17, *(8):* 5–22; *4 (16):* 5–15.

27. Oliveira, B. D. 1958. Selection of coffee types resistant to the *Hemileia* leaf rust. *Coffee Tea Ind. 81:* 112–120.

28. Orillo, F. T. and R. B. Valdez. 1961. Selection of *Coffea* species and varieties resistant to coffee rust in the Philippines. *Philippine Agr. 45:* 223–234.

29. Rayner, R. W. 1960. Rust disease of coffee. *World Crops 12:* 187–190, 222–224, 261–264, 309–312.

30. Rayner, R. W. 1961. Spore liberation and dispersal of coffee rust *Hemileia vastatrix* B. et Br. *Nature 191:* 725.

31. Rayner, R. W. 1961. Germination and penetration studies on coffee rust (*Hemileia vastatrix* B. & Br.). *Ann. Appl. Biol. 49:* 497–505.

32. Rayner, R. W. 1962. The control of coffee rust in Kenya by fungicides. *Ann. Appl. Biol. 50:* 245–261.

33. Reyes, T. T. 1957. Uma nova raca fisiologica de *Hemileia vastatrix* B. et Br. Rev. *Café Português 4:* 12–15.

34. Rodrigues, C. J., Jr. 1956. Nota sobre a resistência de algumas especies de *Coffea* à *Hemileia coffeicola* Maubl. *Rev. Café Português 3:* 48–71.

35. Rodrigues, C. J., Jr., A. J. Bettencourt and L. Rijo. 1975. Races of the pathogen and resistance to coffee rust. *Annu. Rev. Phytopathol. 13:* 49–70.

36. Rodrigues, C. J., Jr., E. F. Medeiros and B. G. Lewis. 1975. Relationship between a phytoalexin-like response in coffee leaves (*Coffea arabica* L.) and compatibility with *Hemileia vastatrix* Berk. & Br. *Physiol. Plant Pathol. 6:* 35–41.

37. Schieber, E. 1972. Economic impact of coffee rust in Latin America. *Annu. Rev. Phytopathol. 10:* 491–510.

38. Schieber, E. 1975. Present status of coffee rust in South America. *Annu. Rev. Phytopathol. 13:* 375–382.

39. Shaw, D. E. 1970. Coffee eradication in a previously coffee rust infected area in Papua. *Papua New Guinea Agr. J. 22:* 59–61.

40. Stevenson, J. A. and R. Beam. 1955. An annotated bibliography of coffee rust (*Hemileia* spp.). *U. S. Dept. Agr. Div. Mycol. Dis. Surv. Spec. Pub. 3.* 80 p.

41. Waller, J. M. 1972. Coffee rust in Latin America. *PANS 18:* 402–408.

42. Wallis, J. A. N. and I. D. Firman. 1962. Spraying Arabic coffee for the control of leaf rust (*Hemileia vastatrix*). *East Afr. Forest. J. 28:* 89–104.

43. Ward, H. M. 1882. The life history of *Hemileia vastatrix*—the fungus of the "coffee-leaf disease." *J. Linn. Soc. (Botany) 19:* 299–335.

44. Wellman, F. L. 1970. The rust *Hemileia vastatrix* now firmly established on coffee in Brazil. *Plant Dis. Reptr. 54:* 539–541.

45. Wellman, F. L. 1970. "Hemileia vastatrix"—*investigaciones presentes y pasadas en la herrumbre del cafe y su importancia en la America tropical.* Organismo Internacional Regional de Sanidad Agropecuaria, San Salvador. 67 p.

Ascochyta leaf blight

A disease of coffee described in 1902 from Brazil has been reported from Guatemala recently and attributed to the same fungus species originally identified in Brazil. Leaf blight and twig dieback symptoms described in 1957 from Ethiopia were quite similar to those noted in Italy some 25 years earlier and more recently in Kenya and Nyasaland.

Necrotic lesions on leaves are tan to brown with distinct dark borders, often with concentric rings, and frequently irregular in shape. Dieback symptoms occur more often on young twigs and may become sufficiently extensive to kill affected plants. Symptoms are easily confused with those of brown blight.

The disease is caused by *Ascochyta tarda* Stewart that produces globose pycnidia 70–110 μm and elliptical and unicellular, hyaline conidia 2–3 × 9–14 μm. A species of *Mycosphaerella* often associated with natural infections (but not proven to be the perfect stage of *A. tarda*) has dark, subglobose peritheica 60–75 × 60–90 μm, clavate, hyaline asci 7.5–12 × 25–45 μm and 2-celled ascospores 4.5–7.5 × 12–15 μm.

The disease is serious only on plants growing in excessive shade and, while plants of all ages are susceptible, seedlings and young trees 3–4 years old are more severely affected. Varietal differences in susceptibility have been reported.

1. Firman, I. D. 1965. Some investigations on a disease of *Coffea arabica* caused by *Ascochyta tarda*. *Trans. Brit. Mycol. Soc. 48:* 161–166.

2. Hennings, P. 1902. Fungi S. Paulensis. *Hedwigia 41:* 295–311.

3. Schieber, E. 1972. Leaf blight incited by *Ascochyta coffeae* on coffee in Guatemala. *Plant Dis. Reptr. 56:* 753–754.

4. Stewart, R. B. 1957. Leaf blight and stem dieback of coffee caused by an undescribed species of *Ascochyta*. *Mycologia 49:* 430–433.

Brown eye spot (Berry blotch, Chasparria)

The fungus that causes symptoms referred to as brown eye spot and berry blotch of coffee was described in 1881. Both brown eye spot and berry blotch are found in Africa, the Americas and in Pacific areas where coffee is grown. Although these diseases are common and widespread now, they are not generally considered of major economic import.

Infections on leaves are first evident as small, brown necrotic spots that enlarge to 4–15 mm and develop light colored centers. The margins of these circular lesions remain brown and are often bordered by chlorotic halos (Fig. 52). Lesions may be numerous, coalesce and cause epinasty and general yellowing of affected leaves that fall prematurely. Defoliation, particularly of lower leaves, may be extensive.

Fruit symptoms appear as brown, sunken spots more commonly found on the upper or exposed surfaces of full size, green or maturing fruit. A purple halo near the color of ripening fruit around each lesion is evident, and, with the associated lesion(s), becomes larger and more obvious as affected fruits ripen. Diseased fruits turn black, shrivel and often fall, but some remain attached to the plant. Such fruits are lighter than healthy fruits and, when mixed with harvested fruits, complicate the depulping process.

Seedlings are quite susceptible to infection and damage can be severe in seedbeds and nurseries. The disease and subsequent dieback may be so serious that plants are unfit for field planting.

The causal fungus is *Cercospora coffeicola* Berk. & Cke. that produces conidiophores (4–6 × 60–210 μm) and conidia (4–5 × 15–300 μm), both of which have been reported to vary considerably in length, possibly because of the influence of different environmental conditions during production.

Cardinal temperatures for growth of the fungus *in vitro* are 8, 24–28 and 36 C and approximately the same for conidial production and germination. Conidial production *in vivo* occurs only when relative humidity is 98

Fig. 52. Brown eye (*Cercospora*) leaf spot of coffee.

percent or above. Conidia are readily dispersed in water and can be windborne, while mycelial penetration of leaves is effected indirectly via stomata. Symptom development is more severe in direct sunlight under conditions of high humidity and noticeably more restricted in shade. Five species of *Coffea* (*C. arabica*, *C. canephora*, *C. eugenioides* S. Moore, *C. liberica* and *C. racemosa* Lour.) and 2 unrelated plants (castor bean and *Leonurus sibiricus* L.) are susceptible to infection. Some differences in severity of symptoms on species of coffee have been noted.

Horticultural practices to reduce humidity in the field, i.e. pruning, weeding and spacing of plants for maximum air drainage, should reduce damage from brown eye spot. Growing coffee plants under shade also should be beneficial. A number of fungicides, including captan, ferbam, maneb, cupric hydroxide and tribasic copper sulfate, have significantly reduced disease incidence.

1. Awatramani, N. A. 1963. *Cercospora* incidence in relation to growth of coffee seedlings in different containers. *Indian Coffee 27:* 339–340.

2. Bianchini P., C. 1959. Sintomas, prevencion y combate de las principales enfermedades del cafe en Costa Rica. *Suelo Tico 11:* 37–48.

3. Bitancourt, A. A. 1958. As manchas da fôlha do cafeeiro. *Biológico 24:* 191–201.

4. Chupp, C. 1953. *A monograph of the fungus genus* Cercospora. Ithaca, N.Y. 667 p.

5. Echandi, E. 1959. La chasparria de los cafetos ausada por el hongo *Cercospora coffeicola* Berk. & Cooke. *Turrialba 9:* 54–67.

6. Gonzalez T., J. A. 1961. Manera de combatir la chasparria del cafe. *Nuestra Tierra 5:* 153–154.

7. Mulder, J. L. and P. Holliday. 1974. *Cercospora coffeicola*. *Commonwealth Mycol. Inst. Descriptions Pathogenic Fungi Bacteria 415.* 2 p.

8. Rodriguez, R. A. and C. A. Sota. 1968. Cupric emulsions for the control of the *Cercospora* disease of coffee. *Phytopathology 58:* 554.

9. Sridhar, T. S. and S. Subramanian. 1968. Studies on the brown-eye-spot disease (*Cercospora coffeicola* Berk. et Cooke) of coffee (*Coffea arabica* L.). *Riv. Patol. Veg. (Ser. 3) 4:* 33–39, 41–47.

10. Sridhar, T. S. and S. Subramanian. 1969. Studies on the brown-eye-spot disease of coffee. *Indian Coffee 33:* 97–99.

11. Wellman, F. L. and T. R. Quesada. 1951. Certain factors limiting *Cercospora* disease on coffee leaves. *Primera Asamblea Latino-Americana Fitoparasitología. Mex. Dept. Agr. Misc. Foletto 4:* 246–251.

12. Wellman, F. L., J. C. Walker, A. Cook and M. Gallegly, Jr. 1951. Effects of temperature on vegetative growth of five coffee disease fungi. *Primera Asamblea Latino-Americana Fitoparasitología. Mex. Dept. Agr. Misc. Foletto 4:* 251–257.

Coffee berry disease (Anthracnose, Brown blight)

A number of common or local names have been applied to diseases of coffee now regarded as a complex caused by related forms of the same fungus. "Brown blight" has been used to describe a disease of ripe or ripening berries only, and "coffee berry disease" (CBD) typically includes first appearance of symptoms on green berries, but flowers and fruits in all stages of development may be infected. The term "anthracnose" has been applied to symptoms of leaves, twigs, flowers and berries that may be induced by forms of the fungus capable of causing either brown blight or coffee berry disease. The fungus that causes these diseases also is often associated with dieback symptoms (see Coffee—Other Diseases), which occur in Kenya only under very wet, high altitude conditions favorable for coffee berry disease infection, but a causal relation has not been established.

Brown blight was recognized in Uganda in 1911 and in India in 1918 and was known to occur in Kenya before coffee berry disease was first reported from there in 1922. Coffee berry disease is confined to Africa, where it is a major disease now in Angola, Cameroon, Congo, Ethiopia, Rwanda and Tanzania, in addition to Kenya, and can cause complete crop losses locally. Anthracnose symptoms were first noted on coffee in Brazil in 1901 and since then have been reported from nearly every country where the crop is grown, including Sri Lanka, Malaysia, Java and Central America, in addition to Africa. Losses can be quite high locally, and anthracnose has been the cause for complete abandonment of coffee production in some areas. Seasonal losses from coffee berry disease in all African producing countries has been estimated at 50 percent, while losses from brown blight were assessed at 4 percent or less.

Green berries in all stages of development are susceptible to coffee berry disease, and infections first appear as a small, dark brown spot that enlarges and becomes slightly sunken. The entire berry later becomes affected as the pulp turns brown and becomes hard and brittle (Fig. 53). The bean under the infection site first begins to show signs of deterioration but, by the time the pulp is involved, both beans have become blackened and shrivelled and of absolutely no commercial value. Berries desiccate and later fall from diseased stalks, but the infection usually does not involve the twig, which, with surrounding leaves, remains healthy.

Fig. 53. Symptoms of coffee berry (anthracnose) disease on coffee fruits. (Courtesy of D. M. Okioga, Coffee Research Foundation, Ruiru.)

Brown blight symptoms on berries consist of sunken, brown lesions which are readily distinguished from other fruit diseases in the field and which result only in superficial blemish to beans from affected berries.

Leaf symptoms of anthracnose consist of small brown spots, often along the margins of leaf laminae, which become progressively darker and 1–2 cm in diameter. Lesions are often delimited by veins and are noticeable from both leaf surfaces. Affected twigs blacken and desiccate, often resulting in growth from axillary buds in diseased seedlings. Berry infections first appear as black spots that enlarge and become slightly sunken on berries in any stage of development. The pulp and seeds later become infected, and the fruit drops. Sporulation by the causal fungus is often seen in leaves, twigs or fruit, either while still on the plant or after they have fallen to the ground.

The fungus that causes these various disease manifestations is *Colletotrichum coffeanum* Noack *(Glomerella cingulata)*. Septate setae in the acervuli measure 85–103 μm and the conidia 4–6 × 13–18 μm. Ascospores 3–5 × 12–23 μm are borne in asci 8–12 × 53–76 μm. Minor cultural and morphological differences have been reported for isolates that cause the different disease manifestations.

Occurrence of the brown blight and anthracnose diseases has been quite general throughout coffee culture for years. Coffee berry disease, in serious proportions, was limited to elevations in Africa in excess of 1675 m until 1961 when the disease appeared in Kenya in lower elevations with devastating effects. Infection of green berries only occurs within a temperature range of 17–26 C when free water is available for 5 hours or more, a condition found in Kenya before 1961 with much greater frequency at higher elevations. Excessive rainfall with lower maximum temperatures first occurred in Kenya in 1961, and the ubiquitous fungus developed quickly in the lower elevations where it has continued to be a serious problem. Strains of the causal fungus, based on gross morphological characteristics, have been recovered from all parts of infected plants, including maturing bark where the fungus persists in saprophytic form. That the fungus can persist in debris of fallen leaves and maturing bark without causing disease symptoms and in close association with form(s) of the fungus that are parasitic but with presumed different pathogenic capabilities has led to persistent confusion about the pathological cycle of the fungus. Conidia, largely dispersed by water, develop on infected twigs and berries and serve to efficiently disseminate the fungus, but the role of the perfect stage of the organism as a pathogen remains to be determined.

Continuous flowering and fruit production complicate disease control, but the fungicides captafol (0.4 percent), benomyl (0.05 percent) and chlorothalonil (0.4 percent) have given satisfactory control applied on a 3-week schedule in Kenya. However, the appearance of biotypes of the pathogen tolerant to benomyl and related fungicides has significantly reduced their usefulness. The varieties Rume Sudan and Geisha 10 have been reported to show resistance in Kenya.

1. Bock, K. R. 1956. Investigations on coffee berry disease—laboratory studies. *East Afr. Agr. J. 22:* 97–103.

2. Bock, K. R. 1963. The control of coffee berry disease in Kenya. *Empire J. Expt. Agr. 31:* 97–107.

3. Boisson, C. 1960. L'anthracnose du caféier. *Rev. Mycol. 25:* 263–292.

4. Clarke, R. T. and J. G. Williamson. 1963. The control of coffee-berry disease in Kenya by seasonal fungicidal sprays. *Empire J. Expt. Agr. 31:* 327–333.

5. Cook, R. T. A. 1975. The future of benzimidazole fungicides for CBD control in Kenya. *Kenya Coffee 40:* 139–144.

6. Fernandez, C. E. 1963. Antracnosis del cafe en Guatemala. *Rev. Cafetalera 27:* 6–8.

7. Firman, I. D. 1964. Screening of coffee for resistance to coffee berry disease. *East Afr. Agr. Forest. J. 29:* 192–194.

8. Foucart, G. 1961. "Shedding" et "coffee berry disease" du caféier d'Arabie. *Parasitica 17:* 22–28.

9. Gibbs, J. N. 1969. Inoculum sources for coffee berry disease. *Ann Appl. Biol. 64:* 515–522.

10. Gibbs, J. N. 1971. Some factors influencing the performance of spray programmes for the control of coffee berry disease. *Ann. Appl. Biol. 67:* 343–356.

11. Gibbs, J. N. 1972. Effects of fungicides on the populations of *Colletotrichum* and other fungi in bark of coffee. *Ann. Appl. Biol. 70:* 35–47.

12. Griffiths, E. and J. N. Gibbs. 1969. Early-season sprays for the control of coffee berry disease. *Ann. Appl. Biol. 64:* 523–532.

13. Griffiths, E., J. N. Gibbs and J. M. Waller. 1971. The control of coffee berry disease. *Ann. Appl. Biol. 67:* 45–74.

14. Griffiths, E. and I. Furtado. 1972. A berry infection technique for assessment of the CBD strain of *Colletotrichum coffeanum* on coffee branchlets. *Trans. Brit. Mycol. Soc. 58:* 313–320.

15. Henrard, P. 1957. *Colletotrichum coffeanum* Noack. *Agricultura (Louvain) 5:* 39–55.

16. Hindorf, H. 1973. *Colletotrichum*-population auf *Coffea arabica* L. in Kenya. *Phytopathol. Z. 77:* 97–116; 216–234; 324–338.

17. Hocking, D. 1966. Brown blight (*Colletotrichum coffeanum* Noack) of arabica coffee in East Africa. *Ann. Appl. Biol. 58:* 409–421.

18. Hocking, D., J. C. Johanns and H. Vermeulen. 1967. Ascospore production, discharge and infection by *Glomerella cingulata* causing coffee berry disease. *Nature 214:* 1144–1145.

19. Huxley, P. A. and A. Turk. 1967. Berry wetness and the incidence of coffee berry disease. *Kenya Coffee 32:* 99–104.

20. McDonald, J. 1926. A preliminary account of a disease of green coffee berries in Kenya colony. *Trans. Brit. Mycol. Soc. 11:* 145–154.

21. Mordue, J. E. M. 1971. *Glomerella cingulata. Commonwealth Mycol. Inst. Descriptions Pathogenic Fungi Bacteria 315.* 2 p.

22. Mulinge, S. K. 1970. Development of coffee berry disease in relation to the stage of berry growth. *Ann. Appl. Biol. 65:* 269–276.

23. Mulinge, S. K. 1971. Effect of altitude on the distribution of the fungus causing coffee berry disease in Kenya. *Ann. Appl. Biol. 67:* 93–98.

24. Mulinge, S. K. 1973. Coffee berry disease. *FAO Plant Protect. Bull. 21:* 85–86.

25. Muller, R. 1961. L'anthracnose des baies du caféier d'arabie (*Coffea arabica*) due à *Colletotrichum coffeanum* Noack au Cameroun. *Café, Cacao, Thé 5:* 227–241.

26. Nutman, J. and F. M. Roberts. 1960–61. Investigations on a disease of *Coffea arabica* caused by a form of *Colletotrichum coffeanum* Noack. *Trans. Brit. Mycol. Soc. 43:* 489–505; 643–659; *44:* 511–521.

27. Nutman, F. J. and F. M. Roberts. 1969. Coffee berry disease: epidemiology in relation to control. *Expt. Agr. 5:* 271–282.

28. Nutman, F. J. and F. M. Roberts. 1969. The effect of fungicidal treatments on sporulating capacity in relation to the control of coffee berry disease. *Ann. Appl. Biol. 64:* 101–112.

29. Nutman, F. J. and F. M. Roberts. 1969. The stimulating effect of some fungicides on *Glomerella cingulata* in relation to the control of coffee berry disease. *Ann. Appl. Biol. 64:* 335–344.

30. Nutman, F. J. and F. M. Roberts. 1969. Climatic conditions in relation to the spread of coffee berry disease since 1962 in the East Rift districts of Kenya. *East Afr. Agr. Forest J. 35:* 118–127.

31. Nutman, F. J., F. M. Roberts and G. H. Freeman. 1969. Seasonal variations in the sporulating capacity of the fungus causing coffee berry disease. *Ann. Appl. Biol. 64:* 85–99.

32. Rayner, R. W. 1952. Coffee berry disease—a survey of investigations carried out up to 1950. *East Afr. Agr. J. 18:* 130–158.

33. Okioga, J. N. 1975. Occurrence of strains of *Colletotrichum coffeanum* resistant to systemic fungicides currently used for control of coffee berry disease in Kenya. *Kenya Coffee 40:* 170–173.

34. Vermeulen, H. 1968. Screening fungicides for control of coffee berry disease in Kenya. Expt. Agr. 4: 255–261.

35. Vine, B. H., P. A. Vine and E. Griffiths. 1973. Evaluation of fungicides for the control of coffee berry disease in Kenya. *Ann. Appl. Biol. 75:* 359–375.

36. Wallis, J. A. N. and I. D. Firman. 1965. Spraying arabica coffee for the control of coffee berry disease. *Ann. Appl. Biol. 55:* 139–148.

37. Woodhead, T. 1968. Micro-meteorological studies of coffee berry disease. *Ann. Appl. Biol. 62:* 451–463.

Fusarium bark disease (Leaf spot, Berry disease)

A bark disease of coffee shoots was reported from Tanganyika (Tanzania) in 1932 and has been found since in various manifestations in Malawi and

Kenya. Incidence of the different conditions may be important locally but generally is of little economic importance and often appears in warm, wet weather following insect, mechanical or herbicide damage or mulching too close to the stem.

Common names have been applied to different syndromes which are distinguished essentially by the part of the plant affected. "Storey's bark disease" appears at the base of a sucker as a sunken, dark lesion surrounded by a yellow halo that gradually enlarges to girdle the sucker and causes an obvious constriction. Leaves on the affected sucker wilt and die but remain attached to the sucker, which is killed. The time course of disease development in suckers of different ages is variable and can range from a week to a month after infection for appearance of symptoms, while young suckers may be killed in a month and old suckers persist for more than 8 months after infection. Sporodochia (pink) develop on the discolored tissue of the lesion, which may extend to the main stem and cause collar rot.

A sunken lesion on the main stem that often extends further downward than upward from the snag where a lateral has been removed by pruning has been described as "scaly bark." The bark over these lesions is noticeably roughened and rubs off easily. Lesions may continue to expand to kill laterals immediately above or below (see above syndrome) or can girdle the stem, killing all plant parts above. Appearance of symptoms first occurs about a month after infection, and disease development generally is slower than with Storey's bark disease.

Plants affected by "collar rot" develop an unthrifty appearance, become chlorotic, wilt and die, but development of first symptoms is considerably slower than the manifestations described above and may take 4 months or more. Affected trees usually exhibit either a stem constriction close to soil level or a cankerous lesion from which extends an indistinct collar just above ground level. Lesion tissue under the bark covering is visibly discolored. Lateral extension of the lesion to girdle the stem is rather slow but results in rapid death of the plant after girdling is complete.

Irregular, brown leaf spots with yellow margins up to 1.2 cm and dark brown, sunken lesions on fruits (berry disease) are caused by the same organism. Sporulation occurs on affected tissues of both leaves and fruits, but neither disease manifestation is of significant import.

The causal fungus for all of the above symptoms is *Fusarium stilboides* Wollenw. (*F. lateritium* var. *longum* Wr.), characterized by cylindrical to fusiform macroconidia variable in size (2–5 × 7–85 μm), which is related to number of septations (0–9), that appear pink to orange *en masse*.

Conidia are largely responsible for dissemination of the fungus and are spread mainly by rain drops but can be carried by insects and on contaminated tools. High humidity, but not free water, is required for germination (optimum at 30 C). Infection, which is dependent on wounds and especially fresh leaf scars, does not extend to the vascular tissue, and viability in infected bark is maintained for a year. The different forms of the disease apparently result from original infection of different plant parts.

Importance of the diseases usually does not warrant special control measures, but careful attention should be given to prevent wounding, and all wounds that do occur should be painted immediately with a fungicidal dressing. Severely diseased plants should be pulled and burned. Reports of varietal resistance remain to be confirmed.

1. Booth, C. and J. M. Waterston. 1964. *Fusarium stilboides. Commonwealth Mycol. Inst. Descriptions Pathogenic Fungi Bacteria 30.* 2 p.

2. Corbett, D. C. M. 1959. *Fusarium* bark diseases of coffee in Nyasaland. *Nyasaland Farmer Forest. 4:* 3–7.

3. Firman, I. D. 1964. *Fusarium* diseases of coffee in Kenya. *Kenya Coffee 29:* 353, 355.

4. Gordon, W. L. 1956. The taxonomy and habitats of the *Fusarium* species of Trinidad, B. W. I. *Can. J. Bot. 34:* 847–864.

5. Siddiqi, M. A. and D. C. M. Corbett. 1963. Coffee bark diseases in Nyasaland. I. Pathogenicity, description and identity of the causal organism. *Trans. Brit. Mycol. Soc. 46:* 91–101.

6. Siddiqi, M. A. and D. C. M. Corbett. 1968. Coffee bark diseases in Malawi. II. Properties of the causal organism and conditions favouring the disease. *Trans. Brit. Mycol. Soc. 51:* 129–135.

7. Storey, H. H. 1932. A bark disease of coffee in East Africa. *Ann. Appl. Biol. 19:* 173–184.

8. Wallace, G. 1939. A non-parasitic disease of Arabica coffee. *East Afr. Agr. J. 4:* 365–368.

9. Wallace, G. and M. M. Wallace. 1955. Bark diseases of coffee. *East Afr. Agr. J. 21:* 25–27.

Tracheomycosis

A wilt disease of coffee was noted in Ivory Coast in 1948 and became serious in Congo in 1950, where up to 30–40 percent of the trees in some areas were killed. Also reported from Central African Republic and Guinea, the disease apparently is restricted to Africa, although 2 wilt diseases of coffee (caused by *Fusarium javanicum* Koorders and *F. solani* var. *minus* Wr.) have been reported from India in 1967 and 1968. The disease can cause rapid and serious loss of trees and has been considered the most important disease of coffee in Guinea (1962).

Leaves on affected plants assume an appearance of general chlorosis and subsequently may become flaccid and curled before turning brown, dry and falling. Branches become brown to black, and numerous vertical or spiral cracks develop in bark on the trunk to reveal blackened tissues underneath. Young wood exhibits black discolored streaks ("ruban noir") which have been considered diagnostic symptoms. Roots are darkly discolored by a wet rot that spreads upward via the tracheids. Incubation varies from a week (trees 2–3 months old) to 6 months on 7–10 year old trees, but trees may be killed in 2–3 weeks after symptoms appear. The crowns of dead trees are completely defoliated.

Tracheomycosis is caused by *Gibberella xylarioides* Heim & Saccas that produces dark perithecia 180–300 × 200–400 μm embedded in dark stromata. Asci are cylindrical, 7–9.5 × 90–110 μm, each with 8 1–3 septate, hyaline ascospores 4.5–6 × 12–14 μm. The unicellular, curved microconidia measure 2.5–3 × 5–10 μm, the falcate, 2–3 septate macroconidia 4–5 × 20–25 μm, and the globose chlamydospores 8–10 × 10–15 μm.

The fungus can be disseminated by ascospores and conidia dispersed by rain or wind from sporulation on the surfaces of diseased plants and by man through distribution of diseased planting materials. Infection occurs through wounds either on aerial plant parts or on superficial roots. Caffeine inhibits growth of the pathogen *in vitro*. The disease is found primarily on *Coffea canephora* and *C. liberica*.

Copper sulfate has been used to treat wounds and prevent infection, but diseased trees should be lifted and burned *in situ* as soon as symptoms are apparent. If sporulation is apparent on diseased or dead trees, they should be sprayed with a disinfectant before being uprooted. Basamid (dazomet) at 150 g/hole has been used to fumigate planting sites in Kenya. Resistant selections have been found in *C. liberica*, but highest resistance occurs in *C. canephora*.

1. Baker, C. J. 1972. *Fusarium solani* associated with a wilt of *Coffea arabica* in Kenya. *East Afr. Agr. J. 38:* 137–140.

2. Booth, C. and J. M. Waterston. 1964. *Gibberella xylarioides. Commonwealth Mycol. Inst. Descriptions Pathogenic Fungi Bacteria 24.* 2 p.

3. Fraselle, J. 1950. Observations preliminaires sur une trachéomycose de *Coffea robusta. Bull. Agr. Congo Belge 41:* 361–372.

4. Fraselle, J. V., G. Vallaeys and O. DeKnop. 1953. La lutte contre la trachéomycose du caféier à Yangambi et le problème que pose actuellement cette maladie en Congo Belge. *Bull. Inform. Inst. Etud. Agron. Congo Belge 2:* 373–394.

5. Govindarajan, T. S. and S. Subramanian. 1968. *Fusarium* wilt of coffee in India. *Indian Coffee 32:* 270–271.

6. Kranz, J. 1962. Coffee disease in Guinea. *FAO Plant Protect. Bull. 10:* 107–109.

7. Rahman, M. U. and S. Subramanian. 1967. A new *Fusarium* wilt of coffee (*Coffea arabica*) in South India. *Plant Dis. Reptr. 51:* 758–759.

8. Saccas, A. M. 1951. La trachéomycose (carbunculariose) des *Coffea excelsa*, neo-arnoldiana et robusta en Oubangui-Chari. *Agron. Trop. (Paris) 6:* 453–506.

Myrothecium stem necrosis and leaf spot (Mancha circular)

A seedling disease characterized by stem necrosis and collapse above soil level was first reported from Colombia (1951) and later Costa Rica, Guatemala and India. Similar symptoms have been described from Brazil and resemble damping-off. Leaf lesions that first are small and appear water-soaked rapidly enlarge to 2 cm or more in diameter and turn brown with noticeable concentric rings.

Sporodochia of *Myrothecium roridum* Tode ex Fries that develop on the lower surfaces of lesions are grayish in appearance and consist of conidiophores (115–165 μm) that produce phialides 1.8–3 × 10–24 μm on which are borne cylindrical, unicellular conidia 1.5 × 5.6–7 μm.

Application of cuprous oxide has been used successfully to control the disease in nurseries where infections are not extensive. Severely affected seedlings should be destroyed, however, because damaged stems often break in high winds when the plants become larger.

1. Abrahão, J. 1954. Estrangulamento da haste do cafeeiro. *Biolǵico 20:* 32–36.
2. Echandi, E. 1961. Mancha circular de la hoja y necrosis del tallo en plantitas de cafe. *Turrialba 11:* 118–119.
3. Fitton, M. and P. Holliday. 1970. *Myrothecium roridum. Commonwealth Mycol. Inst. Descriptions Pathogenic Fungi Bacteria 253.* 2 p.
4. Nag Raj, T. R. and K. V. George. 1958. "Target leaf spot" disease of coffee. I. Occurrence, symptoms and etiology. *Indian Phytopathol. 11:* 153–158.
5. Rossetti, V., E. D. Pinheiro and J. Abrahão. 1960. Estrangulamento da haste do cafeeiro e reproducõ experimental de um dos tipos dessa doença. *Arq. Inst. Biol. 27:* 179–186.
6. Schieber, E. 1962. Estrangulamiento del tallo y mancha de la hoja. *Rev. Cafetalera 1:* 7–8.
7. Schieber, E. and G. A. Zentmyer. 1968. *Myrothecium* stem necrosis and leaf spot: important coffee disease in Guatemala. *Plant Dis Reptr. 52:* 115–117.
8. Silberschmidt, K. M. 1952. Estrangulamento do caule de cafeeiros. *Biológico 18:* 174–175.
9. Urban, O. 1951. Llaga y marchitamiento del cafeto causado por *Myrothecium roridum* Tode. Centro Nacional de Investigaciones de Cafe, Chinchina, Colombia. *Boletin Inform. 2:* 33–45.

Crinkle-leaf

One or more small brown lesions, often along the margins of very young leaves still in the bud, are the first symptoms of crinkle-leaf, a disease of *Coffea arabica* and *C. canephora* recognized in Kenya in 1959. The disease also occurs in Tanganyika, Uganda and the Congo ("brûlure de caféier"). The necrotic areas do not enlarge as the leaf expands, and the severity can range from insignificant to complete destruction. Another characteristic symptom of affected plants is shortened internodes.

Young leaves on plants afflicted with "hot and cold" disease are small with an uneven surface, leathery to the touch and with a marginal yellow band that may extend between the larger veins. Chlorotic tissue may become necrotic later to make affected leaves resemble crinkle-leaf symptoms.

Production from crinkle-leaf trees is reduced, and, although the causative agent has not been determined, sprays of captan with or without copper, but not copper alone or captafol, provide effective control. Best control is achieved when spray is applied at the first appearance of symptoms, but subsequent sprays at intervals of 6 weeks may be needed.

1. Bock, K. R. 1959. Notes on the use of captan sprays in coffee. *Kenya Coffee 24:* 405.

2. Firman, I. D. and T. M. Wormer. 1963. Recommendations for the control of crinkle-leaf. *Kenya Coffee 28:* 463, 465, 467.

3. Tapley, R. G. 1961. Crinkle-leaf of coffee in Tanganyika. *Kenya Coffee 26:* 156–157.

4. Wormer, T. M. and I. D. Firman. 1965. Control of crinkle-leaf of coffee in Kenya. *Expt. Agr. 1:* 1–10.

Dieback (Elgon dieback, Solai dieback)

Wilting of stipules and leaves one or more nodes distant from the ends of twigs followed by death of the leaves involved and necrosis that extends in both directions from the involved node(s) has been termed "dieback." Leaves may drop or remain attached. A bacterium tentatively identified as *Pseudomonas syringae* (personal communication from D. M. Okioga) has been consistently associated with the disease, but a causal relation has not been established nor the prerequisite environmental conditions determined. The variety Geisha has been found resistant to dieback in Kenya where copper (Perenox) foliar sprays also are recommended.

1. Anonymous. 1962. Die-back in coffee. *Planters' Chron. 57:* 330–333.

2. Firman, I. D. 1963. Elgon die-back. *Kenya Coffee 28:* 9, 11–14.

3. Frankel, W. 1963. Unknown die-back in Tanganyika. *Kenya Coffee 28:* 21–24.

4. Mayne, W. W. 1931. Die-back of coffee. *Planters' Chron. 26:* 492–495.

5. Nutman, F. J. and F. M. Roberts. 1960. Investigations on a disease of *Coffea arabica* caused by a form of *Colletotrichum coffeanum* Noack. I. Some factors affecting invasion by the pathogen. *Trans. Brit. Mycol. Soc. 43:* 491–507.

6. Ribeiro, J. 1958. Estudo preliminar sobre a incidencia do "secamento de ponteiros" ou "die back" em algumas variedades de cafeeiro (*Coffea arabica* L.). *Rev. Ceres 10:* 413–421.

7. Thorold, C. A. 1945. Elgon die-back disease of coffee. *East Afr. Agr. J. 10:* 198–206.

Thread blights and Pink disease

Diseases grouped under the general designation "thread blights" are widespread throughout the world where favorable environmental conditions occur, although some confusion has arisen concerning the names for the causal agents. The diseases remain important locally, but generally they are not of major consequence.

"Koleroga" or "black rot" of coffee was described from India in 1876 and has since been recorded from Africa, South America, the Antilles and Caribbean islands as well as the southwest Pacific.

Coarse strands of white or light brown mycelium develop often on the

Fig. 54. Thread blight of coffee. Note dead leaves still hanging but attached only by strands of mycelium.

shaded sides of stems and slowly extend onto the lower surfaces of branches and leaves. The mycelium on the lower leaf surfaces branches into scant, fragile, white growth, with some larger strands, that covers the leaf. Round to irregular lesions may appear on some leaves before they develop black rot symptoms and are completely killed. Dead leaves soon become detached from the twig but typically do not fall and hang vertically, held to the twig by mycelial strands (Fig. 54).

Twigs and small branches also may be killed and berries of any age are susceptible to infection. Sunken, round lesions on fruit may enlarge to 1.5 cm diameter, and the infection progress to blacken the seed and cause the fruit to dry prematurely.

Koleroga, or black rot, symptoms are caused by *Ceratobasidium koleroga*. (See Cacao—Thread blight.)

A number of species of the genus *Marasmius* also may cause thread blights similar to those described, but related species may not be pathogenic. *Rhizoctonia ramicola* Roberts & Weber has been reported to cause "silky thread blight" symptoms on coffee in El Salvador.

Pink disease also is quite widespread on a number of hosts other than coffee and was first described in 1873, also in India. Symptoms often appear on coffee as wilting of the ends of coffee branches accompanied by dieback and dropping of leaves and fruits. Fine, external mycelial strands are often found on the surfaces of leaves and stems, while lesions to 15 cm in length appear in bark. Affected areas become pinkish in coloration, and a crusty surface develops on cankers in the bark.

The pink disease is caused by *Botryobasidium salmonicolor* that produces hyaline, obovate basidiospores 6–7 × 9–12 μm.

Infection may occur first in the fruit, which are penetrated and assume a dark discoloration before falling and from which infection extends via the racemes to the principal branches. Disease development occurs mostly during the rainy season. Incidence of pink disease is favored by mild temperatures but limited by elevation range to areas that include frequent and abundant rainfall and sunny days such as often found at altitudes of 450–900 m.

1. Bianchini, C. L. and F. L. Wellman. 1958. Experimentos en el control de *Pellicularia* del cafe y creitas diferencias en Pelicularias de cineo huespedes. *Turrialba 8:* 73–92.

2. Briton-Jones, H. R. and R. E. D. Baker. 1934. Thread blights in Trinidad. *Trop. Agr. (Trinidad) 11:* 55–67.

3. Burt, E. A. 1918. Corticiums causing *Pellicularia* disease of the coffee plant, hypochnose of pomaceous fruits, and *Rhizoctonia* disease. *Ann. Missouri Bot. Gard. 5:* 119–132.

4. Fawcett, G. L. 1914. *Pellicularia koleroga* on coffee in Porto Rico. *J. Agr. Res. 2:* 231–233.

5. Mathew, K. T. 1954. Studies on the black rot of coffee. *Proc. Indian Acad. Sci., Sect. B, 39:* 133–170; 179–211.

6. Narasimhan, M. J. 1933. Black rot of coffee in Mysore. *Phytopathology 23:* 875–886.

7. Orillo, F. T. and R. B. Valdez. 1958. Four diseases of coffee hitherto undescribed in the Philippines. *Philippine Agr. 42:* 292–302.

8. Paz, M. H. 1966. Mal rosado. *Rev. Cafetalera 64:* 26.

9. Roger, L. 1951. Phytopathologie des pays chauds. Tome I. *Encycl. Mycol. 17:* 948–953; 1087–1091.

10. Rogers, D. P. 1943. The genus *Pellicularia (Thelephoraceae). Farlowia 1:* 95–118.

11. Rorer, J. B. 1910. Pink disease of cacao. *Bull. Dept. Agr. Trinidad Tobago 15:* 86–89.

12. Shaw, D. E. 1962. Diseases of coffee in Papua and New Guinea. *Papua New Guinea Agr. J. 51:* 1–6.

13. Tims, E. C., P. J. Mills and B. Exner. 1954. Thread-blight (*Pellicularia koleroga*) in Louisiana. *Plant Dis. Reptr. 38:* 634–637.

14. Venkatarayan, S. V. 1947. Diseases of coffee. *Mysore Agr. J. 25:* 3–20.

15. Weber, G. F. and L. Abrego. Damping-off and thread blights of coffee in El Salvador. *Plant Dis. Reptr. 42:* 1378–1381.

Mancha mantecosa (Blister spot, Wellman's fruit and leaf spot)

A leaf spot and fruit disease of coffee observed in Costa Rica for several years was reported to be virus induced and transmitted by *Toxoptera aurantiae* Boyer in 1957. Similar symptoms were reported later from Brazil and New Guinea, but attempts to verify the nature of the causal agent were unsuccessful. More recently, the symptoms, which consist of round (2–6 mm diameter), dull, slightly sunken leaf spots often with a noticeable central spot (Fig. 55), and sunken, necrotic fruit lesions, have been attributed to a fungus, *Colletotrichum* sp. Development of symptoms is dependent on high relative humidity.

Control measures have not been devised.

1. Bianchini, C. L. 1960. Report on "mancha mantecosa", "chasparria" and "ojo de gallo", three diseases of coffee in Costa Rica. *Coffee 2:* 29–34.

Fig. 55. Leaf lesions characteristic of mancha mantecosa (blister spot). (Courtesy of E. Vargas, University of Costa Rica, San Jose.)

2. Bitancourt, A. A. 1958. As manchas da folha do cafeeiro. *Biológico 24:* 191–201.

3. Rodriguez, R. A. 1958. La "mancha mantecosa", enfermedad por virus en el cafeto. *Suelo Tico 10:* 94–97.

4. Shaw, D. E. 1968. Wellman's leaf and fruit spot of Arabica coffee in New Guinea. *Papua New Guinea Agr. J. 19:* 152–166.

5. Vargas G., E. and L. C. Gonzalez U. 1972. La mancha mantecosa del café causada por *Colletotrichum* spp. *Turrialba 22:* 129–135.

6. Wellman, F. L. 1957. Blister spot of Arabica coffee from virus in Costa Rica. *Turrialba 7:* 13–15.

Ringspot (virus) disease (Mancha anular)

Spots consisting of concentric necrotic rings to 1.5 cm in diameter have been reported from Brazil (1938) and the Philippines. Symptoms have been observed in *Coffea arabica*, *C. excelsa* Chev., *C. liberica* and *C. robusta* Linden types in seedlings and on leaves, stems and fruits of trees. Transmission has been accomplished by grafting only in Brazil where bacilliform particles have been associated with the disease. Seed transmission of the disease has been described from the Philippines. Similar symptoms have been induced on inoculated leaves of coffee by spotted wilt virus ("vira cabeça"), but identity of the 2 diseases has not been established.

1. Bitancourt, A. A. 1938. A mancha anular, uma nova doença do cafeeiro. *Biológico 4:* 404–405.

2. Bitancourt, A. A. 1939. Lesoes nas frutas da mancha anular do cafeeiro. *Biológico 5:* 33–34.

3. Costa, A. S., D. M. Silva and A. M. B. Carvalho. 1960. Infeccão de cafeeiros com o virus de vira-cabeça. *Bragantia 19:* 47–52.

4. De Ingunza S., M. A. 1966. Virus que atacan al cafeta. *Vida Agr. 42:* 605–607, 609, 611–613, 615–616.

5. Reyes, T. T. 1959. Ring spot disease of coffee. *FAO Plant Protect. Bull. 8:* 11–12.

6. Reyes, T. T. 1961. Seed transmission of coffee ring spot by excelsa coffee (*Coffea excelsa*). *Plant Dis. Reptr. 45:* 185.

7. Silberschmidt, K. 1941. A transmissäo experimental da mancha anular do cafeeiro. *Biológico 7:* 93–99.

8. Silberschmidt, K. M. and A. A. Bitancourt. 1965. Las enfermedades de virus en el cafeto. *Cafe (Lima) 6:* 1–7.

Mineral deficiency symptoms

Lack of *nitrogen* may become apparent within a month after deficiency occurs. Leaves on lower branches are the first to become chlorotic and sometimes develop apical necrosis. General chlorosis progresses upward to eventually include all of the foliage, and plants become stunted and asymmetrical in habit. Leaves produced on deficient plants are smaller in size, and defoliation of old leaves is often more extensive, particularly during periods of hot weather. Chlorosis of leaves caused by nitrogen deficiency is accentuated by high light intensity. Different forms of nitrogen can be added to the soil to alleviate this deficiency, but urea is most readily available to plants. For very acid soils, sodium nitrate (or urea) is recommended, while ammonium sulfate should be considered for less acid soils. Urea also can be used as a foliar spray but should be applied in low concentration for either foliar or soil application because of possible (biuret) toxicity.

Phosphorus deficiency symptoms, infrequent in nature, develop as mottling in the oldest leaves accompanied by irregular spotting, often chlorotic but tinged red or brown. Soil application of phosphate fertilizer only in large quantities has given significant response, but both manure and bone meal also have been used to correct phosphorus deficiency.

Deficiency in *potassium* is evident as apical necrosis that extends as necrotic spotting and later confluent, red-brown necrosis back along both leaf margins from the leaf tip. Extension of necrosis continues until the leaf is completely dead and falls. Older leaves are affected first, but eventually all but the very youngest foliage exhibit symptoms, and defoliation is rapid. Flowering is not affected, but fruit set may be reduced. Coffee trees respond readily to soil application of potassium fertilizer.

Evidence of *calcium* deficiency is first seen in the youngest leaves on lateral branches and the apical part of the plant as a loss of green color along the leaf margins and between the main veins. This condition advances until affected leaves have yellow laminae with green midribs. These symptoms later result in dieback of apical growth including secondary axillary growth. Growth of trees deficient in calcium is retarded and flowering and fruit set noticeably reduced. Soil amendments with calcium-containing materials have been less than completely satisfactory, but changes in soil pH (by liming) may prove beneficial.

The oldest leaves on plants deficient in *magnesium* develop small, chlorotic or necrotic spots that may coalesce along the margins and apices of laminae. Some loss of green color progresses into interveinal areas, but the midrib and main veins of affected leaves turn yellow and remain bordered by narrow strips of green tissue. Symptoms progress to older leaves of lateral branches and affected leaves soon fall. Continual deficiency results in dieback of branch tips, reduction of flowering, absence of fruit set and may result in death of affected plants. Addition of materials containing magnesium in regular fertilizer applications may be sufficient to alleviate deficiency of this element, but supplemental soil amendments may be required. Utilization of magnesium is inhibited by high concentrations of potassium, but association has been found between uptake of phosphorus and magnesium.

Chlorosis in the youngest leaves accompanied by light green veins is characteristic of insufficient *sulfur*. Growth may be somewhat retarded, but no effect on flowering has been noted.

Veins of the youngest leaves remain noticeably green in distinctly chlorotic laminae on plants deficient in *iron*. Defoliation is usually minor and flowering and fruit set affected little. It has been reported that iron deficiency results from excessive phosphorus or soluble manganese. Soil application of metal chelates has been used to correct iron deficiency induced by manganese toxicity. Deficiency of iron is often associated with deficiency of potassium and is corrected by application of potassium fertilizer. A condition known as "amber beans," which causes problems in processing, has been associated with iron deficiency.

Coffee plants deficient in *zinc* characteristically develop short internodes and abnormally small, lanceolate leaves with interveinal chlorosis. Leaves are concentrated at the ends of branches with noticeably fewer leaves in the middles. Defoliation may be severe and accompanied by dieback. Foliar applications(s) of zinc sulfate (0.25 percent) will give prompt improvement.

The youngest leaves on plants deficient in *manganese* develop interveinal chlorosis, with narrow bands on either side of main veins and the midrib remaining dark green. If the condition continues, the young leaf pairs assume a uniform light green or yellow color in shade, but a uniform bright lemon color with white, interveinal flecks develops on plants grown in the sun. Inconsistent results have been obtained with soil applications of manganese salts to correct deficiency, but foliar sprays with manganese compounds (0.2–0.8 kg/100 ℓ water with sticker added) can be recommended. Caution against excessive use of manganese should be noted to prevent toxicity.

Boron deficiency is first evident in young leaves that are small, distorted and may have necrotic spots in the laminae. Dieback of twigs and the growing point is followed by weak growth from axillary buds. Remedial treatments can be made either by soil application (15–125 g boron compound/year) or 1 or more foliar sprays (0.2 percent solution with sticker) each year. Boron is rapidly taken up by coffee plants, and the range of concentrations between deficiency and toxicity is relatively narrow, which warrants careful attention to concentration of applied materials.

1. Ananth, B. R. and N. G. Chokkanna. 1962. Zinc deficiency in arabica coffee. *World Crops 14:* 89–91.

2. Cibes, H. and G. Samuels. 1955. Mineral-deficiency symptoms displayed by coffee trees grown under controlled conditions. *Univ. Puerto Rico Agr. Expt. Sta. Tech. Paper 14.* 21 p.

3. Fernández, C. E. 1961. The effect of triiodobenzoic acid, urea and iron in correcting iron chlorosis in coffee. *Proc. Am. Soc. Hort. Sci. 77:* 236–239.

4. Fernändez, C. E. 1963. Deficiencia de manganesa en cafetos de Guatemala. *Rev. Cafetalera 22:* 11–12.

5. Muller, L. 1958. Mineral nutrition: detection and control of minor element deficiencies. *Coffee Tea Ind. 81:* 71–74, 76–77.

6. Ochavillo, E. A. 1956. Responses of coffee to deficiency of various elements in the culture medium. Philippine Agr. 40: 315–334.

7. Perez, V. M. 1958. Algunas deficiencias minerales del cafeto en Costa Rica. *Rev. Cafe 14:* 7–8, 10.

8. Robinson, J. B. D. 1959. Iron deficiency control. *Kenya Coffee 24:* . 16–19.

9. Robinson, J. B. D. 1960. Amber beans. *Kenya Coffee 25:* 91–93.

10. Robinson, J. B. D. and E. M. Chenery. 1958. Magnesium deficiency in coffee with special reference to mulching. *Empire J. Expt. Agr. 26:* 259–272.

11. Southern, P. J. and G. Hart. 1969. Nutritional studies of coffee in the Territory of Papua and New Guinea. *Dept. Agr. Stock Fisheries Port Moresby Res. Bull. 5.* 91 p.

OTHER DISEASES

Bacterial leaf blight [*Pseudomonas cichorii* (Swingle) Stapp]

1. Robbs, C. F., O. Kimura, R. de L. Ribeiro and L. C. Oyadomari. 1974. "Crestamento bacteriano das folhas": nova enfermedade do cafeeiro (*Coffea arabica* L.) incitada por *Pseudomonas cichorii* (Swingle) Stapp. *Arq. Univ. Fed. Rural Rio de Janeiro 4:* 1–5.

Canker (*Nectria dodgei* Heiser)

1. Flores, M. A., R. Pazos and F. J. LeBeau. 1957. A canker disease of coffee in Guatemala. *Phytopathology 47:* 11.

Collar canker (*Helicobasidium compactum* Boed.)

1. Schieber, E. and G. A. Zentmyer. 1967. Collar canker of coffee trees in Guatemala. *Plant Dis. Reptr. 51:* 267–269.

Damping-off (*Rhizoctonia solani*), Collar rot, (*Macrophomina phaseolina =Rhizoctonia bataticola*) and *Botryobasidium rolfsii*

1. Holliday, P. and E. Punithalingam. 1970. *Macrophomina phaseolina. Commonwealth Mycol. Inst. Descriptions Pathogenic Fungi Bacteria 275.* 2 p.

2. Saccas, A. M. 1966. La pourriture du collet des plantules de caféiers en pépinière, due à *Rhizoctonia bataticola* (Taub.) Butler. *Café, Cacao, Thé 10:* 28–43.

3. Valdez, R. B. and J. R. Acedo. 1963. An evaluation of fungicides for the control of damping-off of coffee seedlings. *Plant Dis. Reptr. 47:* 176–179.

4. Weber, G. F. and L. Abrego. 1958. Damping-off and thread blights of coffee in El Salvador. *Plant Dis. Reptr. 42:* 1378–1381.

Diffuse yellow leaf spot

1. Shaw, D. E. 1966. Diffuse yellow leaf spot of arabica coffee in Papua and New Guinea. *Papua New Guinea Agr. J. 18:* 120–121.

Envassouramento (Witches' broom)

1. Da Silva, D. M. 1955. Envassouramento do "*Coffea congensis*". *Bol. Superintendencia Serv. Cafe, São Paulo 30:* 8–11.

Fruit rot (*Trachysphaera fructigena*)

Dark brown discoloration of coffee fruits is later covered by a pink-brown, crusty surface of noticeably echinulate conidia 35 μm in diameter of the fungus *T. fructigena*. Oogonia are 24 × 40 μm, relatively thick-walled and characteristically have irregular outgrowths over their surfaces. Antheridia are amphigynous.

1. Holliday, P. 1970. *Trachysphaera fructigena. Commonwealth Mycol. Inst. Descriptions Pathogenic Fungi Bacteria 229.* 2 p.

2. Tabor, R. J. and R. H. Bunting. 1923. On a disease of cocoa and coffee fruits caused by a fungus hitherto undescribed. *Ann. Bot. 37:* 153–157.

Phloem necrosis and Wilt

A flagellate, *Phytomonas leptovasorum* Stahel, has been found consistently associated with diseased *Coffea liberica* plants in Surinam, but a causal relation remains to be established.

1. Bitancourt, A. A. 1935. Un protozoario parasita do cafeeiro. *Rev. Inst. Café, São Paulo, 19:* 2486–2490.

2. Ingunza S., A. D. 1966. Protozoarios que atacan al cafeto. *Rev. Cafetalera 56:* 21–24.

3. Stahel, G. 1931. Zur kenntnis der siebrohrenkrankheit (phloëmnekrose) des kaffeebaumes in Surinam. I. *Phytopathol. Z. 1:* 65–82.

4. Stahel, G. 1954. Die siebröhrenkrankheit (phloëmnekrose, flagellatose) des kaffeebaumes. *Neth. J. Agr. Sci. 4:* 260–264.

5. Vermeulen, H. 1963. A wilt of *Coffea liberica* in Surinam and its association with a flagellate, *Phytomonas leptovasorum* Stahel. *J. Protozoology 10:* 216–222.

6. Vermeulen, H. 1968. Investigations into the cause of the phloem necrosis disease of *Coffea liberica* in Surinam, South America. *Neth. J. Plant Pathol. 74:* 202–218.

Phthiriose

1. Lavabre, E. M. 1964. La "phthiriose" du caféier. *Café, Cacao, Thé 8:* 275–292.

Phyllosticta leaf spot (Derrite) (*Phyllosticta coffeicola* Speg.)

Scattered, spherical pycnidia 67–114 μm in diameter and unicellular, hyaline pycnidiospores 4–6 × 13–17 μm were found in dark, scattered, irregular leaf spots in India.

1. Rao, V. G. 1964. The genus *Phyllosticta* in Bombay–Maharashtra. IV. *Mycopathol. Mycol. Appl. 22:* 157–166.

2. Rodriguez, R. A., C. A. Bianchini and C. A. Soto. 1957. Studies on the control of "derrite" disease of coffee caused by *Phyllosticta coffeicola* Del. *Plant Dis. Reptr. 41:* 560–563.

Root rots

Cylindrocarpon tenue Bugn.

1. Subramanian, S. and T. S. Govindarajan. 1968. A new root rot of coffee in India. *Plant Dis. Reptr. 52:* 773–774.

Marasmius viegasii Sing.

1. Viégas, A. P. 1957. Podridão das raizes do cafeeiro. *Bol. Superintendencia Serv. Cafe, São Paulo, 32:* 7–16, 10–19.

Clitocybe tabescens (Fr.) Bres.

1. Dadant, R. 1957. Le pourridié du caféier à Madagascar. *Café, Cacao, Thé 1:* 126–131.

Fusarium oxysporum Schl. f. *coffeae* (Garcia) Wellman

1. Alvarez Garcia, L. A. 1945. Studies on coffee root rot disease in Puerto Rico. 1. A coffee *Fusarium* wilt. *J. Agr. Univ. Puerto Rico 29:* 1–29.

2. Dhaliwal, T. S., J. H. Lopez Rosa, G. Steiner, L. Igaravidez and A. T. Sepulveda. 1963. Studies on coffee root rot and horticultural practices for its amelioration. *Univ. Puerto Rico Agr. Expt. Sta. Tech. Paper 36.* 30 p.

Rhizoctonia choussii Crandall & Arillaga (Black root rot, Podredumbre negra)

1. Crandall, B. S. and J. G. Arillaga. 1955. A new *Rhizoctonia* from El Salvador associated with root rot of coffee. *Mycologia 47:* 403–407.

Rosellinia sp. (Llaga negra)

1. Bonilla, E. 1958. La "llaga negra" del cafeto y su combate. *Rev. Cafetera Colombia 14:* 35–38.

2. Castaño, J. J. 1953. Algunas observaciones sobre la "llaga negra" radicular del cafeto. *Agr. Trop. (Colombia) 9:* 41–47.

3. Lopez-Duque, S. and O. Fernandez-Borrero. 1966. Llagas radicales necra (*Rosellinia bunodes*) y estrellada (*Rosellinia pepo*) del cafeto. II. Efecto de la humedad y pH del suelo en el desarrollo micelial e infeccion. *Cenicafe 17:* 61–69.

4. Roger, L. 1953. Phytopathologie des pays chauds. II. *Encycl. Mycol. 17:* 1286–1297.

5. Saccas, A. M. 1956. Les *Rosellinia* des caféiers en Oubangui-Chari. *Agron. Trop. (Nogent) 11:* 687–706.

Sclerotium disease (*Sclerotium coffeicola* Stahel)

Serious losses in Surinam were incurred in 1917 by "sclerotium (coremium) disease" which is now known to occur also in Guyana, Trinidad, Tobago, Sierra Leone and India.

Mostly round leaf spots to 2.5 cm in diameter with alternating concentric rings of light and dark brown necrotic tissue with a distinct dark brown outer margin are typical. On the lower surfaces of lesions, white external mycelium may cover the leaf and form dense spots or aggregates that darken with age and form easily detached "bristles" and, later, sclerotia. Infection may spread by way of rhizomorphs or detached bristles and may infect flowers and berries.

High humidity is required for infection and disease development. The fungus is infectious also on *Commelina nudiflora* L., *Cecropia peltata* L. and *Vitis sicyoides* L.

Pruning to remove infected plant parts, spraying with Bordeaux mixture (2 percent), and planting to permit better aeration have been recommended for control.

1. Martyn, E. B. 1929. The *Sclerotium* disease of coffee and its occurrence in this colony. *Agr. J. Brit. Guiana 2:* 7–10.

2. Martyn, E. B. 1930. The *Sclerotium* disease of coffee. Some notes on the origin of the disease, its outbreak and control. *Agr. J. Brit. Guiana 3:* 28–34.

3. Nag Raj, T. R. and K. V. George. 1960. A disease of coffee hithe 'o undescribed from India. *Indian Coffee 24:* 120–122.

Stalk rot of cherries and leaves (New malady dieback) (*Colletotrichum gloeosporioides*)

Death of leaves 1 or 2 nodes below the apical end of a branch followed by twig dieback was described from India in 1960.

1. Anonymous. 1960. Die-back disease in Seklespur and North Coorg areas. *Indian Coffee 24:* 29, 44.

2. Anath, K. C., K. V. George, R. L. Narasimhaswamy and N. G. Chokkanna. 1960. Die-back of coffee. Outlook and observations. *Indian Coffee 24:* 238–245.

3. Hindorf, H. and B. N. Muthappa. 1974. A comparison of *Colletotrichum coffeanum* Noack from South India and Kenya. *Phytopathol. Z. 80:* 9–12.

4. Mordue, J. E. M. 1971. *Glomerella cingulata. Commonwealth Mycol. Inst. Descriptions Pathogenic Fungi Bacteria 315.* 2 p.

5. Muthappa, B. N. 1972. Studies on *Colletotrichum coffeanum* in India. I. Stalk-rot of green berries and leaves of *Coffea arabica* during the wet weather of monsoon. *J. Coffee Res. 2:* 16–22.

Stem pitting

1. Firman, I. D. 1962. The symptoms and incidence of stem pitting of coffee in Kenya. *Trop. Agr. (Trinidad) 39:* 95–102.

2. Sheffield, F. M. L. 1959. Stem-pitting condition in coffee. *Kenya Coffee 24:* 265–266.

Stem-wasting ("Kondlie") disease

A constriction of the main stem of seedlings 8–11 months old in the vicinity of the first node that results in chlorotic foliage and easily broken stems has been reported from India. Axillary growth that develops from buds below the constricted area after the main stem has broken is usually normal. No pathogenic agent has been identified with this disease, which is more prevalent on plants sprayed with Bordeaux mixture.

1. George, K. V. 1959. Stem-wasting "Kondlie" disease of coffee. I. Occurrence and symptoms. *Indian Coffee 23:* 495–496, 502.

2. Sheriff, M., T. R. Nag Raj and K. V. George. 1960. Stem-wasting "Kondlie" disease of coffee. II. Preliminary report on investigations. *Indian Coffee 24:* 160–167.

Sudden death (Morte subita)

1. Arviza, J. M. 1960. La "morte subita." *Ager 10:* 11–12.

2. De Medina, A. E. R. 1959. A morte súbita dos cafeeiros de Angola. *Gaz. Agr. Angola 4:* 187–191.

3. Ferrão, A. D. F. and F. J. D. Serafim. 1960. A "morte súbita" dos cafeeiros em Angola. *Rev. Cafe Portugues 7:* 17–27.

Zonal leaf spot

Irregular, gray-brown spots from 1–6 cm, often with incomplete dark brown, concentric rings and, on the lower surfaces of the spots, gray-white powdery growth of loose, hyaline mycelium, were first observed in Puerto Rico in 1915 and later from Africa, Costa Rica, Peru, the Philippines and India. The causal fungus, *Cephalosporium zonatum* Sawada (*C. eichorniae* Padwick), produces hyaline conidiophores 44–77 μm and acrogenous conidia 1.8–2.5 × 4.2–8.4 μm.

1. Abbott, E. V. 1929. Diseases of economic plants in Peru. *Phytopathology 19:* 645–656.

2. Bitancourt, A. A. 1954. As doenças do caffeiro na Costa do Marfim (Africa Occidental Francesa). *Biológico 20:* 205–222.

3. Nag Raj, T. R. and K. V. George. 1962. Occurrence of zonal leaf spot of coffee caused by *Cephalosporium zonatum* in India. *Current Sci. 31:* 104–105.

4. Orillo, F. T. and R. B. Valdez. 1958. Four diseases of coffee hitherto undescribed in the Philippines. *Philippine Agr. 42:* 292–302.

5. Salas, A. and E. Echandi. 1967. Mancha zonal de lo hoja de los cafetos (*Coffea arabica* L.) causada por el hongo *Cephalosporium zonatum*. *Turrialba 17:* 292–295.

CORIANDER: *Coriandrum sativum* L.

Bacterial blight

Brown, elliptical spots develop on all plant parts, but symptoms are most characteristic on inflorescences, which are often malformed. Foliage of affected plants, which may wilt, becomes darkened, and growth is arrested. Blight has been found in India, Poland, Rumania and Russia. The causal organism has been identified as *Xanthomonas corianderi* Srinivaṣan et al., which is also infectious to fennel.

1. Godlewska-Lipowa, W. 1966. Investigations on the etiology of bacterial necrosis of coriander (*Coriandrum sativum*) inflorescences. *Acta Agrobot. 19:* 59–91.

2. Németh, J., L. Paizs and Z. Klement. 1969. The flowerstand blight and seed decay of coriander. *Acta Phytopathol. Acad. Sci. Hung. 4:* 57–62.

3. Srinivasan, M. C., M. K. Patel and M. J. Thirumalachar. 1961. A bacterial blight disease of coriander. *Proc. Indian Acad. Sci., Sect. B, 53:* 298–301.

Stem gall

The disease of coriander known as stem gall or tumor was recognized in India in 1918 but was not thought to cause much damage at first. Later estimates, however, included an average of more than 20 percent of the plants affected and an average loss per plant of approximately 15 percent in areas where the disease occurs. Stem gall disease is not known to occur outside India.

Elongated swellings to 3–5 × 9–12 mm develop on the veins of leaves, petioles, stems and flower stalks and vary in size depending on the plant part infected. Hyperplastic growth results in swollen distortion of affected tissues.

Mycelium of *Protomyces macrosporus* Unger, the fungus that causes gall disease, is irregularly branched and intercellular. Individual, multinucleate, mycelial cells swell to form globose, thick-walled, resting sporangia (chlamydospores) 50–60 × 60–70 μm. The sporangia germinate by extruding cytoplasm and nuclei from the exospore, which remains surrounded by the sac-like inner wall, and the nuceli migrate to the periphery, leaving a large, central vacuole. The cytoplasm is divided by cleavage into small, uninucleate cells that each divide twice to form 4 spores that later mass at the center of the sporangium and are released in a ball with the slimy contents of the vacuole. Spores may multiply by budding but, upon contact with the host, pair by a conjugation tube in which the 2 nuclei fuse and, after diploidization, parasitic mycelium develops.

Resting sporangia are liberated into soil from debris of infected plants and undergo a dormant period of 4–6 months before they are capable of germination. Sporangia may remain viable in soil for 5 years, but continual seasonal planting assures a plentiful supply of primary inoculum from germinating sporangia. Infection may occur from either soil- or seed-borne inoculum. Maximum germination of sporangia occurs from 17–22 C at 75 percent (soil) moisture, and disease development is most severe in slightly basic soils.

A number of materials (sulfathiazole, streptomycin, perenox), each at 250–1000 ppm, inhibit germination of sporangia to varying degrees in the laboratory, but combined seed (0.25 kg/100 kg seed) and soil (16 kg/ha) treatments with thiram have given the best control of the disease in the field. Crop rotation and/or sanitation should prove beneficial for disease control.

1. Gupta, J. S. 1954. Disease appraisal of stem-gall of *Coriandrum sativum* L. *Indian Phytopathol. 7:* 53–60.

2. Gupta, J. S. 1958. Laboratory assay of certain fungicides, antibiotics and sulpha drugs on the germination of chlamydospores of *Protomyces macrosporus* Ung. *Agra Univ. J. Res. Sci. 7:* 105–112.

3. Mathur, S. B. 1962. Soil conditions in relation to some diseases of crops. *Agra Univ. J. Res. Sci. 11:* 105–107.

4. Mathur, S. B. 1962. Soil pH in relation to stem gall disease of coriander caused by *Protomyces macrosporus* Unger. *Indian Phytopathol. 15:* 75–76.

5. Narula, P. N. 1962. Efficacy of different fungicides against stem-gall disease of *Coriandrum sativum* L. *Sci. Cult. 28:* 481–482.

6. Nene, Y. L., I. A. Siddiqui and P. D. Kharbanda. 1966. Control of stem-gall of coriander by fungicides. *Mycopathol. Mycol. Appl. 29:* 142–144.

7. Pavgi, M. S. and A. N. Mukhopadhyay. 1969. Transmission of *Protomyces macrosporus* Unger, the incitant of stem gall of coriander. *Ann. Phytopathol. Soc. Japan 35:* 265–270.

8. Srivastava, H. C. 1955. Studies on tumour disease of *Coriandrum sativum* L. caused by *Protomyces macrosporus* Unger; Viability of chlamydospores of *Protomyces macrosporus* Unger. *Proc. 42 Indian Sci. Congr., Part 3 (Abstracts):* 218.

Alternaria blight

Small, brown, necrotic lesions on succulent tissues that quickly developed into a severe blight were reported from the Poona area (India) in 1962. Disease development was favored by heavy rainfall. *Alternaria poonensis* Raghunath, with simple conidiophores 6.5–8.6 × 25.8–80 μm and muriform, clavate conidia 8.3–19.3 × 47–204 μm and long, septate beaks, was identified as the causal fungus.

1. Raghunath, T. 1963. A new species of *Alternaria* causing blight of *Coriandrum sativum* L. *Mycopathol. Mycol. Appl. 21:* 315–317.

2. Raghunath, T. 1963. A new *Alternaria* blight of *Coriandrum sativum* from India. *Plant Dis. Reptr. 47:* 259–260.

Fusarium wilt

Widespread occurrence of coriander wilt in India has been recognized since 1936, and frequent losses of 50 percent have been noted with some instances of complete crop loss. The disease has been reported as the limiting factor for production in the Guna district.

Affected plants become generally chlorotic and develop progressive wilt symptoms. Growth is arrested, and most plants eventually die. Vascular discoloration appears only in the roots. Plants in all stages of growth are susceptible, but those infected late often do not set seeds, or seeds set before symptoms develop do not mature.

The causal fungus has been identified as *Fusarium oxysporum* f. *corianderii* Kulkarni et al. that produces both unicellular, ovoid microconidia and falcate, 1–4 septate macroconidia 2.6–4.8 × 13.8–37μm. Cardinal temperatures for growth *in vitro* are 12, 20, and 35 C.

Control practices have not been developed, but some indication of differences in varietal susceptibility has been noted.

1. Chattopadhyay, S. B. 1967. *Diseases of plants yielding drugs, dyes and spices.* Indian Council of Agricultural Research, New Delhi. 100 p.

2. Nikam, B. G., S. N. Kulkarni and F. K. Pawar. 1959. Coriander and its 3 diseases. *Indian Farming 9:* 12–13.

3. Srivastava, U. S. 1969. Effect of inoculum potential on wilt development of coriander caused by *Fusarium oxysporum* f. *corianderii. Indian Phytopathol. 22:* 406–408.

4. Srivastava, U. S. 1972. Effect of interaction of factors on wilt of coriander caused by *Fusarium oxysporum* Schlecht ex. Fr. f. *corianderii* Kulkarni, Nikam & Joshi. *Indian J. Agr. Sci. 42:* 618–621.

OTHER DISEASES

Anthracnose (*Glomerella cingulata* = *Colletotrichum gloeosporioides*)

Affected flowers turn pale and droop as the infection rapidly spreads to unopened flowers that are killed and to leaves in severe cases. Formation of seeds may be completely prevented, or late infection can result in shrivelled seeds. Conidia may be seed-borne and cause damping-off of seedlings. Unicellular, hyaline conidia 3.5–5 × 3.5–18 μm and setae (52–270 μm) occur in acervuli (330 μm). *C. capsici* also causes anthracnose symptoms in India.

1. Mordue, J. E. M. 1971. *Glomerella cingulata. Commonwealth Mycol. Inst. Descriptions Pathogenic Fungi Bacteria 315.* 2 p.

2. Moses, G. J. and P. G. Rao. 1969. Coriander anthracnose—caused by *Glomerella cingulata* a new record from India. *Indian Phytopathol. 22:* 145–146.

3. Sehgal, S. P., A. L. Deshpande and U. S. Srivastava. 1965. A new disease of coriander caused by *Colletotrichum capsici* (Syd.) Butler and Bisby from India. *Current Sci. 34:* 89–90.

4. Thomas, C. A. 1949. Observations on diseases of some special crops. *Plant Dis. Reptr. 33:* 453–454.

Foot rot (*Sclerotinia sclerotiorum*)

Light brown lesions with white, external mycelium develop on stems at ground level and kill affected plants. Disease development was found to be more rapid at 19 C than 24 C, and plants were not killed at 29 C and 34 C. The fungus can persist in the soil as a saprophyte.

1. Gupta, J. S. 1963. Effect of soil temperature on the foot rot disease of coriander [*Sclerotinia sclerotiorum* (Lib.) de Bary]. *Indian Phytopathol. 16:* 210–213.

2. Mehta, P. R., B. Singh and S. K. Bose. 1946. Some new hosts of *Sclerotinia sclerotiorum* (Lib.) de Bary. *Current Sci. 15:* 171–172.

3. Tanrikut, S. and E. K. Vaughan. 1951. Studies on the physiology of *Sclerotinia sclerotiorum*. *Phytopathology 41:* 1099–1103.

Leaf rot [*Erwinia aroideae* (Townsend) Holland]

1. Chakravarti, B. P. and M. Rangarajan. 1966. Occurrence of *Erwinia aroideae* on two new hosts in India. *Plant Dis. Reptr. 50:* 701–702.

Powdery mildew (*Erysiphe polygoni*)

1. Nikam, B. G., S. N. Kulkarni and F. K. Pawar. 1959. Coriander and its 3 diseases. *Indian Farming 9:* 12–13.

COWPEA: *Vigna sinensis* (Torner) Savi [*V. unguiculata* (L.) Walp.]; CATJANG: *Vigna catjang* Walp.; ASPARAGUS BEAN: *Vigna sesquipedalis* Wight

Bacterial canker

Cowpea plants affected by bacterial canker were collected in July and August, 1942 in Oklahoma and Texas (United States), respectively, although there were unpublished observations of the disease more than 10 years earlier. The first published record of the occurrence of canker (in Texas) ap-

Fig. 56. Bacterial canker (small, dark) lesions on cowpea leaflet. (Courtesy of R. J. Williams, International Institute of Tropical Agriculture, Ibadan.)

peared a year later, and the disease has been found since in Illinois and Florida, as well as in India. Losses have been extensive in some fields where it has been estimated that 60 percent of the plants were killed, while in other instances, the crop has been assessed as almost a total loss.

Lesions appear on leaves as small, light colored spots about a mm in diameter but soon change to 1–2 cm and develop brown, necrotic centers that frequently drop causing "shot hole" symptoms (Fig. 56). A halo often remains around the hole from which necrotic tissue has dropped. Leaf lesions may vary in size and shape, and severely affected leaves sometimes wilt and drop. Stem lesions develop at the bases of affected leaves as well as elsewhere along the stems and may extend rapidly along the entire length of the stem. Plants with deep, open stem cankers often break over and die, although some plants may survive and even produce seed.

The causal agent of canker has been identified as *Xanthomonas vignicola* Burkholder, characterized by gram negative rods 0.5–0.9 × 1.0–2.8 μm with single, polar flagella. Cardinal temperatures for growth *in vitro* are approximately 9, 28 and 37 C.

Infection of seedlings was readily obtained under controlled conditions by dipping germinated seeds in a bacterial suspension or spraying inoculum on uninjured leaves, but infection also resulted when leaves, stems or roots were inoculated after wounding. Incubation lasted for 1–3 weeks, depending on method of inoculation, although leaf lesions often appeared earlier than stem cankers. Disease development was faster and more extensive at 30 C than at lower temperatures (20 and 25 C), but the relative susceptible/resistant responses of individual cultivars of cowpea were not affected noticeably. Resistance was found to be dominant and to result from a single genetic factor, but 2 pathotypes of the bacterium were distinguished in later studies.

Only seeds harvested from healthy plants should be used for planting purposes and infected plants rogued when symptoms first appear. The cultivars Brabham, Buff, Iron, Suwanee and Victor have been found resistant to both pathotypes of the bacterium, while cvs. Groit and New Era were susceptible only to pathotype 2.

1. Boewe, G. H. 1948. Bacterial canker of cowpea in Illinois. *Plant Dis. Reptr. 32:* 275.

2. Burkholder, W. H. 1944. *Xanthomonas vignicola* sp. nov. pathogenic on cowpeas and beans. *Phytopathology 34:* 430–432.

3. Dunlap, A. A. 1943. Some notes on newly observed or unusual diseases. *Plant Dis. Reptr. 27:* 274.

4. Hoffmaster, D. E. 1944. Bacterial canker of cowpeas. *Phytopathology 34:* 439–441.

5. Larsh, H. W. 1943. Bacterial crown canker of cowpea in Oklahoma. *Plant Dis. Reptr. 27:* 555.

6. Lefebvre, C. L. and H. S. Sherwin. 1950. Inheritance of resistance to bacterial canker in cowpea, *Vigna sinensis. Phytopathology 40:* 17–18.

7. Patel, M. K. and N. D. Diwan. 1950. Bacterial blight of cowpea. *Indian Phytopthol. 3:* 75–80.

8. Patel, P. N. and J. K. Jindal. 1970. Studies on resistance to bacterial diseases in India. II. Resistance in cowpea to bacterial blight (canker) disease. *Indian Phytopathol. Soc. Bull. 6:* 28–34.

9. Sherwin, H. S. and C. L. Lefebvre. 1951. Reaction of cowpea varieties to bacterial canker. *Plant Dis. Reptr. 35:* 303–317.

10. Watkins, G. M. 1943. Further note on a cowpea bacterial canker in Texas. *Plant Dis. Reptr. 27:* 556.

Bacterial spot

The organism that causes bacterial spot symptoms of cowpea is also responsible for the bacterial spot disease of lima bean, but it was first noted as a pathogen of bean in 1892. The diseases of cowpea and lima bean both were described in 1923 from the United States (Indiana), although they had been observed earlier, and the cowpea disease soon was noted from several additional states including Florida and Kansas. The pathogen is worldwide in occurrence and may cause serious defoliation and reduction in seed harvest.

Spots on the leaves are first small, water-soaked, sunken and circular but enlarge to become mostly round or angular, brown lesions 1–4 mm in diameter. Many lesions are larger yet with light colored, necrotic centers surrounded by red-brown borders. The centers of such lesions may dry, crack and fall. Lesions, which are not delimited by veins, often follow veins for considerable distances. Young leaves are quite susceptible to infection, which results in serious tearing, curling or distortion of laminae, often accompanied by red-brown coloration of marginal tissues.

Lesions on stems and petioles are typically sunken and elongated, while lesions on pods are often red to brown and also sunken, but more nearly circular and 1–8 mm in diameter. Water-soaked tissues often surround lesions on petioles, stems and pods, but pods frequently are constricted at infection sites and may be noticeably bent at these locations. Lesions often completely penetrate the pod wall, and seeds directly under large lesions often are discolored and shrivelled. Seedlings from such infected seeds frequently develop large, cotyledonary lesions from which hypocotyl lesions or systemic infections, characterized by red-brown vascular streaks, often result.

Bacterial spot is caused by *Pseudomonas syringae*, which produces gram negative cells 0.5–0.7 × 0.8–2.2 μm with polar flagella. A green, fluorescent pigment is produced *in vitro* on King's B medium.

The pathogen is readily carried with seeds from infected pods and often results in cotyledonary infection and subsequent systemic infection. Seedlings and young plants are susceptible to infection from stomatal penetration without wounding. Incubation continues for 2–4 days, while bacterial invasion remains largely intercellular until vascular tissue becomes affected. Disease development is usually more severe under conditions of high rainfall, which also contributes to field spread of the disease.

Only seeds from healthy plants should be selected for planting. Storage of seeds for more than a year before planting has been reported to reduce viable, seed-borne bacteria.

1. Beach, W. S. 1892. Some bean diseases. *N. Y. State Agr. Expt. Sta. Bull. 48:* 305–333.

2. Gardner, M. W. and J. B. Kendrick. 1923. Bacterial spot of cowpea. *Science 57:* 275.

3. Gardner, M. W. and J. B. Kendrick. 1925. Bacterial spot of cowpea and lima bean. *J. Agr. Res. 31:* 841–863.

4. Hayward, A. C. and J. M. Waterston. 1965. *Pseudomonas syringae. Commonwealth Mycol. Inst. Descriptions Pathogenic Fungi Bacteria 46.* 2 p.

5. Tisdale, W. B. and M. M. Williamson. 1921. Bacterial leaf spot of lima bean. *Phytopathology 11:* 52.

6. Zaumeyer, W. J. and H. R. Thomas. 1956. A monographic study of bean diseases and methods for their control. *U. S. Dept. Agr. Tech. Bull. 868.* 255 p.

Red stem canker and (*Phytophthora*) Stem rot

Two diseases of cowpea, caused by distinct species of the same fungus genus, have been described. Red stem canker was observed in the United States

(Georgia) in 1944 but not reported until 5 years later. It was subsequently found in South Carolina but apparently remains confined to the United States, where it is not generally considered of major importance. Stem rot disease of cowpea was reported from Queensland (Australia) in 1953, having been noted earlier from New South Wales. Serious losses were caused in both locations.

Elongated stem lesions that first appear water-soaked but soon turn brown to deep red are the most characteristic symptoms of red stem canker disease. Lesions may develop on any part of the stem or petioles and may extend from the soil line to the growing tip. The centers of lesions, which may split lengthwise when the tissues dry, are often lighter and vary in color but usually are some shade of dark red. Seedlings are sometimes girdled and killed, although infection may be confined to only a single side of the stem and result in unilateral plant development.

Plants affected by stem rot are noticeable because they turn yellow and quickly wilt and die. However, such plants usually develop sunken, brown stem lesions that extend from ground level upward, often reaching the tops of the plants. The margins of lesions, which may girdle the stems, appear water-soaked where there may be sporangial production on scant mycelial growth during moist weather. Aerial stem lesions also may develop during periods of high humidity, while the root systems of affected plants are progressively destroyed, but there is a notable absence of red coloration associated with stem lesions.

The fungus that causes red stem canker is *Phytophthora cactorum* (Leb. & Cohn) Schroet. which produces hyphae usually 4–4.5 μm in diameter, but sometimes much broader, *in vitro* with branched sporangiophores (including secondary sporangiophores) quite variable in length (50–1000 μm). Oval to elliptical sporangia 25–50 μm in diameter germinate by formation of zoospores, and the nearly round oogonia measure 21–31 $\times$ 18–30 μm with paragynous or amphigynous antheridia. Oospores are mostly spherical and 14–29 $\times$ 14–27 μm.

Sporangiophores of *P. vignae* Purss (stem rot fungus) are mostly simple and undifferentiated and bear ovoid to pyriform sporangia 15–54 $\times$ 24–72 μm that also germinate by production of zoospores. Antheridia are nearly round and essentially all amphigynous, 9–18 $\times$ 12–27 μm, and the smooth, almost spherical oogonia 24–46 $\times$ 27–42 μm. Oospores are light brown, spherical, with smooth walls and measure 21–30 μm in diameter.

Cardinal temperatures for growth *in vitro* of both fungi are approximately 10, 28, and 35 C, although *P. vignae* causes the most serious damage to plants from 20–28 C. Infection has been shown to occur without wounding and may result either from mycelial infection of the stem under ground or zoospores liberated from sporangia on the stems above ground. Sporangia, which are easily dislodged and wind-borne, are responsible for much of the in-field spread of both diseases, but surface water, particularly in drainage ditches, is also effective in dissemination. It has been demonstrated that successive cropping to cowpea serves dramatically to increase losses to *P. vignae*, particularly when soil moisture levels are high. However, severe losses may be sustained with a fairly dry soil environment if the soil inoculum level is high. There is no evidence of seed transmission of either fungus.

Control procedures have not been developed, but annual rotation of cowpea plantings utilizing only well-drained soils should help to preclude serious losses. Varietal differences in susceptibility to *P. vignae* have been investigated, and cultivars with specific resistances developed. Four races of the fungus are presently recognized.

1. Armstrong, G. M. 1951. *Phytophthora cactorum* on cowpeas in South Carolina. *Plant Dis. Reptr. 35:* 418.
2. Purss, G. S. 1953. Stem rot of cowpeas. *Queensland Agr. J. 77:* 139–142.
3. Purss, G. S. 1957. Stem rot: a disease of cowpeas caused by an undescribed species of *Phytophthora. Queensland J. Agr. Sci. 14:* 125–154.
4. Purss, G. S. 1958. Studies on varietal resistance to stem rot (*Phytophthora vignae* Purss) in the cowpea. *Queensland J. Agr. Sci. 15:* 1–14.
5. Purss, G. S. 1963. Caloona—stem rot resistant cowpea. *Queensland Agr. J. 89:* 756–759.
6. Purss, G. S. 1972. Pathogenic specialization in *Phytophthora vignae. Australian J. Agr. Res. 23:* 453–456.
7. Purss, G. S. 1975. A new stem-rot resistant cowpea cultivar. *Queensland Agr. J. 101:* 551.
8. Purss, G. S. and G. D. Bowen. 1952. A stem rot of cowpeas. *Australian Plant Dis. Recorder 4:* 43.
9. Waterhouse, G. M. and J. M. Waterston. 1966. *Phytophthora cactorum. Commonwealth Mycol. Inst. Descriptions Pathogenic Fungi Bacteria 111.* 2 p.
10. Weimer, J. L. 1949. Red stem canker of cowpea, caused by *Phytophthora cactorum. J. Agr. Res. 78:* 65–75.

Cercospora leaf spot

A number of leaf spot diseases of cowpea and catjang, all attributed to species of *Cercospora*, have been reported and described since 1880. Mild disagreement among authors concerning nomenclatural identity of the pathogen(s) involved has prompted their inclusion herein as a single species. The disease probably occurs wherever cowpea is grown but has been reported from India and the United States. Yield losses attributable to leaf spot have not been determined but would be proportional to defoliation estimated to reach 10 percent or more.

Lesions restricted mostly to leaf laminae are first seen as small, brown spots with chlorotic halos but soon become irregular (1–16 mm), red-brown and necrotic (Fig. 57). Coloration in lesions is often more noticeable around

Fig. 57. *Cercospora* leaf spot of cowpea. (Courtesy of R. J. Williams, International Institute of Tropical Agriculture, Ibadan.)

the margins than in the centers. Spots may coalesce to cause yellowing of leaves and premature defoliation.

Various names have been given the causal agent(s) of *Cercospora* leaf spot, herein attributed to *C. cruenta*, that produces lightly colored, occasionally branched conidiophores with 0–2 septae that measure 2.5–5 × 10–75 μm. Subhyaline to brown, cylindrical to obclavate, multiseptate conidia are 2–5 × 25–150 μm and straight to slightly curved. There remains some question whether this fungus may also infect other plants, although infection of hyacinth bean has been consistently reported. Epidemiological observations have revealed that the disease often appears in late season under conditions of mild mean temperature (26 C) and relative humidity (60–80 percent).

Control procedures have not been reported, although differences in varietal susceptibility have been noted and a number of resistant types identified.

1. Chandrasekaran, S. and G. Rangaswami. 1960. Studies on *Cercospora cruenta* occurring on *Vigna catjang*. *Indian Phytopathol. 13:* 96–99.

2. Chupp, C. 1953. *A monograph of the fungus genus* Cercospora. Ithaca, New York. 667 p.

3. Latham, D. H. 1934. Life history of a *Cercospora* leafspot fungus of cowpea. *Mycologia 26:* 516–527.

4. Fery, R. L., P. D. Dukes and F. P. Cuthbert, Jr. 1976. The inheritance of *Cercospora* leaf spot resistance in the southernpea [*Vigna unguiculata* (L.) Walp.]. *J. Am. Soc. Hort. Sci. 101:* 148–149.

5. Mulder, J. L. and P. Holliday. 1969. *Cercospora cruenta. Commonwealth Mycol. Inst. Descriptions Pathogenic Fungi Bacteria 463.* 2 p.

6. Schneider, R. W., R. J. Williams and J. B. Sinclair. 1976. *Cercospora* leaf spot of cowpea: models for estimating yield loss. *Phytopathology 66:* 384–388.

7. Verma, P. R. and P. N. Patel. 1969. Host range, varietal resistance and epidemiological observations on *Cercospora* leaf spot disease of cowpea. *Indian Phytopathol. 22:* 61–66.

Choanephora pod rot

Pod rot of cowpea caused by *Choanephora cucurbitarum* (sporangia 40–80 μm, sporangiospores 10–16 × 17–37 μm and conidia 8–16 × 16–37 μm) was reported as a minor disease in North Carolina in 1920 but caused from 1–5 percent losses in Georgia in 1937 and 1965. Mature pods show water-soaked spots that develop into a soft rot on which the fungus fruits abundantly (Fig.

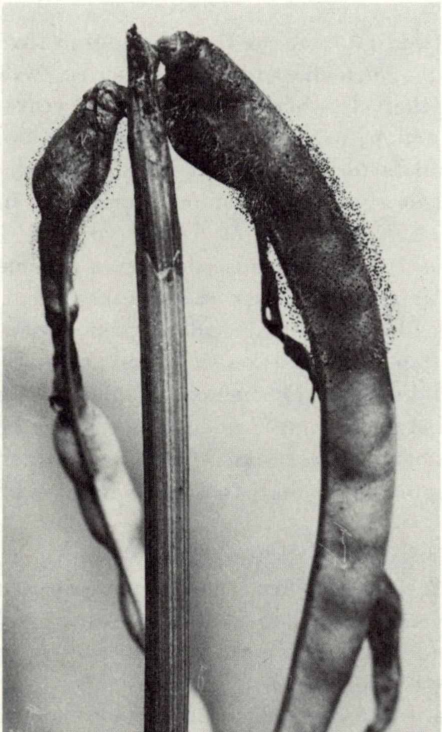

Fig. 58. *Choanephora* pod rot of cowpea. (Courtesy of R. J. Williams, International Institute of Tropical Agriculture, Ibadan.)

58). Peduncles and senescent corollas also are affected, although less frequently, but immature pods, leaves and flowers are not infected. The disease has occurred only during extended periods of high humidity and moderate daily temperatures.

Incidence of the disease has been quite erratic and usually is of little consequence. Varietal differences in susceptibility have been reported, but no variety has been identified as immune to infection.

Rot of insect damaged pods in cool, wet weather has been attributed to *C. infundibulifera* (Currey) Sacc. in Nigeria.

1. Lefebvre, C. L. and J. L. Weimer. 1939. *Choanephora cucurbitarum* attacking cowpeas. *Phytopathology 29:* 898–901.

2. Toler, R. W. and P. D. Dukes. 1965. *Choanephora* pod rot of cowpeas. *Plant Dis. Reptr. 49:* 347–350.

3. Williams, R. J. 1975. Diseases of cowpea [*Vigna unguiculata* (L) Walp.] in Nigeria. *PANS 21:* 253–267.

4. Wolf, F. A. and S. G. Lehman. 1920. Notes on new or little known plant diseases in North Carolina in 1920. *N. C. Agr. Expt. Sta. Ann. Rep. 43:* 55–58.

Fusarium wilt

The wilt disease of cowpea was described first from the United States in 1899 but now is known to occur also in Asia, Australia and South America. Wilt was more serious in the United States than elsewhere and, for several years, was a major factor in cowpea production, particularly in California and some of the southeastern states. Satisfactory control has now been developed, and wilt has been relegated to a position of minor importance in cowpea production.

Infected plants are more frequently noted because the leaves become flaccid and chlorotic before falling prematurely. Closer examination reveals the lower stem may be swollen before chlorosis appears, and necrotic vascular tissue in the stems and roots often is more extensive than might be expected from foliar symptoms. A rapid wilt develops in young plants, which are usually killed, but plants infected at a later stage in development may be stunted with slower progression of foliar chlorosis and wilt. Some infected plants may never show external symptoms, yet the vascular tissue may be severely disrupted and discolored.

Wilt is caused by *Fusarium oxysporum* Schlecht. ex Fries f. sp. *tracheiphilum* (E. F. Smith) Snyder & Hansen. Three races of the pathogen have been distinguished, i.e. race 1 has been isolated from both cowpea and soybean; race 2 from some cowpea cultivars only; and race 3 from cowpea cv. Arlington, which is resistant to the other 2 races.

The fungus that causes wilt tolerates an environment with low oxygen and high carbon dioxide concentrations, which may relate to its long persistence in soil. While seeds are not known to be internally infected, it has been demonstrated that inoculum can be carried on the seed surface, from which infection can readily result even in seeds more than a year old. The severity of wilt infection from infested soil or artificial inoculation has been seriously exacerbated by the presence of root-knot nematodes (*Meloidogyne* sp.) in controlled experiments. Races of the fungus also have been distinguished on the basis of varietal susceptibility with cv. Iron resistant to all 3 races, Arlington resistant to races 1 and 2 but killed by race 3 (also isolated from infected soybean), Groit killed by race 1 but not race 2 and Extra Early Blackeye killed by race 2 but not by race 1. Genetic resistance from cv. Iron has been found to result from a single dominant factor for race 1 of the causal fungus and 2 dominant factors each for races 2 and 3.

Satisfactory control has been achieved (in the United States) through use of resistant varieties, viz. Brown Sugar Crowder S-1, Grant and Mississippi Crowder.

1. Armstrong, G. M. and J. K. Armstrong. 1950. Biological races of the *Fusarium* causing wilt of cowpeas and soybeans. *Phytopathology 40:* 181–193.

2. Erwin, D. C. and I. J. Thomason. 1956. Wilt resistant blackeye beans. *Calif. Agr. 10:* 6.

3. Hare, W. W. 1952. Reaction of cowpea varieties to *Fusarium* wilt. *Phytopathology 42:* 283.

4. Hare, W. W. 1953. A new race of *Fusarium* causing wilt of cowpea. *Phytopathology 43:* 291.

5. Hare, W. W. 1955. Resistance to *Fusarium* wilt in Brown Sugar Crowder cowpeas. *Phytopathology 45:* 347.

6. Hare, W. W. 1957. Inheritance of resistance of *Fusarium* wilt in cowpeas. *Phytopathology 47:* 312.

7. Hare, W. W. 1957. Mississippi Crowder, a new disease-resistant cowpea. *Phytopathology 47:* 565–566.

8. Holliday, P. 1970. *Fusarium oxysporum* f. sp. *tracheiphilum. Commonwealth Mycol. Inst. Descriptions Pathogenic Fungi Bacteria 220.* 1 p.

9. Kendrick, J. B. 1931. Seed transmission of cowpea *Fusarium* wilt. *Phytopathology 21:* 979–983.

10. Kendrick, J. B. 1936. A cowpea resistant to *Fusarium* wilt and nematode root knot. *Phytopathology 26:* 98.

11. Mackie, W. W. 1946. Blackeye beans in California. *Calif. Agr. Expt. Sta. Bull. 696.* 56 p.

12. Orton, W. A. 1902. The wilt disease of the cowpea and its control. *U. S. Dept. Agr. Bur. Plant Ind. Bull. 17:* 9–20.

13. Oyekan, P. O. 1975. Occurrence of cowpea wilt caused by *Fusarium oxysporum* f. sp. *tracheiphilum* in Nigeria. *Plant Dis. Reptr. 59:* 488–490.

14. Singh, R. S. 1954. Wilt of lobia in Uttar Pradesh. *Sci. Cult. 19:* 454–456.

15. Smith, E. F. 1899. Wilt disease of cotton, watermelon and cowpea (*Neocosmospora* nov. gen.). *U. S. Dept. Agr. Div. Veg. Physiol. Pathol. Bull. 17.* 72 p.

16. Thomason, I. J. 1958. The effect of the root-knot nematode, *Meloidogyne javanica*, on blackeye bean wilt. *Phytopathology 48:* 398.

17. Thomason, I. J., D. C. Erwin and M. J. Garber. 1959. The relationship of the root-knot nematode, *Meloidogyne javanica*, to *Fusarium* wilt of cowpea. *Phytopathology 49:* 602–606.

Target spot

Leaf spot symptoms on cowpea were described from Japan in 1931 and later (1945) reported to occur in several locations in southeastern United States where localized damage was serious a year earlier. The disease is not known to occur elsewhere, nor is it considered a major disease of cowpea now.

Small, reddish spots on cowpea leaves enlarge to become mostly round lesions, typically with prominent red-brown, concentric zonations on lighter brown, necrotic tissue. A red-brown spot is often evident in the centers of lesions 3–20 mm, which sometimes coalesce to form extensive necrotic areas in which tissues rupture and fall, imparting a ragged appearance to affected leaves. Similar spots, but less evident and without zonation, may develop on leaves of soybean, and reddish spots and streaks may develop on the mature petioles and stems of severely affected cowpea plants.

Helminthosporium vignicola (Kawamura) Olive (syn. *Cercospora vignicola* Kawamura) has been identified as the causal fungus. It produces 1–20 septate conidiophores 6–11 × 44–380 μm, either single or in clumps of 2–6, mostly on the zonate rings in leaf lesions. Brown, elongate, tapered (from the base) conidia with thickened walls are 3–20 septate, measure 8–19 × 40–270 μm and may be straight or curved and catenulate in chains of 2–5. Conidia are formed mostly in acropetal fashion. Two pathotypes of the fungus have been distinguished, one isolated from cowpea that causes serious symptoms on cowpea leaves but only minor leaf spotting on soybean, and another isolated from soybean that causes a minor leaf spot disease on soybean and only a fleck on cowpea leaves.

Disease incidence has not warranted development of control measures for target spot, which typically occurs in late season.

1. Olive, L. S. 1949. Target spot of cowpea and soybean. *Mycologia 41:* 355.

2. Olive, L. S., D. C. Bain and C. L. Lefebvre. 1945. A leaf spot of cowpea and soybean caused by an undescribed species of *Helminthosporium*. *Phytopathology 35:* 822–831.

Cowpea virus diseases

Cowpea is widely grown throughout the world for purposes other than human consumption, and many varieties and cultivars have resulted from planned breeding programs as well as incidental local selection. The natural growth habit of the cowpea plant makes it a convenient test host, especially for virus studies. As a result, the cowpea has been used extensively in virus research, and some references made to cowpea do not clearly distinguish natural infection of this host as opposed to infection under controlled conditions. Thus, agricultural research literature may not always accurately reflect the natural disease producing capabilities of causal agents referred to in some instances as pathogens of cowpea.

Aphid-borne cowpea mosaic

A virus disease affecting cowpea in Florida (United States) was designated "blackeye cowpea mosaic" in 1955. A similar disease was reported later from Italy as "cowpea aphid-borne mosaic," which now is known to occur in Asia, Africa and the southwest Pacific. Economic effects of the disease have not been critically examined, but there have been serious local infections.

Infected plants develop a severe mottling with dark green vein banding and interveinal chlorosis (Fig. 59). There can be considerable distortion and blistering of affected leaves, while plants are stunted and yield is markedly reduced.

The virus that causes aphid-borne mosaic is characterized by flexuous particles 750 nm in length. Virus infectivity is lost in sap from diseased cowpea by dilution greater than 10^{-4}, exposure for 10 minutes to temperatures above 60 C or storage *in vitro* for more than 3 days, although there has been some disagreement about these criteria. A number of aphids, including *Myzus persicae*, *Aphis medicaginis*, *A. craccivora*, and *A. gossypii*, readily transmit the virus in non-persistent fashion. The virus also is transmissible mechanically and is seed-borne in cowpea. Several apparent strains of the virus, as determined by differing host ranges and physical properties, have been described, but most are infectious to a sizeable number of leguminous plants and species in other families as well, i.e. *Amaranthaceae*, *Chenopodiaceae*, *Cucurbitaceae*, *Labiatae* and *Solanaceae*. Some strains have been shown to have distant serological relation to bean common mosaic.

A virus described from India (catjang mosaic), which also is seed-borne and transmissible by aphids (*A. medicaginis* was the most efficient vector), was characterized as having quite different physical properties (inactivated by exposure to 90 C, dilution to 10^{-5} or storage *in vitro* for 4 days). Despite sizeable disparities and closer conformity with some physical properties of southern bean mosaic, it was suggested that the virus from India was closely related to those viruses described from the United States and China (i.e. aphid-borne mosaic).

Little attention has been given to control of aphid-borne mosaic, but reported differences in severity of field symptoms on cowpea cultivars offer some promise of useful genetic resistance. Seeds for planting should be harvested only from healthy plants and new fields isolated from existing or previously infected fields of cowpea. The immediate area surrounding cowpea fields should be kept free of plants known to be susceptible to aphid-borne mosaic and application of insecticide to reduce and keep aphid populations as low as practicable can be recommended.

1. Anderson, C. W. 1955. *Vigna* and *Crotalaria* viruses in Florida. II. Notations concerning cowpea mosaic virus *(Marmor vignae)*. *Plant. Dis. Reptr. 39:* 349–353.

2. Bock, K. R. 1973. East African strains of cowpea aphid-borne mosaic virus. *Ann. Appl. Biol. 74:* 75–83.

Fig. 59. Leaf symptoms of aphid-borne cowpea mosaic (top) (Courtesy of F. W. Zettler, University of Florida, Gainesville) and appearance of infected (left) and healthy cowpea plants in the field (Courtesy of W. J. Kaiser, East African Agriculture and Forestry Research Organization, Nairobi).

3. Bock, K. R. and M. Conti. 1974. Cowpea aphid-borne mosaic virus. *Commonwealth Mycol. Inst. Descriptions Plant Viruses 134.* 4 p.

4. Capoor, S. P. and P. M. Varma. 1956. Studies on a mosaic disease of *Vigna cylindrica* Skeels. *Indian J. Agr. Sci. 26:* 95–103.

5. Capoor, S. P., P. M. Varma and B. N. Uppal. 1947. A mosaic disease of *Vigna catjang* Walp. *Current Sci. 16:* 151.

6. Fischer, H. U. and B. E. Lockhart. 1976. Strain of cowpea aphid-borne mosaic virus isolated from cowpeas in Morocco. *Phytopathol. Z. 85:* 43–48.

7. Kaiser, W. J., D. Danesh and M. Okhovat. 1968. Diseases of pulse crops (edible legumes) in Iran. *Plant Dis. Reptr. 52:* 687–691.

8. Kaiser, W. J. and G. H. Mossahebi. 1975. Studies with cowpea aphid-borne mosaic virus and its effect on cowpea in Iran. *FAO Plant Protect. Bull. 23:* 33–39.

9. Kuhn, C. W., B. B. Brantley and G. Sowell, Jr. 1966. Southern pea viruses: identification, symptomatology, and sources of resistance. *Georgia Agr. Expt. Sta. Bull., n.s. 157.* 22 p.

10. Lovisolo, O. and M. Conti. 1966. Identification of an aphid-transmitted cowpea mosaic virus. *Neth. J. Plant Pathol. 72:* 265–269.

11. Martyn, E. B. 1968. Plant virus names. An annotated list of names and synonyms of plant viruses and diseases. *Commonwealth Mycol. Inst. Phytopathol. Paper 9.* 204 p.

12. McLean, D. M. 1941. Studies on mosaic of cowpeas, *Vigna sinensis. Phytopathology 31:* 420–430.

13. Padma, R. and A. S. Summanwar. 1973. *Chenopodium murale*—a differential host for cowpea mosaic viruses. *Current Sci. 42:* 620–621.

14. Raheja, A. K. and O. I. Leleji. 1974. An aphid-borne virus disease of irrigated cowpea in northern Nigeria. *Plant Dis. Reptr. 58:* 1080–1084.

15. Snyder, W. C. 1942. A seed-borne mosaic of asparagus bean, *Vigna sesquipedalis. Phytopathology 32:* 518–523.

16. Tsuchizaki, T., K. Yora and H. Asuyama. 1970. The viruses causing mosaic of cowpea and azuki beans, and their transmissibility through seeds. I. Seed transmission of viruses in cowpea and azuki bean plants. II. Factors influencing seed transmission. *Ann. Phytopathol. Soc. Japan 36:* 112–120; 121–126.

17. Van Velsen, R. J. 1962. Cowpea mosaic, a virus disease of *Vigna sinensis* in New Guinea. *Papua New Guinea Agr. J. 14:* 153–161.

18. Verma, V. S. 1971. Effect of heat on seed transmission of mosaic disease of cowpea (*Vigna sinensis* Savi). *Acta Microbiol. Pol., Ser. B., 3:* 163–166.

19. Vidano, C. and M. Conti. 1965. Transmissioni con afidi di un "cowpea mosaic virus" isolato da *Vigna sinensis* Endl. in Italia. *Atti Accad. Sci., Torino 99:* 1041–1050.

20. Yu, T. F. 1946. A mosaic disease of cowpea (*Vigna sinensis* Endl.). *Ann. Appl. Biol. 33:* 450–454.

21. Zettler, F. W. and I. R. Evans. 1972. Blackeye cowpea mosaic virus in Florida: host range and incidence in certified cowpea seed. *Proc. Florida State Hort. Soc. 85:* 99–101.

Beetle-borne cowpea mosaic

The name "yellow mosaic" was applied to a disease of cowpea first described from Nigeria in 1959, but later studies revealed that probably the same disease, now termed "cowpea mosaic," had been reported from the United States 25 years earlier. Another disease, presently considered the same as yellow mosaic but for which no common name was designated, was reported from Trinidad in 1949. The same disease was later reported from the United States (Arkansas) in 1963, from Venezuela in 1967 and from Cuba in 1970. Losses attributable to cowpea mosaic have not been determined, but it has been characterized as the "most important disease affecting cowpeas . . . in Western Nigeria, sometimes reducing seed yields to nothing."

Inoculated leaves of cowpea varieties susceptible to beetle-borne mosaic develop chlorotic (ring or ringspot) lesions 2–4 mm in diameter, but systemic infection appears in the next leaf above as vein clearing. Leaves formed later develop an irregular, yellow-green mottle and may be distorted. Growth of infected plants is noticeably retarded and internodes shortened. However, these symptoms do not distinguish beetle-borne mosaic infection from symptoms of other cowpea virus diseases.

Different symptoms on primary leaves have been found indicative of plant response to beetle-borne cowpea mosaic inoculation. Plants immune to infection develop no symptoms on inoculated, primary leaves, while resistant plants develop necrotic lesions (with no systemic infection) and susceptible plants develop chlorotic lesions followed by systemic infection.

Beetle-borne cowpea mosaic disease is caused by a (polyhedral) virus with particles variously reported as 25–33 nm in diameter. Physical properties in sap from infected cowpea include dilution end-point of about 10^{-5}, thermal inactivation between 65–70 C and longevity *in vitro* of 7–20 days. The virus is seed-borne (10 percent), mechanically transmissible and transmitted by beetles [galerucid beetle *Ootheca mutabilis* Sahlb., chrysomelid beetle *Paraluperodes quaternus* (Fairm.), curculionid beetle *Nematocerus acerbus* Fst.], 2 grasshoppers, *Cantatops spissus spissus* Wlk. and *Zonocerus variegatus* F.; 2 thrips, *Sericothrips occipitals* Hood and *Taeniothrips sjostedti* (Tryb.) in Nigeria; the leaf beetle *Cerotoma ruficornis* in Trinidad and the bean leaf beetle *C. trifurcata* in the United States. Host plants for the viruses in the 3 locations (Arkansas, Trinidad and Nigeria) are not identical, but the virus from Arkansas has by far the broadest range of infectivity that includes systemic infection of cucumber, tobacco, *N. glutinosa*, pigeon pea and pea, and localized, chlorotic infection of jimson weed, globe amaranth and *Chenopodium amaranticolor*. The virus from Arkansas also causes necrotic, localized infections on *Cassia occidentalis* and *Sesbania exaltata* (Raf.) Cory. The viruses from Trinidad and Nigeria cause localized infections on bean, Bengal bean (*Mucuna aterrima* Holland) and *Chenopodium amaranticolor*. The viruses from all 3 areas have been found serologically related, but only the Trinidad and Arkansas viruses were found serologically related to bean pod mottle virus.

Dissemination of the disease is readily accomplished by seed from infected plants and the feeding of beetles. These factors, without the added possibility of ready mechanical transmission, provide adequate means for broad distribution and severe local incidence of cowpea mosaic.

Varieties which have been reported immune to infection by beetle-borne cowpea mosaic include Arlington, Blackeye, Brabham, Buff, Louisiana Purchase, Six Weeks Alabama, Variegated, Victor and Whippoorwill. Other varieties, including Brown Sugar Crowder, Clay Buff, Early Red and Paraguay 5, have been found resistant and, with the immune varieties, provide satisfactory control.

1. Capoor, S. P., P. M. Varma and B. W. Uppal. 1947. A mosaic disease of *Vigna catjang* Walp. *Current Sci. 16:* 151.

2. Chant, S. R. 1959. Viruses of cowpea, *Vigna unguiculata* (L.) Walp., in Nigeria. *Ann. Appl. Biol. 47:* 565–572.

3. Chant, S. R. 1960. The effect of infection with tobacco-mosaic and cowpea yellow-mosaic viruses on the growth rate and yield of cowpea in Nigeria. *Empire J. Expt. Agr. 28:* 114–120.

4. Chant, S. R. 1961. The use of *Chenopodium amaranticolor* in the study of Nigerian cowpea yellow mosaic virus. *Phytopathology 51:* 332–333.

5. Chant, S. R. 1962. Further studies on the host range and properties of Trinidad cowpea mosaic virus. *Ann. Appl. Biol. 50:* 159–162.

6. Dale, W. T. 1949. Observations on a virus disease of cowpea in Trinidad. *Ann. Appl. Biol. 36:* 327–333.

7. Dale, W. T. 1953. The transmission of plant viruses by biting insects, with particular reference to cowpea mosaic. *Ann. Appl. Biol. 40:* 384–392.

8. Debrot C., E. and C. E. B. de Rojas. 1967. El virus del mosaico del frijol, *Vigna sinensis* Endl., (cowpea mosaic virus) en Venezuela. *Agron. Trop. 17:* 3–16.

9. Kvíčala, B. A., J. Smrž and N. Blanco. 1973. A beetle-transmitted virus disease of cowpea in Cuba. *FAO Plant Protect. Bull. 21:* 27–29, 39.

10. Martyn, E. B. 1968. Plant virus names. An annotated list of names and synonyms of plant viruses and diseases. *Commonwealth Mycol. Inst. Phytopathol. Paper 9.* 204 p.

11. Robertson, D. G. 1963. Further studies on the host range of cowpea yellow mosaic virus. *Trop. Agr. (Trinidad) 40:* 319–324.

12. Robertson, D. G. 1965. The local lesion reaction for recognizing cowpea varieties immune from and resistant to cowpea yellow mosaic virus. *Phytopathology 55:* 923–925.

13. Shepherd, R. J. 1963. Serological relationship between bean pod mottle virus and cowpea mosaic viruses from Arkansas and Trinidad. *Phytopathology 53:* 865–866.

14. Shepherd, R. J. 1964. Properties of a mosaic virus of cowpea and its relationship to the bean pod mottle virus. *Phytopathology 54:* 466–473.

15. Shepherd, R. J. and R. W. Fulton. 1962. Identity of a seed borne virus of cowpea. *Phytopathology 52:* 489–493.

16. Smith, C. E. 1924. Transmission of cowpea mosaic by bean leaf-beetle. *Science 60:* 268.

17. Van Kammen, A. 1971. Cowpea mosaic virus. *Commonwealth Mycol. Inst. Descriptions Plant Viruses 47.* 4 p.

18. Wells, D. G. and R. Deba. 1961. Sources of resistance to the cowpea yellow mosaic virus. *Plant Dis. Reptr. 45:* 878–881.

19. Whitney, W. K. and R. M. Gilmer. 1974. Insect vectors of cowpea mosaic in Nigeria. *Ann. Appl. Biol. 77:* 17–21.

Chlorotic mottle

Chlorotic mottle disease of cowpea was identified and reported from the United States (Georgia) in 1964. The disease remains confined to the United

Fig. 60. Symptoms of chlorotic mottle in cowpea leaflets. (Courtesy of C. W. Kuhn, University of Georgia, Athens.)

States, but no information concerning effect on yield has been reported.

Affected plants usually develop a bright yellow mottle that may range from distinct chlorotic rings to patchy coloration to complete and uniform yellowing of leaflets (Fig. 60). Some cultivars which are fully susceptible exhibit only very mild symptoms, however.

Particles of the virus that causes chlorotic mottle disease are spherical, 25 nm in diameter, inactivated (in cowpea sap) by exposure for 10 minutes to temperatures between 65–70 C, dilution greater than 10^{-5} or storage *in vitro* for more than 24–48 hours. Two additional strains (*Desmodium* and soybean strains) with somewhat different physical properties have been distinguished by reduced symptom severity on cowpea. The virus is readily transmissible mechanically and is infectious to several legumes, causing systemic symptoms in *Vigna cylindrica* L. and asparagus bean, local lesions in *Cassia tora*, vein necrosis in *C. occidentalis*, hyacinth bean and several varieties of bean, and symptomless infection in broad bean, *Lupinus albus*, pea and peanut. Local lesions also are induced on *Chenopodium album*. Cowpea chlorotic mottle virus and bean yellow stipple virus, which has been reported from Central America, are related serologically.

The virus infects all parts of susceptible cowpea plants, including roots, petals, sepals, stamens and pistils, but not pollen. Virus can be recovered from the seedcoat but not the cotyledons or embryos of immature seeds, and no evidence of transmission in mature, dry seed has been obtained. Mixed infection with southern bean mosaic causes a synergistic reduction of plant growth and yield.

Eight cultivars resistant to chlorotic mottle have been identified (P. I. 147652, 186458, 186465, 194203, 255788, 255811, 297561, and 297562), and resistance in one, P. I. 255811, has been found to be conditioned by a single, recessive allele.

1. Bancroft, J. B. 1971. Cowpea chlorotic mottle virus. *Commonwealth Mycol. Inst. Descriptions Plant Viruses 49.* 4 p.

2. Bancroft, J. B., E. Hiebert, M. W. Rees and R. Markham. 1968. Properties of cowpea chlorotic mottle virus, its protein and nucleic acid. *Virology 34:* 224–239.

3. Bliss, F. A. and D. G. Robertson, 1971. Genetics of host reaction in cowpea to cowpea yellow mosaic virus and cowpea mottle virus. *Crop Sci. 11:* 258–262.

4. Dawson, W. O. and C. W. Kuhn. 1972. Enhancement of cowpea chlorotic mottle virus biosynthesis and *in vivo* infectivity by 2-thiouracil. *Virology 47:* 21–29.

5. Dawson, W. O. and C. W. Kuhn. 1974. Kinetics of multiplication, inactivation, and particle-breakdown of cowpea chlorotic mottle virus in cowpea. *Phytopathology 64:* 951–957.

6. Fulton, J. P., R. Gamez and H. A. Scott. 1975. Cowpea chlorotic mottle and bean yellow stipple viruses. *Phytopathology 65:* 741–742.

7. Gay, J. D. 1969. Effect of seed maturation on the infectivity of cowpea chlorotic mottle virus. *Phytopathology 59:* 802–804.

8. Gay, J. D. and C. W. Kuhn. 1968. Specific infectivity of cowpea chlorotic mottle virus from five hosts. *Phytopathology 58:* 1609–1615.

9. Kuhn, C. W. 1964. Separation of cowpea virus mixtures. *Phytopathology 54:* 739–740.

10. Kuhn, C. W. 1964. Purification, serology, and properties of a new cowpea virus. *Phytopathology 54:* 853–857.

11. Kuhn, C. W. 1965. Decline of specific infectivity of cowpea chlorotic virus *in vivo. Virology 25:* 9–14.

12. Kuhn, C. W. 1968. Identification and specific infectivity of a soybean strain of cowpea chlorotic mottle virus. *Phytopathology 58:* 1441–1442.

13. Kuhn, C. W. and W. O. Dawson. 1973. Multiplication and pathogenesis of cowpea chlorotic mottle virus and southern bean mosaic virus in single and double infections in cowpea. *Phytopathology 63:* 1380–1385.

14. Rogers, K. M., J. D. Norton and O. L. Chambliss. 1973. Inheritance of resistance to cowpea chlorotic mottle virus in southern pea, *Vigna sinensis. J. Am. Soc. Hort. Sci. 98:* 62–63.

15. Sowell, G., Jr., C. W. Kuhn and B. B. Brantley. 1965. Resistance of southern pea, *Vigna sinensis,* to cowpea chlorotic mottle virus. *Proc. Am. Soc. Hort. Sci. 86:* 487–490.

Cucumber mosaic

The disease now referred to as cucumber mosaic was described in 1916 simultaneously by 2 men who worked independently in the United States. The first report that cucumber mosaic could cause a "systemic disease" in cowpea appeared in 1934 from controlled inoculations, but natural occurrence of the disease in cowpea was not established until 1955. Now, cucumber mosaic is worldwide, particularly in temperate regions, and causes diseases in a wide assortment of plants, including banana and passion fruit, as well as cucurbits and solanaceous plants. Limited information concerning influence of the disease on productivity in cowpea provides no evidence of serious reduction in yield (under controlled conditions) despite often severe foliar symptoms.

Symptoms of cucumber mosaic infection in cowpea may vary depending on the cultivar of host and isolate or strain of virus involved. Susceptible combinations of host and virus typically result in early vein clearing on the first leaves systemically infected, followed by severe mottling of leaves subsequently formed (Fig. 61). Affected plants are seriously stunted and size of distorted leaves often noticeably reduced.

Many combinations of virus isolates, or strains, and specific cowpea cultivars result in only localized, necrotic infection on inoculated leaves with no systemic infection.

Cucumber mosaic disease is caused by a virus characterized by isometric particles 30 nm in diameter that are rendered noninfectious by exposure (in tobacco sap) to about 70 C, dilution greater than 10^{-4} or storage *in vitro* for more than 3–6 days. The virus is readily transmissible mechanically and

Fig. 61. Symptoms (on left) of cucumber mosaic in cowpea. (Courtesy of C. W. Kuhn, University of Georgia, Athens.)

by more than 60 species of aphids, including *Aphis gossypii* and *Myzus persicae*, some of which can acquire the virus by feeding for less than a minute and transmit in a like period of time. Cucumber mosaic virus is also transmissible by seed (up to 28 percent with some strains) and several species of dodder (*Cuscuta* sp.) but is poorly immunogenic. A number of strain variants of the virus have been reported and variously characterized, but not all afford uniform cross-protection, and it has been demonstrated that at least 2 strains can multiply simultaneously in the same plant.

Cucumber mosaic infection of cowpea has prompted considerable attention because of the variety of different virus strains that have been described and the localized or systemic infection responses of identifiable cowpea cultivars. Infection does not seriously reduce yield, even though the disease can be seed-borne in cowpea, which probably accounts for the absence of a concerted effort to control the disease. However, cucumber mosaic is infectious to a large number of other crop plants, and planting only seeds harvested from healthy plants or cultivar(s) resistant to the (local) isolate of the virus is recommended.

1. Anderson, C. W. 1955. *Vigna* and *Crotalaria* viruses in Florida. I. Preliminary report on a strain of cucumber mosaic virus obtained from cowpea plants. *Plant Dis. Reptr. 39:* 346–348.

2. Anderson, C. W. 1957. Seed transmission of three viruses in cowpea. *Phytopathology 47:* 515.

3. Anderson, C. W. 1959. *Vigna* and *Crotalaria* viruses in Florida. V. Comparative transmission tests with aphids and beetles. *Phytopathology 49:* 117–118.

4. Brantley, B. B., C. W. Kuhn and G. Sowell, Jr. 1965. Effect of cucumber mosaic virus on southern pea (*Vigna sinensis*). *Proc. Am. Soc. Hort. Sci. 87:* 355–358.

5. Fulton, J. P. 1950. Studies on strains of cucumber virus 1 from spinach. *Phytopathology 40:* 729–736.

6. Gibbs, A. J. and B. D. Harrison. 1970. Cucumber mosaic virus. *Commonwealth Mycol. Inst. Descriptions Plant Viruses 1.* 4 p.

7. Klesser, P. J. 1960. Virus diseases of cowpeas. *Bothalia 7:* 233–251.

8. Price, W. C. 1934. Isolation and study of some yellow strains of cucumber mosaic. *Phytopathology 24:* 743–761.

9. Sill, W. H., Jr. and J. C. Walker. 1952. A strain of cowpea susceptible to systemic infection by various strains of common cucumber mosaic virus. *Phytopathology 42:* 442.

10. Sill, W. H., Jr. and J. C. Walker. 1952. Cowpea as an assay host for cucumber virus 1. *Phytopathology 42:* 328–330.

11. Sinclair, J. B. and J. C. Walker. 1955. Inheritance of resistance to cucumber mosaic virus in cowpea. *Phytopathology 45:* 563–564.

12. Sinclair, J. B. and J. C. Walker. 1956. Extent of cross protection among strains of cucumber mosaic virus in cucumber and cowpea. *Phytopathology 46:* 367–371.

13. Welkie, G. W., S. F. Yang and G. W. Miller. 1967. Metabolite changes induced by cucumber mosaic virus in resistant and susceptible strains of cowpeas. *Phytopathology 57:* 472–475.

14. Whipple, O. C. and J. C. Walker. 1941. Strains of cucumber mosaic virus pathogenic on bean and pea. *J. Agr. Res. 62:* 27–60.

Mild mottle

A disease of cowpea in Ghana was described in 1973 as mild mottle. Differences noted in symptom severity on cowpea plants made economic assessment of the disease uncertain, but it was considered a definite threat to successful production of other crops.

Naturally affected cowpea plants exhibit either a mild systemic mottle or no symptom, but inoculated plants of intolerant cultivars develop local (necrotic) lesions on inoculated leaves and severe systemic chlorosis and necrosis.

Particles of the causal virus measure 13 × 650 nm and sap from infected soybean is rendered noninfectious by dilution to 10^{-4} or exposure to 70 C for 10 minutes. Sap remains infectious after storage at 20 C for 8 days (20 days at 2 C). Sap lyophilized in 7 percent peptone and dextrose and stored at 20 C remained infective for 3 years. Mild mosaic virus is serologically related to carnation latent virus and readily transmissible mechanically, but not by aphids. Seed transmission was greater than 90 percent in soybean and cowpea but 15 percent or less in 3 varieties of bean.

The virus is infectious to and induces severe symptoms in peanut, jackbean [*Canavalia ensiformis* (L.) DC] and pigeon pea in addition to soybean. Less severe systemic infection develops in plants of lentil, winged bean [*Psophocarpus tetragonolobus* (L.) DC], lima bean and *P. coccineus*. Plants of chickpea, pea, *Mucuna deeringianum* (Bort) Merr. and broad bean are not infected, nor are plants of pepper, *Celosia argentea* L., cucumber, jimson weed, *Nicandra physalodes* Gaertn. *Nicotiana glutinosa*, *Ocimum basilicum* L., nasturtium, *Vinca rosea* L. or zinnia. However, the mild mottle virus is infectious to *Nicotiana clevelandii* and cacao and caused chlorotic or necrotic localized lesions on leaves of 6 species of *Chenopodiaceae*, including *C. quinoa*, *C. amaranticolor* and spinach.

Control procedures have not been devised, but selection of seeds from healthy plants should prove beneficial. Seed lots suspected of containing infected seeds should not be planted near peanuts, pigeon pea, jackbean or cacao.

1. Brunt, A. A. and R. H. Kenten. 1973. Cowpea mild mottle, a newly recognized virus infecting cowpeas (*Vigna unguiculata*) in Ghana. *Ann. Appl. Biol. 74:* 67–74.

2. Brunt, A. A. and R. H. Kenten. 1974. Cowpea mild mottle virus. *Commonwealth Mycol. Inst. Descriptions Plant Viruses 140.* 4 p.

Southern bean mosaic

A seed-transmitted disease of cowpea first recognized in greenhouse-grown plants was reported from the United States (Wisconsin) in 1962. The cowpea disease is now known to occur in the field in the United States (Georgia) and

Africa (Ghana) and commands some attention because of potential import.

Seedlings from infected seed usually develop vein clearing in the primary leaves, followed by mottling in trifoliate leaves. Inoculated (primary) leaves often develop no symptoms, although sometimes chlorotic lesions appear and may form ring spots. Vein clearing appears in the first trifoliate leaves formed after inoculation, and this symptom is soon followed by a coarse mottle in leaves formed subsequently (Fig. 62). At least 3 other responses to inoculation, including no symptoms whatever, necrotic local lesions on the inoculated leaves only, and necrotic local lesions on the inoculated leaves followed by systemic necrosis, have been noted in different cultivars.

The disease of cowpea is caused by a strain of southern bean mosaic virus characterized by isometric particles 25 nm in diameter inactivated (in cowpea sap) by dilution to 10^{-6}, heating (for 10 minutes) to 90 C or storage *in vitro* for 20 days. Despite slight differences in identifying characteristics of the cowpea isolates of the virus, identity with southern bean mosaic has been established serologically. Infectivity is confined to the *Leguminosae*, but marked differences in specific hosts and cultivars has been reported. The virus originally described from cowpea (grown in the greenhouse in Wisconsin) systemically infected pea but was not infectious to bean, while the virus reported from Ghana did not infect pea, and bean cultivars varied from immune to susceptible to localized infection (none were systemically infected). Transmission of the virus has been demonstrated through seed from infected plants and by the bean leaf beetle.

Fig. 62. Symptoms (on left) of southern bean mosaic in cowpea. (Courtesy of C. W. Kuhn, University of Georgia, Athens.)

Seed from infected plants has been found to result in 3–4 percent diseased seedlings, which provides for widespread dissemination of the disease. Mixed infection with cowpea chlorotic mottle was found to increase seed transmission to 12–30 percent. Spread of infection within fields is accomplished efficiently by the bean leaf beetle, which can remain infective for more than 2 weeks after feeding on diseased plants.

Cowpea varieties resistant to the cowpea strain of southern bean mosaic that occurs in the United States (Black, Brabham, Clay, Climax, Iron, Groit, Potomac and Texas Cream 40) and the virus strain in Ghana (Brown Crowder, Blue Goose, Cream Crowder, Running Acre and Tardy, as well as Dixie Cream, Georgia 21, Long Pod Purple Hull and White Sugar Crowder) offer the best method for field control of the disease. When susceptible varieties are to be grown, seeds for planting should be harvested only from healthy plants.

1. Brantley, B. B. and C. W. Kuhn. 1970. Inheritance of resistance to southern bean mosaic virus in southern pea, *Vigna sinensis*. *J. Am. Soc. Hort. Sci. 95:* 155–158.

2. Capoor, S. P., P. M. Varma and B. N. Uppal. 1947. A mosaic disease of *Vigna catjang* Walp. *Current Sci. 16:* 151.

3. Kuhn, C. W. 1963. Field occurrence and properties of the cowpea strain of southern bean mosaic virus. *Phytopathology 53:* 732–733.

4. Kuhn, C. W. and B. B. Brantley. 1963. Cowpea resistance to the cowpea strain of southern bean mosaic virus. *Plant Dis. Reptr. 47:* 1094–1096.

5. Kuhn, C. W. and W. O. Dawson. 1973. Multiplication and pathogenesis of cowpea chlorotic mottle virus and southern bean mosaic virus in single and double infections in cowpea. *Phytopathology 63:* 1380–1385.

6. Lamptey, P. N. L. and R. I. Hamilton. 1974. A new cowpea strain of southern bean mosaic virus from Ghana. *Phytopathology 64:* 1100–1104.

7. Shepherd, R. J. 1971. Southern bean mosaic virus. *Commonwealth Mycol. Inst. Descriptions Plant Viruses 57.* 4 p.

8. Shepherd, R. J. and R. W. Fulton. 1962. Identity of a seed-borne virus of cowpea. *Phytopathology 52:* 489–493.

9. Walters, H. J. and D. G. Henry. 1970. Bean leaf beetle as a vector of the cowpea strain of southern bean mosaic virus. *Phytopathology 60:* 177–178.

Tobacco mosaic

A disease of cowpea caused by a virus serologically identical to tobacco mosaic was reported from Nigeria in 1955. Particle morphology of the virus, production of local lesions on *Nicotiana glutinosa* and systemic mottle symptoms in tobacco also conformed to identity with tobacco mosaic virus. The virus that caused varying degrees of mottle severity in cowpea and Bengal bean also was systemically infectious to several varieties of bean which responded to inoculation with another (common) strain of tobacco mosaic with localized, necrotic lesions only. The disease is not known to occur elsewhere but was presumed to have been present in Nigeria since 1941.

1. Chant, S. R. 1959. Viruses of cowpea, *Vigna unguiculata* (L.) Walp., in Nigeria. *Ann. Appl. Biol. 47:* 565–572.

2. Lister, R. M. and J. M. Thresh. 1955. A mosaic disease of leguminous plants caused by a strain of tobacco mosaic virus. *Nature 175:* 1047–1048.

3. Zaitlin, M. and H. W. Israel. 1975. Tobacco mosaic virus (type strain). *Commonwealth Mycol. Inst. Descriptions Plant Viruses 151.* 5 p.

OTHER DISEASES

Bacterial pustule (*Xanthomonas* sp.)

1. Williams, R. J. 1975. Diseases of cowpea [*Vigna unguiculata* (L.) Walp.] in Nigeria. *PANS 21:* 253–267.

Bacterial wilt (*Pseudomonas solanacearum*) (see Potato—Bacterial wilt)

1. Hayward, A. C. and J. M. Waterston. 1964. *Pseudomonas solanacearum. Commonwealth Mycol. Inst. Descriptions Pathogenic Fungi Bacteria 15.* 2 p.

2. Ponte, J. J. and A. A. Santos. 1974. Bacterial wilt of cowpea (*Vigna sinensis* Endl.). *Fitopatologia 9:* 65.

Powdery mildew (*Erysiphe polygoni*)

Foliar spray of benomyl (2.4 g/4 ℓ water with surfactant added) provided control in Arkansas (United States).

1. Arif, A. and M. J. Goode. 1970. Systemic chemical control of cowpea powdery mildew. *Plant Dis. Reptr. 54:* 346.

Pythium stem rot (*Pythium aphanidermatum*)

1. Onourah, P. E. 1973. *Pythium* seed decay and stem rot of cowpea [*Vigna sinensis* (Linn.) Savi] in southern Nigeria. *Plant Soil 39:* 187–191.

2. Waterhouse, G. M. and J. M. Waterston. 1964. *Pythium apanidermatum. Commonwealth Mycol. Inst. Descriptions Pathogenic Fungi Bacteria 36.* 2 p.

3. Williams, R. J. and A. Ayanaba. 1975. Increased incidence of *Pythium* stem rot in cowpeas treated with benomyl and related fungicides. *Phytopathology 65:* 217–218.

Smut (*Entyloma vignea* Bat., Bezerra, Pontes & Vasconcelos)

1. Ponte, J. J. 1974. Cowpea varieties resistant to smut (*Entyloma vignae*). *Fitopatologia 9:* 65.

2. Ponte, J. J., A. A. Santos and J. M. F. Chagas. 1974. The occurrence of cowpea smut in the states of Piaui and Rio Grande do Norte. *Fitopatologia 9:* 65.

Anthracnose (*Colletotrichum lindemuthianum*)

Incidence of anthracnose (see Bean—Anthracnose) was first noted in Nigeria in 1969 and was more severe in cool, wet weather. Seed-borne inoculum was found to account for as much as 40 percent infection. The pathotype of the fungus that infects cowpea was not infectious to 3 varieties of bean, i.e. Black Valentine, Dark Red Kidney and Michelite, that are differentially susceptible to several forms of the fungus pathogenic on bean.

Control was achieved with weekly applications of benomyl or mancozeb (0.2 percent), and differences in varietal susceptibility have been noted.

1. Leakey, C. L. A. and M. Simbwa-Bunnya. 1972. Races of *Colletotrichum lindemuthianum* and implications for bean breeding in Uganda. *Ann. Appl. Biol. 70:* 25–34.

2. Mordue, J. E. M. 1971. *Colletotrichum lindemuthianum. Commonwealth Mycol. Inst. Descriptions Pathogenic Fungi Bacteria 316.* 2 p.

3. Onesirosan, P. T.and L. N. Barker. 1971. Stem anthracnose of cowpeas in Nigeria. *Plant Dis. Reptr. 55:* 820–822.

4. Williams, R. J. 1975. Diseases of cowpea [*Vigna unguiculata* (L.) Walp.] in Nigeria. *PANS 21:* 253–267.

Corynespora leaf spot (*Corynespora cassiicola*)

Zonate leaf spot (*Dactuliophora tarrii* Leakey)

1. Ellis, M. B. and P. Holliday. 1971. *Corynespora cassiicola. Commonwealth Mycol. Inst. Descriptions Pathogenic Fungi Bacteria 303.* 2 p.

2. Williams, R. J. 1975. Diseases of cowpea [*Vigna unguiculata* (L.) Walp.] in Nigeria. *PANS 21:* 253–267.

Cladosporium spot (*Cladosporium vignae* Gardner)

1. Strider, D. L. 1960. Control of *Cladosporium* spot of southern pea. *Plant Dis. Reptr. 44:* 955.

Macrophomina stem canker, Ashy stem blight, Charcoal rot (*Macrophomina phaseolina*)

Mostly underground stem cankers are surrounded by brown borders with light colored centers and dark pycnidia.

1. Holliday, P. and E. Punithalingam. 1970. *Macrophomina phaseolina. Commonwealth Mycol. Inst. Descriptions Pathogenic Fungi Bacteria 275.* 2 p.

2. Luttrell, E. S. and J. L. Weimer. 1952. *Macrophomina* stem canker and ashy stem blight of cowpea. *Plant Dis. Reptr. 36:* 194–195.

3. Sackston, W. E. 1969. *Sclerotium bataticola* on seeds of cowpea (*Vigna sinensis*). *Plant Dis. Reptr. 53:* 438–439.

4. Tompkins, C. M. and M. W. Gardner. 1935. Relation of temperature to infection of bean and cowpea seedlings by *Rhizoctonia bataticola*. *Hilgardia 9:* 219–230.

5. Weimer, J. L. and E. S. Luttrell. 1945. A canker of cowpea and *Macrophomina* leaf spot of cowpea and snap bean. *Plant. Dis. Reptr. 29:* 127–129.

Septoria leaf spot (*Septoria vignicola* Rao)

1. Rao, V. G. 1963. A new species of *Septoria* on an economic host. *Current Sci. 32:* 367–368.

Verticillium wilt (*Verticillium albo-atrum*)

1. Hawksworth, D. L. and P. W. Talboys. 1970. *Verticillium albo-atrum. Commonwealth Mycol. Inst. Descriptions Fungi Bacteria 255.* 2 p.

2. Strobel, J. W. 1961. *Verticillium* wilt of okra and southern pea in southern Florida. *Proc. Florida State Hort. Soc. 74:* 171–175.

Rust [*Uromyces phaseoli* (Perx.) Wint. var. *vignae* (Barcl.) Arth.]

Southern blight (*Botryobasidium rolfsii*) (Fig. 63)

Stem canker and Web blight (*Thanatephorus cucumeris =Rhizoctonia solani*) (Fig. 64)

(*Diaporthe*) Stem rot [*Diaporthe phaseolorum* (Cke. & Ell.) Sacc.]

Tip-burn (*Alternaria tenuis* Auct.)

1. Onesirosan, P. T. 1975. Seedborne and weedborne inoculum in web blight of cowpea. *Plant Dis. Reptr. 59:* 338–339.

2. Toler, R. W., S. S. Thompson and J. M. Barber. 1963. Cowpea (southern pea) diseases in Georgia, 1961–1962. *Plant Dis. Reptr. 47:* 746–747.

Fig. 63. Southern blight of cowpea. Note sclerotia on stem at soil line. (Courtesy of R. J. Williams, International Institute of Tropical Agriculture, Ibadan.)

Fig. 64. Lesions typical of web blight disease on cowpea leaflet. (Courtesy of R. J. Williams, International Institute of Tropical Agriculture, Ibadan.)

3. Williams, R. J. 1975. Diseases of cowpea [*Vigna unguiculata* (L.) Walp.] in Nigeria. *PANS 21:* 253–267.

Alfalfa mosaic

Local lesions only were induced on inoculated leaves of 9 cowpea cultivars with 26 virus strains.

1. Bos, L. and E. M. J. Jaspars. 1971. Alfalfa mosaic virus. *Commonwealth Mycol. Inst. Descriptions Plant Viruses 46.* 4 p.

2. Milbrath, J. A. and F. P. McWhorter. 1954. Response of cowpea varieties to strains of alfalfa mosaic virus. *Phytopathology 44:* 498.

Bean common mosaic

This virus was reported to infect cowpea in India in 1967.

1. Bos, L. 1971. Bean common mosaic virus. *Commonwealth Mycol. Inst. Descriptions Plant Viruses 73.* 4 p.

2. Sachchidananda, J., S. Singh, N. Prakash and V. S. Verma. 1973. Bean common mosaic virus on cowpea in India. *Z. Pflanzenkrankh. Pflanzenschutz 80:* 88–91.

Bean yellow mosaic

A strain of bean yellow mosaic virus systemically infectious to cowpea (Fig. 65) was identified from Florida (United States) in 1955 and reported again from New York in 1962. Common strains of the virus do not infect cowpea.

1. Anderson, C. W. 1955. *Vigna* and *Crotalaria* viruses in Florida. II. Notations concerning cowpea mosaic virus (*Marmor vignae*). *Plant Dis. Reptr. 39:* 349–353.

2. Bos, L. 1970. Bean yellow mosaic virus. *Commonwealth Mycol. Inst. Descriptions Plant Viruses 40.* 4 p.

3. Brierly, P. and F. F. Smith. 1962. Three cowpea mosaic viruses from gladiolus. *Plant Dis. Reptr. 46:* 335–337.

4. Harrison, A. N. and R. T. Gudauskas. 1968. Effects of some viruses on growth and seed production of two cowpea cultivars. *Plant Dis. Reptr. 52:* 509–511.

5. Kuhn, C. W., B. B. Brantley and G. Sowell, Jr. 1965. Immunity to bean yellow mosaic virus in cowpea. *Plant Dis. Reptr. 49:* 879–881.

Coconut wilt disease

Severe distortion in the first trifoliate leaves was reported following inoculation on primary leaves of cowpea used as an assay host for coconut wilt disease.

Fig. 65. Cowpea leaflets with symptoms of bean yellow mosaic virus infection (healthy, upper left). (Courtesy of C. W. Kuhn, University of Georgia, Athens.)

1. Holmes, F. O., S. B. Lal and P. Shanta. 1964. Cowpea inoculation test for diagnosis of coconut wilt disease in India. *FAO Plant Protect. Bull. 13:* 30–34.

2. Shanta, P. and K. P. V. Menon. 1960. Cowpea (*Vigna sinensis* Endl.), an indicator plant for the coconut wilt virus. *Virology 12:* 309–310.

Cowpea mottle

Mottling and distortion of leaves caused by a mechanically transmissible virus were reported from Nigeria in 1963. Tolerant varieties included Alabunch, Bechuana, Dixilee Selection and New Era.

1. Bliss, F. A. and D. G. Robertson. 1971. Genetics of host reaction in cowpea to cowpea yellow mosaic virus and cowpea mottle virus. *Crop. Sci. 11:* 258–262.

Tobacco ringspot

Hypersensitive resistance (necrotic, localized lesions) found in the susceptible variety Black was reported to be inherited as a single, dominant factor linked to cucumber mosaic virus susceptibility.

1. deZeeuw, D. J. and J. C. Ballard. 1958. A test strain of Black cowpea. *Plant Dis. Reptr. 42:* 898.

2. deZeeuw, D. J. and J. C. Ballard. 1959. Inheritance in cowpea of resistance to tobacco ringspot virus. *Phytopathology 49:* 332–334.

3. deZeeuw, D. J. and R. A. Crum. 1963. Inheritance of resistance to tobacco ringspot and cucumber mosaic viruses in Black cowpea crosses. *Phytopathology 53:* 337–340.

4. Stace-Smith, R. 1970. Tobacco ringspot virus. *Commonwealth Mycol. Inst. Descriptions Plant Viruses 17.* 4 p.

CUMIN: *Cuminum cyminum* L.

Powdery mildew

Importance of the mildew disease on cumin in India has varied since it was first observed in 1921. The disease was severe in several states of that country in 1953 and caused complete failure in some crops, while losses in yield elsewhere have been estimated at 5–15 percent. Mildew is known to occur only in India and Bulgaria.

Symptoms appear first on the lower leaves as small white specks that gradually enlarge and coalesce to cover the entire leaf surface with superficial, external mycelium. Infection may spread upward rapidly to involve young leaves, flowers and stems, and influence on fruit yield varies from only a slight reduction in weight to complete loss because of failure to mature. Sporulation on affected plant parts imparts an obvious powdery, white appearance.

Mildew disease is caused by *Erysiphe polygoni*, characterized by septate, superficial hyphae 3.5–6.5 μm in diameter with lobed haustoria. Hyaline, unicellular, erect conidiophores 103–170 μm bear hyaline, unicellular ovoid conidia 10–16 × 28–40 μm in chains, but conidia tend to fall when mature, so continuous chains of conidia are seldom seen. The perithecial stage of *E. polygoni* has not been found on cumin in India (although it has been found on other hosts) but has been reported from Bulgaria.

Disease development is enhanced by warm, moist conditions, and intensity of the disease is severely reduced when maximum temperatures fall below 26 C. Cardinal temperatures for conidial germination are 16, 20–22 and 30 C, and mature conidia readily germinate in water but will not germinate when dry and maintain viability only for a short period. The fungus survives from one season to another as dormant mycelium on seeds.

Control has been achieved by dusting sulfur (13.8–23 kg/ha) over the plants about flowering, but a second dusting at half the original rate may be required under environmental conditions favorable for disease development. Perenox as a foliar spray, applied before disease development has become extensive, has provided satisfactory control.

1. Chattopadhyay, S. B. 1967. *Diseases of plants yielding drugs, dyes and spices.* Indian Council of Agricultural Research, New Delhi. 100 p.

2. Joshi, N. C. 1955. Notes on two diseases of *Cumin cyminum* L. hitherto unreported from Ajmer State. *Sci. Cult. 21:* 101–102.

3. Uppal, B. N. and M. K. Desai. 1939. Cumin powdery mildew in Bombay. *Bombay Dept. Agr. Bull. 169.* 22 p.

Alternaria blight

Blight disease was described from India in 1938, where losses incurred usually were light but were estimated in some years as high as 70 percent.

Symptoms first appear on all plant parts as light colored, necrotic lesions that turn purple before changing to brown and later black. Lesions typically are scattered but under favorable conditions enlarge considerably. Affected plant parts, particularly leaves and flowers, are killed.

Conidia of the causal fungus, *Alternaria burnsii* Uppal et al., are beaked and have 1–9 transverse septa and measure 7–25 × 14–55 μm, excluding the beak. Conidiophores are straight, branched and geniculate and measure 2.0–3.7 × 15–52. μm.

Cardinal temperatures for growth *in vitro* are 5, 26–30 and 35 C for the causal fungus, which persists overwinter on plant debris left in the field.

Two foliar applications of either ziram (0.1 percent) or zineb (0.2 percent) at monthly intervals have been used for satisfactory control of blight. The first spray should be applied soon after first appearance of the disease.

1. Gemawat, P. D. and N. Prasad. 1969. Efficacy of different fungicides for the control of *Alternaria* blight of *Cuminum cyminum. Indian Phytopathol. 22:* 49–52.

2. Joshi, N. C. 1955. Notes on two diseases of *Cumin cyminum* L. hitherto unreported from Ajmer State. *Sci. Cult. 21:* 101–102.

3. Uppal, B. N., M. K. Patel and M. N. Kamat. 1938. *Alternaria* blight of cumin. *Indian J. Agr. Sci. 8:* 49–62.

Fusarium wilt

Wilt disease was first reported from India in 1949, where incidence of plant infection has been estimated as high as 70 percent. In the state of Rajasthan, wilt has been the most serious disease of cumin, and losses have approached 60 percent and caused the crop to be abandoned in some areas. Losses to 96 percent have been reported from the state of Gujarat.

Plants of all ages are susceptible to infection, characterized in seedlings by a drooping of the growing tips and leaves, followed quickly by wilting and death. Lower leaves on plants affected at an older age first turn from green to yellow, and this change in coloration progresses upward, but usually the plant wilts and dies before the youngest leaves have turned yellow. The plants gradually become dry and brittle after wilting begins, and the root systems of affected plants often show some signs of deterioration. Infection at flowering either completely prevents seed formation, or only small, shriveled seeds are formed, while plants infected later may produce both large, healthy seeds as well as some small, shriveled seeds.

The disease is caused by *Fusarium oxysporum* f. *cumini* Prasad & Patel, which produces hyaline, ovoid, aseptate microconidia 2–5 × 5–12 μm and mostly 2–3 septate, slightly curved macroconidia that measure 2–8 × 20–35 μm. Chlamydospores, both terminal and intercalary, are 6–10 μm in diameter.

Internal infection of plants by the fungus is confined to the vascular tissues but extends from the roots into the above-ground parts of diseased plants and into seeds, which may be invaded internally as well as contaminated externally. While dissemination may be achieved through diseased seeds or plant debris, the fungus apparently persists readily in soil, and continuous culture of the crop often results in "wilt-sick" soil. A number of morphological and pathological variants that differ in virulence have been distinguished, but the importance of individual pathotypes in nature has not been determined. No cultural practice other than an extended crop rotation program has been found that will effectively alter disease incidence.

Careful selection of seed materials to avoid obvious inclusion of diseased seeds is essential. Addition of sodium dimethyldithiocarbamate (50 percent solution) to soil 20 days before seeding effectively reduced wilt incidence in pot tests. Seed infection can be greatly reduced but not eliminated by treatment in hot water for 10 minutes at 54 C. Exposure to 58 C for 10 minutes may eliminate seed infection, but germination may also be seriously reduced. Rotation of cumin with other crops is recommended, following as long a schedule as is practicable.

1. Gaur, M. M. 1949. Diseases of cumin and fennel. *Plant Protect. Bull. 1:* 20–21.

2. Joshi, N. C. and J. P. Agnihotri. 1958. Studies on the wilt disease of cumin (*Cuminum cyminum* L.) in Ajmer State, India. *Lloydia 21:* 29–33.

3. Mathur, B. L. and R. L. Mathur. 1965. Metabolites of *Fusarium* oxysporum f. *cumini* in relation to cumin wilt. *Indian Phytopathol. 18:* 335–339.

4. Mathur, B. L. and R. L. Mathur. 1967. Effect of SMDC on germination, plant growth, and *Fusarium* wilt incidence of cumin (*Cuminum cyminum*). *Plant Dis. Reptr. 51:* 629–631.

5. Mathur, B. L. and N. Prasad. 1964. Studies on wilt disease of cumin caused by *Fusarium oxysporum* f. *cumini. Indian J. Agr. Sci. 34:* 131–137.

6. Mathur, B. L., H. C. Sankhla and R. L. Mathur. 1967. Influence of cultural practices on cumin wilt incidence. *Indian Phytopathol. 20:* 32–35.

7. Patel, P. N. and N. Prasad. 1963. *Fusarium* wilt of cumin (*Cuminum cyminum*) in Gujarat State, India. *Plant Dis. Reptr. 47:* 528–531.

8. Patel, P. N., N. Prasad, R. L. Mathur and B. L. Mathur. 1957. *Fusarium* wilt of cumin. *Current Sci. 26:* 181–182.

9. Singh, R. D., S. L. Choudhary and K. G. Patel. 1972. Seed transmission and control of *Fusarium* wilt of cumin. *Phytopathol. Mediter. 11:* 19–24.

F

FENNEL (FINOCCHIO): *Foeniculum vulgare* Mill.

Cercospora spot

Occurrence of spot disease in India was recorded in 1929, and a preliminary account appeared in 1950, but severe incidence of the disease soon thereafter prompted a more detailed study that was published in 1957. Extensive infection can cause complete crop loss.

Spots light brown in color develop on fruits, stems and leaves, on which they enlarge to cause complete necrosis that makes the leaves dark brown. Lesions on stems, branches and peduncles elongate as necrotic streaks. Severe disease incidence may cause entire fields to appear brown and necrotic. Seed formation may be completely prevented, or seeds may be shriveled and blackened.

Conidiophores of the causal fungus, *Cercospora foeniculi* Magnus, measure 4.5–7 $\times$ 10–70 μm, are septate but not branched and occur in fascicles. Cylindrical, subhyaline conidia are 1–5 septate and 4–6.5 $\times$ 15–75 μm.

Disease development is enhanced by temperatures between 25–35 C accompanied by some rainfall. Another fungus species (*Passalora foeniculi* Kamal & Khan) has been described as the causal agent in West Pakistan.

Four applications of Bordeaux mixture at intervals of about a month effectively reduced disease incidence in field trials.

1. Anonymous. 1950. Blight disease of fennel by *Cercospora foeniculi* Magn. *Plant Protect. Bull. 1:* 28–30.

2. Joshi, N. C. 1957. Studies on the diseases of *Foeniculum vulgare.* II. Symptoms, morphology, phenological relations and the efficacy of the Bordeaux mixture against *Cercospora foeniculi* Magn. *Agra Univ. J. Res. Sci. 6:* 89–91.

3. Kamal, M. and S. A. Khan. 1962. Identity of *Cercospora foeniculi* P. Magnus causing

infection of *Foeniculum vulgare* in Indo-Pakistan sub-continent. *Biologia (Lahore) 8:* 61–62.

4. Saccardo, P. A. 1931. *Sylloge fungorum 25:* 892.

Powdery mildew

Small, white lesions on leaves of fennel that became distorted and dry were described from Ajmer (India) in 1958. Lesions were found also on flowers, fruits and stems, and affected plants were stunted and produced small fruits with a bitter taste. Yield was noticeably reduced.

The fungus associated with the disease was identified as a species of *Oidium* with stout conidiophores 115–180 μm in length that developed from external mycelium and on which were borne chains of unicellular, hyaline conidia 13–18 × 36–40 μm.

Conidia germinated throughout a temperature range of 17–31 C, but highest percentage germination occurred at 25–27 C. Disease development also was more severe during periods when the maximum daily temperature remained above 25 C.

Approximately 23 kg sulfur/ha applied as dust in 3 applications beginning just prior to flowering and completed at the time of seed formation effectively controlled the disease in experimental plots.

1. Joshi, N. C. 1958. Studies on powdery mildew of *Foeniculum vulgare* from Ajmer. *Agra Univ. J. Res. Sci. 7:* 87–91.

Phytophthora leaf disease

Water-soaked discoloration of leaf laminae and petioles, soon developing into leaf lesions that girdled and constricted affected leaves, was noted in Italy in 1958. The disease was sufficiently extensive in some instances to destroy all the leaves on selected plants. Hyphae of the causal fungus, *Phytophthora syringae* (Kleb.) Kleb., varied from 5–11 μm in diameter, with ellipsoid sporangia that ranged in size from 16–53 × 25–193 μm and germinated either directly (when held at 18 C or higher) or indirectly (when kept below 15 C). Oogonia measured 22–48 μm in diameter and oospores 12–44 μm. Cardinal temperatures for mycelial growth *in vitro* were about 1, 21 and 26 C.

The disease was more severe in cool, humid weather. Four foliar applications of Bordeaux mixture (1 percent) gave best control in limited experiments.

1. Noviello, C. and W. C. Snyder. 1962. A *Phytophthora* disease of fennel. *Phytopathol. Z. 46:* 139–163.

2. Waterhouse, G. M. and J. M. Waterston. 1964. *Phytophthora syringae. Commonwealth Mycol. Inst. Descriptions Pathogenic Fungi Bacteria 32.* 2 p.

FENUGREEK: *Trigonella foenum-graecum* L.

Leaf spot

The first description of leaf spot on fenugreek was from Italy in 1904, but the disease has been found since that time in nearly all of the countries in southern Asia from Burma to Turkey. Leaf spot also has been reported to occur in Rumania and Hungary, Jordan, Israel, Egypt and Ethiopia, Brazil and the United States. Yield losses sustained have varied from those attributable to a "rather destructive disease" to "heavy," but probably the greatest potential danger from leaf spot is the consistent dissemination of the pathogen in seed.

Symptoms usually appear first about the time of flowering on leaves and stems as large, dark spots in which sporulation occurs (Fig. 66) and provides inoculum for further spread of the disease, including infection of pods and seeds.

Cercospora traversiana Sacc. produces conidiophores 3.5–4.5 × 30–40 μm with visible scars where 5–6 conidia (2.2–4.5 × 50–140 μm with 3–7 septa) have developed.

Infected seeds are not readily noticeable and often pass undetected through quarantine inspections. Fungus growth begins soon after seed germination, and abundant sporulation on the seedcoat provides inoculum for seedling infection. Systemic mycelial infection persists for an extended incubation period until near anthesis, at which time symptoms appear. Seed infection commonly results soon thereafter. Widespread dissemination of the fungus in recent years has been directly related to distribution of infected seed.

High frequency of seed-borne infection and failure of seed treatments to provide effective elimination of the pathogen necessitate selection of healthy seed for planting. Foliar spray applications of Benlate, Brestan and Cuprosan (1.0, 2.8 and 5.5 kg/ha, respectively) provided complete disease control in recent experimental tests in Hungary.

1. Chupp, C. 1953. *A monograph of the fungus genus* Cercospora. Ithaca, N. Y. 667 p.

2. Leppik, E. E. 1959. World distribution of *Cercospora traversiana. FAO Plant Protect. Bull. 8:* 19–21.

3. Leppik, E. E. 1960. *Cercospora traversiana* and some other pathogens of fenugreek new to North America. *Plant Dis. Reptr. 44:* 40–44.

4. Maublanc, M. 1920. Contribution à l'étude de la flore mycologique brésilienne. *Bull. Trimestr. Soc. Mycol. France 36:* 33–43.

5. Minz, G. and Z. Solel. 1959. New records of field crop diseases in Israel. *Plant Dis. Reptr. 43:* 1051.

6. Saccardo, P. A. 1904. Notae mycologicae, Ser. IV. *Ann. Mycol. 2:* 18.

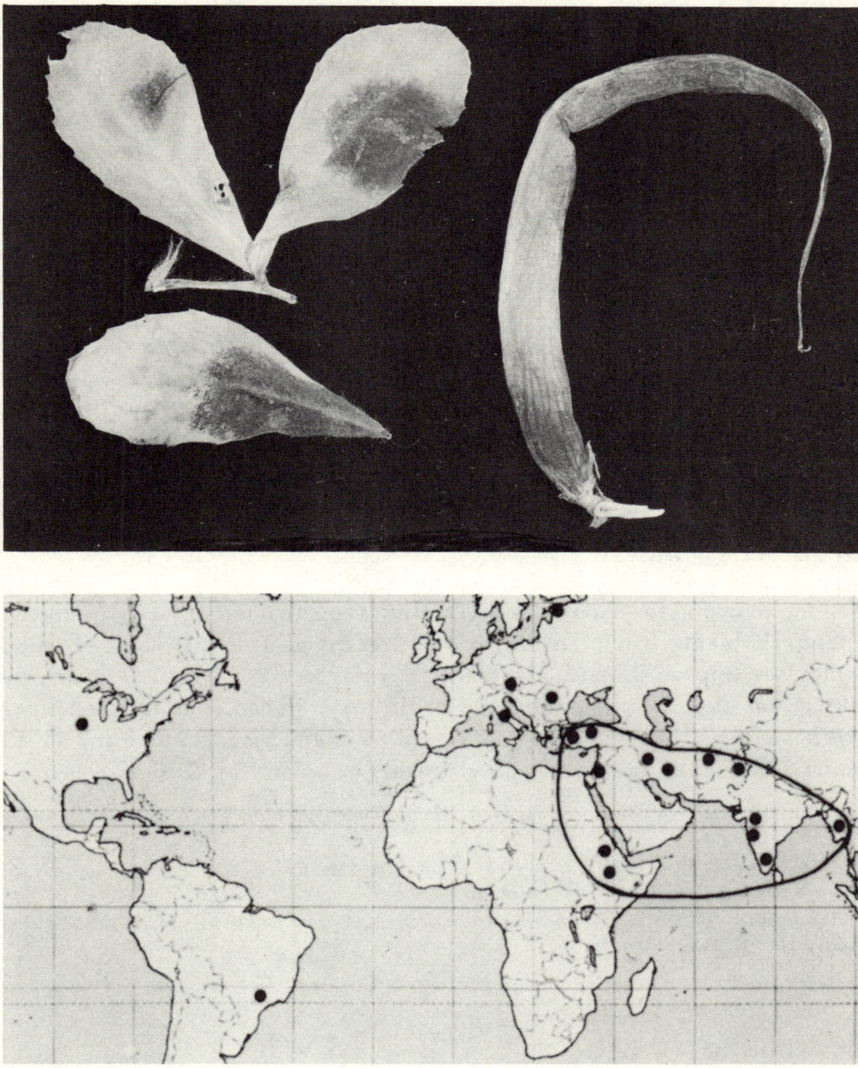

Fig. 66. Leaf and pod symptoms and world distribution of (*Cercospora*) leaf spot of fenugreek. (Courtesy of E. E. Leppik, U. S. Department of Agriculture, Beltsville.)

7. Săvulescu, T. and C. Sandu-ville. 1933. Beiträge zur kenntnis der micromyceten Rumäniens. *Hedwigia 73:* 71–132.

8. Vörös, J. and F. Nagy. 1972. *Cercospora traversiana* Sacc., a new destructive pathogen of fenugreek in Hungary. *Acta Phytopathol. Acad. Sci. Hung.* 7: 71–76.

Mildews: *Leveillula taurica* (Lev.) Armand; *Peronospora trigonella* Gaüm.

Reported only from India.

1. Chattopadhyay, S. B. 1967. *Diseases of plants yielding drugs, dyes and spices.* Indian council of Agricultural Research, New Delhi. 100 p.

2. Mukerji, K. G. 1968. *Leveillula taurica. Commonwealth Mycol. Inst. Descriptions Pathogenic Fungi Bacteria 182.* 2 p.

Rust: *Uromyces anthyllidis* (Grev.) Schroet.

Reported only from India.

1. Chattopadhyay, S. B. 1967. *Diseases of plants yielding drugs, dyes and spices.* Indian Council of Agricultural Research, New Delhi. 100 p.

Bean yellow mosaic

Yellow vein mosaic symptoms were first reported on naturally infected fenugreek plants from Canada in 1969. Plant growth was stopped when symptoms appeared, and plants (with vascular discoloration) promptly wilted and died. Dilution end-point of the virus in sap from infected fenugreek was 10^{-4}, inactivation temperature 54 C and longevity *in vitro* 48–72 hours. The disease was transmitted mechanically and by *Myzus persicae.*

1. Bos, L. 1970. Bean yellow mosaic virus. *Commonwealth Mycol. Inst. Descriptions Plant Viruses 40.* 4 p.

2. Grogan, R. G. and J. C. Walker. 1948. A pod distorting strain of the yellow mosaic virus of bean. *J. Agr. Res. 77:* 301–314.

3. Singh, R. P. 1969. Natural infection of fenugreek by two legume viruses. *Plant Dis. Reptr. 53:* 194–198.

Pea mosaic

Symptoms consisting of green mosaic and rugosity of leaves on naturally infected plants were reported also from Canada in 1969. The virus, transmissible mechanically and by *Myzus persicae,* remained infectious *in vitro* less than 24 hours and was inactivated by exposure to temperature higher than 58 C (for 10 minutes) or dilution of sap from infected fenugreek plants more than 10^{-3}.

1. Hagedorn, D. J. 1948. Three viruses of canning pea. *Phytopathology 38:* 11.

2. Singh, R. P. 1969. Natural infection of fenugreek by two legume viruses. *Plant Dis. Reptr. 53:* 194–198.

Cucumber mosaic

Pea enation mosaic

Red clover vein mosaic

Wisconsin pea streak

1. Gibbs, A. J. and B. D. Harrison. 1970. Cucumber mosaic virus. *Commonwealth Mycol. Inst. Descriptions Plant Viruses 1.* 4 p.

2. Hagedorn, D. J. 1952. Cross-protection tests with viruses affecting canning peas in Wisconsin. *Phytopathology 42:* 10.

3. Shepherd, R. J. 1970. Pea enation mosaic virus. *Commonwealth Mycol. Inst. Descriptions Plant Viruses 25.* 4 p.

4. Varma, A. 1970. Red clover vein mosaic virus. *Commonwealth Mycol. Inst. Descriptions Plant Viruses 22.* 4 p.

Subterranean clover vein mosaic

1. Aitken, Y. and B. J. Grieve. 1943. A mosaic virus of subterranean clover. *J. Australian Inst. Agr. Sci. 9:* 81–82.

G

GINGER: *Zingiber officinale* Roscoe

Bacterial wilt

Bacterial wilt disease of ginger was reported from Mauritius in 1953 and by 1965 had been reported from Hawaii, Malaya and Australia. Losses to wilt have not been determined, but the disease has been described as "serious" in Hawaii and a "considerable threat" in Australia.

Affected plants first exhibit yellowing and wilting of the lower leaves, often on only one pseudostem. These symptoms quickly extend upward to

Fig. 67. Bacterial wilt of ginger. (Courtesy of K. G. Pegg, Queensland Department of Primary Industries, Indooroopilly.)

involve the entire pseudostem, which may wilt without turning yellow (Fig. 67). Young shoots that are affected typically become soft and water-soaked at the bases and easily break away from the rhizomes. Discoloration of the vascular tissues in diseased shoots extends into the rhizomes and may be localized or uniform throughout but often is more pronounced where cortical tissues border the vascular elements. Eventually the entire rhizome is discolored, and white, milky bacterial exudate appears from the vascular area of rhizomes cut transversely.

Pseudomonas solanacearum causes bacterial wilt of ginger and forms small gram negative rods $0.5–0.6 \times 0.8–1.2\ \mu m$, usually with a single polar flagella. Biotypes of the bacterium have been distinguished and the form pathogenic on tomato identified as the type responsible for wilt of ginger in Mauritius. The disease in Hawaii, Malaya and Australia is caused by another biotype but one that also causes wilting in tomato. (See Potato—Bacterial wilt.)

Considerable variation between isolates of the bacterium has been noted *in vitro*, but cardinal temperatures for growth of all are approximately 12, 33 and 38 C. Disease development is enhanced by heavy rainfall and high soil moisture, but outbreaks of bacterial wilt may occur either on soils that are well-drained or have poor drainage. Ability of the bacterium to persist in soil may vary according to environmental and soil conditions, but it usually does not remain viable for more than a few months in infected plant debris. Dissemination of the bacterium and the wilt disease of ginger is largely through infected rhizomes and seed-pieces.

No satisfactory control for bacterial wilt of ginger has been developed other than strict attention to selection of healthy planting materials and abandonment of infested land. Ginger should not be planted following crops susceptible to the wilt bacterium, particularly tomatoes, nor in locations which will receive surface water from neighboring fields where wilt has occurred.

1. Buddenhagen, I. and A. Kelman. 1964. Biological and physiological aspects of bacterial wilt caused by *Pseudomonas solanacearum*. *Ann. Phytopathol. 2:* 203–230.

2. Hayward, A. C., M. L. Moffett and K. G. Pegg. 1967. Bacterial wilt of ginger in Queensland. *Queensland J. Agr. Animal Sci. 24:* 1–5.

3. Hayward, A. C. and J. M. Waterston. 1964. *Pseudomonas solanacearum. Commonwealth Mycol. Inst. Descriptions Pathogenic Fungi Bacteria 15.* 2 p.

4. Ishii, M. and M. Aragaki. 1963. Ginger wilt caused by *Pseudomonas solanacearum* E. F. Smith. *Plant Dis. Reptr. 47:* 710–713.

5. Kelman, A. 1953. The bacterial wilt caused by *Pseudomonas solanacearum. N. C. Agr. Expt. Sta. Tech. Bull. 99.* 194 p.

6. Pegg, K. G., M. L. Moffett and R. C. Colbran. 1974. Diseases of ginger in Queensland. *Queensland Agr. J. 100:* 611–618.

7. Quinon, V. L., M. Aragaki and M. Ishii. 1964. Pathogenicity and serological relationship of three strains of *Pseudomonas solanacearum* in Hawaii. *Phytopathology 54:* 1096–1099.

8. Trujillo, E. E. 1964. Diseases of ginger (*Zingiber officinale*) in Hawaii. *Hawaii Agr. Expt. Sta. Circ. 62.* 14 p.

Fusarium wilt

Occurrence of a severe wilt disease of ginger in southern Queensland, Australia, was reported in 1956, and the cause identified as a species of *Fusarium*, but the disease was not described until 1965. A similar disease, although less severe, was reported from Hawaii in 1963. *Fusarium* wilt also has been found recently in one location in India. Although losses from *Fusarium* wilt have not been established, the disease has been widespread in Hawaii and Queensland and, coupled with the longevity of the fungus, must be considered a serious threat to continued successful production.

The lower leaves on diseased plants become yellow, and this change in coloration progresses to involve all the leaves on plants, which then wilt and die. The vascular tissues in rhizomes from infected plants are discolored light yellowish also, and, as the infection extends outward, sunken areas develop in the cortical tissues of rhizomes (Fig. 68).

The causal fungus is *Fusarium oxysporum* f. *zingiberi* Trujillo, which

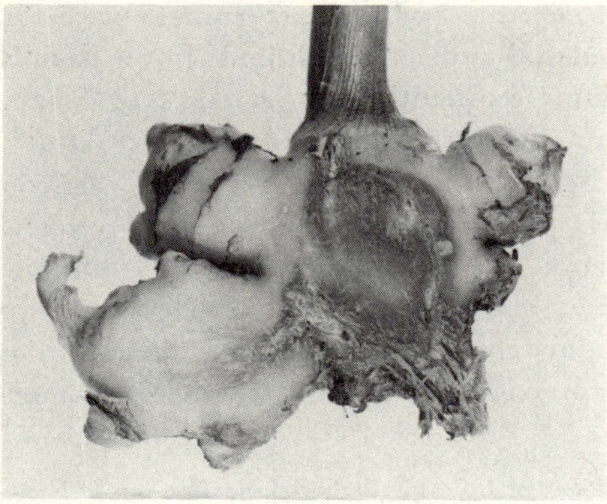

Fig. 68. *Fusarium* (wilt) rhizome rot of ginger. (Courtesy of K. G. Pegg, Queensland Department of Primary Industries, Indooroopilly.)

produces vascular discoloration and both inter- and intracellular mycelial invasion of cortical tissues. Mycelial invasion in vascular elements is limited to the discolored vessels. The fungus is not pathogenic on a number of other hosts, including cowpea, cucumber, passion fruit (*Passiflora edulis* Sims), tomato or watermelon (*Citrullus vulgaris* Schrad.) nor is ginger susceptible to isolates of *Fusarium* from other hosts. [Another species of *Fusarium* (*F. roseum*) has been reported to cause a rot of stored ginger rhizomes in India.]

Infection may result from soil-borne inoculum, and, although no factual information has been reported, the organism presumably remains viable and pathogenic in the soil for a number of years. Inoculum may be introduced and disseminated by infected rhizomes.

Seed materials always should be selected from healthy rhizomes, and seed-borne inoculum can be controlled by immersing rhizomes for 10 minutes in benomyl (250 g/100 ℓ). Annual crop rotation should be practiced strictly so that ginger will never be planted on soils which bore ginger or a solanaceous crop the previous season. Care should be exercised to avoid damage to plants during cultivation, by sunburn or insects.

1. Haware, M. P. and L. K. Joshi. 1973. Yellow disease of ginger (*Zingiber officinale*) from Madhya Pradesh. *Indian Phytopathol. 26:* 754–755.

2. Mehrotra, B. S. 1952, *Fusarium roseum* Link and *Sclerotium rolfsii* Sacc. on ginger rhizomes. *Indian Phytopathol. 5:* 52–54.

3. Oxenham, B. L. 1956. Stem rot of ginger. *Australian Plant Dis. Record 8:* 16.

4. Pegg, K. G., M. L. Moffett and R. C. Colbran. 1974. Diseases of ginger in Queensland. *Queensland Agr. J. 100:* 611–618.

5. Teakle, D. S. 1965. *Fusarium* rhizome rot of ginger in Queensland. *Queensland J. Agr. Animal Sci. 22:* 263–272.

6. Trujillo, E. E. 1963. *Fusarium* yellows and rhizome rot of common ginger. *Phytopathology 53:* 1370–1371.

7. Trujillo, E. E. 1964. Diseases of ginger (*Zingiber officinale*) in Hawaii. *Hawaii Agr. Expt. Sta. Circ. 62.* 14 p.

OTHER DISEASES

Soft rots

Pythium sp.

1. Bhagwat, V. Y. 1960. Control of soft rot of ginger. *Poona Agr. Coll. Mag. 51:* 47–49.

2. Venkatakrishniah, N. S. 1958. Koleroga or soft rot of ginger (*Zingiber officinale*). *Mysore Agr. Cal. 1957/1958:* 185–187.

Erwinia carotovora (Jones) Holland

1. Nakazawa, M. 1953. Control of bacterial rot (*Bacillus carotovorus*) of ginger. *Jap. J. Plant Protect. 44:* 87–92. (In Japanese)

2. Nakazawa, M. and K. Kato. 1953. Control method of bacterial rot of ginger. II. *Jap. J. Plant Protect. 45:* 133–137. (In Japanese)

3. Pegg, K. G., M. L. Moffett and R. C. Colbran. 1974. Diseases of ginger in Queensland. *Queensland Agr. J. 100:* 611–618.

4. Trujillo, E. E. 1964. Diseases of ginger (*Zingiber officinale*) in Hawaii. *Hawaii Agr. Expt. Sta. Circ. 62.* 14 p.

Rhizome, root and basal stem rots (*Armillariella mellea; Sclerotium rolfsii*) (Fig. 69)

1. Mordue, J. E. M. 1974. *Corticium rolfsii. Commonwealth Mycol. Inst. Descriptions Pathogenic Fungi Bacteria 410.* 2 p.

2. Pegler, D. N. and I. A. S. Gibson. 1972. *Armillariella mellea. Commonwealth Mycol. Inst. Descriptions Pathogenic Fungi Bacteria 321.* 2 p.

Big bud (Tomato big bud mycoplasm) (Fig. 70)

1. Pegg, K. G., M. L. Moffett and R. C. Colbran. 1974. Diseases of ginger in Queensland. *Queensland Agr. J. 100:* 611–618.

Phyllosticta leaf spot (*Phyllosticta zingiberi* Ramakrishnan)

Round to elongate leaf spots 3–4 × 9–10 mm with white centers surrounded by dark brown margins and yellowish halos were first noted in India in 1938. The fungus that caused the spots produced ostiolate, subglobose, dark pyc-

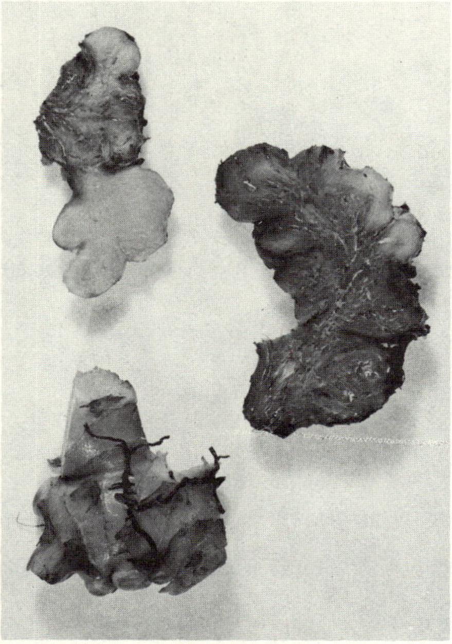

Fig. 69. *Armillaria* rhizome rot of ginger. (Courtesy of K. G. Pegg, Queensland Department of Primary Industries, Indooroopilly.)

Fig. 70. Ginger infected with the (tomato) big bud disease. (Courtesy of K. G. Pegg, Queensland Department of Primary Industries, Indooroopilly.)

nidia 78–150 μm in diameter and hyaline, oblong, unicellular conidia 1.2–2.5 × 3.7–7.4 μm. Incidence did not usually justify control measures, but 2 applications of Bordeaux mixture (1 percent), the first before the disease appeared, provided ample protection.

1. Ramakrishnan, T. S. 1942. A leaf spot disease of *Zingiber officinale* caused by *Phyllosticta zingiberi* n. sp. Proc. Indian Acad. Sci. 15 B: 167–171.

Root rot (*Pythium zingiberum* Takahashi)

Reported from Oosaka Prefecture (Japan) in 1954, mycelium of the causal fungus measured 4–8 μm in width and zoospores (10–20 in each zoosporangium) were irregularly elliptical and 20–31 μm in diameter. Terminal oogonia measured 24–43 μm and were globose. The globose oospores were 23–35 μm in diameter.

1. Waterhouse, G. M. 1968. The genus *Pythium* Pringsheim. *Commonwealth Mycol. Inst. Mycol. Paper 110.* 71 p.

GUAR: *Cyamopsis tetragonoloba (L.) Taub.*

Bacterial blight

The disease of guar now referred to as bacterial blight was originally described in 1953 as a leaf spot disease first seen in India. In another report 5 years later, also from India, bacterial blight symptoms were attributed to a different bacterium, but it was established 4 years later that only one pathogen was responsible for the entire complex of symptoms. The common name "bacterial blight" has been retained for this disease, which can be so severe under favorable environmental conditions that plantings may be completely destroyed.

Leaf symptoms consist of round, dark, oily spots, which are from 0.5–2 mm on the undersurfaces of leaves but appear tan to olive on upper leaf surfaces. Infections along leaf margins result in characteristic light colored, angular (cuneiform) lesions delimited by veins. Interveinal leaf spots may coalesce and also involve veins, from which systemic infection through the petioles into the stem results. Long, dark streaks develop in stems of affected plants, which soon wilt and exhibit splitting and cracking of the stems that later break over. Flowers and fruits have not been found to show symptoms. Infected seedlings develop yellow cotyledons and blackened hypocotyls and are usually killed.

Xanthomonas cyamopsidis Patel et al., the causal bacterium, produces gram negative rods 0.8 × 1.5 μm with single, polar flagella. Optimum temperature for growth *in vitro* is 25–31 C, thermal death point is 52 C, and no growth occurs below 13 C. Colonies on agar range from butyrous to sulfur to mustard yellow, depending on the medium.

The blight pathogen is consistently seed-borne, from which seedling infection readily occurs. Dissemination of the disease in the field is dependent on rainfall or surface water with concomitant high humidity and temperature conducive for bacterial growth. Infection probably occurs via stomata, including hydathodes, along leaf margins. The existence of 2 distinguishable pathotypes (races) of the bacterium was established in controlled inoculations following observation of severe blight symptoms on varieties previously resistant in Texas and Oklahoma (United States). Antigenic differences between the 2 bacterial pathotypes were also demonstrated, as were serotypes within 1 of the pathotypes. Disease severity in 1 variety (Brooks) was directly correlated with increase in nitrogen fertilization under 16 hour (1,500 ft. candle illumination, 27 C) day/8 hour (22 C) night conditions and ranged from resistant to susceptible.

Seeds for planting should be harvested only from healthy plants and supplemented with treatment of seeds in hot water (56 C) for 10 minutes, which has been shown to eliminate seed infection. Varieties with satisfactory (field) resistance to 1 pathotype (race 0) include Brooks, Hall and Mills, but they are susceptible to another pathotype (race 1). Avoid high nitrogen fertilization, which has been shown to enhance disease development, particularly during long day/short night seasons with moderate temperatures.

1. Orellana, R. G. and M. L. Kinman. 1970. A new virulent race of *Xanthomonas cyamopsidis*, bacterial blight of guar. *Plant Dis. Reptr. 54:* 111–113.

2. Orellana, R. G. and C. A. Thomas. 1968. Light and nitrogen affect reaction of guar to bacterial blight caused by *Xanthomonas cyamopsidis. Phytopathology 58:* 250–251.

3. Orellana, R. G., C. A. Thomas and M. L. Kinman. 1965. A bacterial blight of guar in the United States. *FAO Plant Protect. Bull. 13:* 9–13.

4. Orellana, R. G. and D. F. Weber. 1971. Immunodiffusion analysis of isolates of *Xanthomonas cyamopsidis. Appl. Microbiol. 22:* 622–624.

5. Patel, M. K., G. W. Dhande and Y. S. Kulkarni. 1953. Bacterial leaf-spot of *Cyamopsis tetragonoloba* (L.) Taub. *Current Sci. 22:* 183.

6. Patel, A. J. and M. K. Patel. 1958. A new bacterial blight of *Cyamopsis tetragonoloba* (L.) Taub. *Current Sci. 27:* 258–259.

7. Srivastava, D. N. and Y. P. Rao. 1963. Bacterial blight of guar. *Indian Phytopathol. 16:* 69–73.

8. Srivastava, D. N. and Y. P. Rao. 1963. Seed transmission and control of bacterial blight of guar [*Cyamopsis tetragonoloba* (L.) Taub.]. *Indian Phytopathol. 16:* 390–391.

Bacterial leaf spot

Leaf spot symptoms caused by a bacterium were reported from Australia in 1965 and the United States in 1967, but in neither instance was there indication of disease severity.

Symptoms are confined to the leaves and consist of round or irregular lesions 1–5 mm, first water-soaked but quickly turning light brown and necrotic (Fig. 71). Spots typically have narrow dark margins and light green halos, and severely affected leaves fall, but infection does not become systemic, and diseased plants usually survive.

Pseudomonas syringae, characterized by gram negative rods each with a single polar flagellum, causes leaf spot and is pathogenic on a number of other hosts. Another species of bacteria, *P. cyamopsicola* Rangaswami & Sanne Gowda, has been reported on guar from India.

Incidence of the disease has not warranted development of control measures.

1. Hayward, A. C. and J. M. Waterston. 1965. *Pseudomonas syringae. Commonwealth Mycol. Descriptions Pathogenic Fungi Bacteria 46.* 2 p.

2. Moffett, M. L. and J. L. Alcorn. 1965. Bacterial leaf spot of guar. *Queensland J. Agr. Animal Sci. 22:* 351–352.

3. Orellana, R. G. 1967. Leaf spot of guar caused by *Pseudomonas syringae* in the United States. *Plant Dis. Reptr. 51:* 182–184.

4. Rangaswami, G. and S. Sanne Gowda. 1973. On some bacterial diseases of ornamentals and vegetables in Madras State. *Indian Phytopathol. 16:* 74–85.

Fig. 71. Bacterial leaf spot lesions on leaflets of guar. (Courtesy of R. G. Orellana, U. S. Department of Agriculture, Beltsville.)

Alternaria blight

Diseases of guar caused by *Alternaria* sp. have been recognized since 1951, when the first description of symptoms from Georgia (United States) appeared. Subsequently, *Alternaria* blight has been found in other locations in the United States (Oklahoma, Arizona) and in India. The disease can cause considerable damage under conditions favorable for disease development, but blight usually is not considered a major problem in production.

Brown, nearly round lesions to 10 mm with concentric zonations, and sometimes with chlorotic halos, develop on leaves (Fig. 72). Lesions also develop on petioles, and both leaf and petiole infections can cause defoliation. Symptoms on stems are black, round or elongated (to 15 mm) lesions that are sunken, while pod spots usually are no more than 5 mm in diameter and may be either sunken or superficial.

Identification of the causal agent as a species of *Alternaria* has never been in dispute, but the specific epithet has been more controversial. The most recent combination is *A. cucumerina* (Ell. & Ev.) Elliott var. *cyamopsidis* (Rangaswami & Rao) Simmons (syn. *A. cyamopsidis* Rangaswami & Rao). Dimensions of conidia reported from India (12–19 × 62–146 μm) are somewhat different from the measurements of conidia from the United States (Georgia), 13–24 × 81–199 μm. However, there is general agreement that conidia are beaked and muriform with 0–5 vertical and 6–11 transverse septa.

Factors affecting incidence of blight have not been determined, but infection has been obtained under controlled conditions on plants held in a moist chamber for 48 hours at 25 C. Typical symptoms were evident in 8 days.

Fig. 72. Leaf spot symptoms of *Alternaria* blight on guar. (Courtesy of R. G. Orellana, U. S. Department of Agriculture, Beltsville.)

Attempts at fungicidal control have not been reported, but application of carbamate fungicides should prove beneficial. The varieties Groehler and Texsel have been found susceptible, while the variety Brooks was resistant in field tests in the United States.

1. Hymowitz, T. 1963. Annotated bibliography on the diseases of guar. *Oklahoma Agr. Expt. Sta. Proc. Ser. P-468.* 23 p.

2. Luttrell, E. S. 1951. Diseases of guar in Georgia. *Plant Dis. Reptr. 35:* 166.

3. Matlock, R. S. and D. C. Aepli. 1948. Growth and diseases of guar. *Arizona Agr. Expt. Sta. Bull. 216.* 42 p.

4. Orellana, R. G. and E. G. Simmons. 1966. *Alternaria* blight of guar in the United States. *Mycopathol. Mycol. Appl. 29:* 129–133.

5. Rangaswami, G. and A. Venkata Rao. 1957. *Alternaria* blight of clusterbeans. *Indian Phytopathol. 10:* 18–25.

6. Sowell, G., Jr. 1965. *Alternaria* leafspot of guar. *Plant Dis. Reptr. 49:* 605–607.

Anthracnose

Diseases described as anthracnose have been reported from India (1955, 1956) and observed in the United States (Georgia, 1960). Damage was "severe" in both locations, but no quantitative assessments of losses were attempted. Herein, the disease is attributed to a single pathogen, although there remains some question that the diseases reported from the 2 countries are caused by identical agents.

Guar plants infected under controlled conditions develop bright yellow spots 0.5–3 mm in diameter that are round and have a spot of dark tissue 0.5 mm in diameter in the center of each lesion (Fig. 73). (Spots on leaves in India are gray colored and 4–6 mm in diameter). Lesions on stems are less well defined with more subdued coloration, and spots on fruits are often sunken and round. Black acervuli in lesions on all plant parts appear as black dots. Conidia are borne abundantly and appear as gray-white, gelatinous masses.

The causal fungus (in the United States) has been identified as *Colletotrichum dematium* f. *truncata* (Schw.) v. Arx with conidia 2–3 × 26–33 μm. (Tentative identity of the pathogen in India, with similar conidial dimensions of 1.5–4.6 × 15–29 μm, was *C. capsici.*)

Inoculum is readily seed-borne and seedlings, which are quite susceptible under favorable conditions, may be killed within a week after infection. Disease development is more severe under high rainfall conditions.

No routine for control of anthracnose on guar has been developed, but seeds for planting should be harvested from healthy plants. Treatment of seeds with an approved fungicide before planting, i.e., thiram, and maneb or zineb foliar sprays should prove beneficial. Guar should not be planted the following year in locations where anthracnose has been present.

Fig. 73. Anthracnose lesions on guar leaflets. (Courtesy of G. Sowell, Jr., U. S. Department of Agriculture, Experiment.)

1. Desai, M. V. and N. Prasad. 1955. A new *Colletotrichum* from India. *Indian Phytopathol. 8:* 52–57.

2. Rao, P. G. and D. K. Rao. 1956. An anthracnose disease on *Cymopsis tetragonoloba* Taub. *Sci. Cult. 21:* 457–458.

3. Sowell, G., Jr. 1965. Anthracnose of guar. *Plant Dis. Reptr. 49:* 607–609.

4. von Arx, J. A. 1957. Die arten der gattung *Colletotrichum* Cda. *Phytopathol. Z. 29:* 413–468.

Tobacco ringspot

Top necrosis of guar was first observed in the United States (Oklahoma), in 1944, where the number of affected plants increased from less than 1 percent to 90 percent in less than 3 months in an experimental planting. Losses were estimated to be "at least 75 percent." A similar disease was noted in Arizona the same year, but the disease in guar apparently disappeared until nearly 20 years later, when it reappeared in Texas and Oklahoma. It is not known to occur outside the United States.

Symptoms in leaves include vein clearing followed by faint line patterns and necrosis in the top of the plant. Some plants recover partially and remain noticeably stunted, but the usual result of infection is death.

The cause of the top necrosis disease is tobacco ringspot virus, characterized by isometric particles about 29 nm in diameter, thermal inactivation at 65 C, dilution end-point of about 10^{-4} and 6–10 days longevity *in vitro*. Numerous strains of the virus have been described, and vectors include the nematode *Xiphenema americanum* Cobb, spider mites (*Tetranchyus* sp.), grasshoppers (*Melanoplus* sp.), *Thrips tabaci* Lind. and the tobacco flea beetle [*Epitrix hirtipennis* (Melsheimer)]. Host range of the virus is quite wide and includes members of more than 35 genera in 17 families. The virus is seed-borne in a number of hosts.

Factors contributing to the disease in guar have not been identified, but the causal virus has been found to be seed-borne (0.2 percent) in guar. Numerous vectors and the wide host range of the virus do not permit optimism for control of the disease in guar by cultural methods. The possibilities for varietal resistance have not been promising but need to be further investigated. Planting only seeds from healthy guar plants in locations isolated from other crop plants susceptible to tobacco ringspot virus offers the best promise for avoiding the top necrosis disease at this time.

1. Chester, K. S. and W. E. Cooper, 1944. A new virus disease of guar. *Phytopathology 34:* 998.

2. Cooper, W. E. 1949. Top necrosis, a virus disease of guar. *Phytopathology 39:* 347–358.

3. McLean, D. M. 1960. Diseases caused by tobacco ringspot virus in the Lower Rio Grande Valley of Texas. *Plant Dis. Reptr. 44:* 738–741.

4. Orellana, R. G. 1966. A new occurrence of tobacco ringspot of guar in the United States. *Plant Dis. Reptr. 50:* 7–10.

5. Orellana, R. G. 1967. Reaction of guar to strains of tobacco ringspot virus. *Phytopathology 57:* 791–792.

6. Stace-Smith, R. 1970. Tobacco ringspot virus. *Commonwealth Mycol. Inst. Descriptions Plant Viruses 17.* 4 p.

7. Streets, R. B. 1944. Diseases of guar (*Cyamopsis tetragonoloba*). *Arizona Agr. Expt. Sta. Ann. Rep. 55:* 79.

OTHER DISEASES

Fusarium blight [*Fusarium moniliforme* (Sheld.) Snyder & Hansen]

1. Desai, M. V. and N. Prasad. 1955. *Fusarium* blight of guar. *Indian Phytopathol. 8:* 112–123.

2. Prasad, N. and M. V. Desai. 1952. *Fusarium* blight of cluster beans. *Current Sci. 21:* 17–18.

Myrothecium leaf spot (*Myrothecium roridum*)

Brown or black zonate lesions 10–15 mm in diameter that develop on leaves, stems and pods are characteristic.

1. Arya, H. D. 1956. On a new leaf spot disease of gawar (*Cyamopsis tetragonoloba* Taub.) caused by *Myrothecium roridum* Tode ex. Fr. *Indian Phytopathol. 9:* 174–181.

2. Fitton, M. and P. Holliday. 1970. *Myrothecium roridum. Commonwealth Mycol. Inst. Descriptions Pathogenic Fungi Bacteria 253.* 2 p.

Potato S (virus) disease

Small, necrotic lesions develop on guar following inoculation with potato S virus.

1. Wetter, C. 1971. Potato virus S. *Commonwealth Mycol. Inst. Descriptions Plant Viruses 60.* 3 p.

2. Yarwood, C. E. and A. H. Gold. 1955. Guar as a local lesion host of potato virus S. *Plant Dis. Reptr. 39:* 622.

Powdery mildew (*Leveillula taurica*)

1. Mukerji, K. G. 1968. *Leveillula taurica. Commonwealth Mycol. Inst. Descriptions Pathogenic Fungi Bacteria 182.* 2 p.

Rhizoctonia (leaf) blight (*Rhizoctonia solani*)

1. Singh, R. S. 1953. A leaf blight of guar. *Sci. Cult. 19:* 155–156.

Root rot [*Sclerotium rolfsii; Phymatotrichum omnivorum* (Shear) Dug.]

1. Luttrell, E. S. 1951. Diseases of guar in Georgia. *Plant Dis. Reptr. 35:* 166.

2. Mathur, S. B. and S. Sinha. 1968. Disease development in guar (*Cyamopsis psoraloides* DC) and gram (*Cicer arietinum* L.) attacked with *Sclerotium rolfsii* under different soil pH conditions. *Phytopathol. Z. 62:* 319–322.

3. Mordue, J. E. M. 1974. *Corticium rolfsii. Commonwealth Mycol. Inst. Descriptions Pathogenic Fungi Bacteria 410.* 2 p.

4. Singh, R. S. 1951. Root-rot of guar. *Sci. Cult. 17:* 131–134.

5. Streets, R. B. 1948. Diseases of guar (*Cyamopsis psoraloides*). *Phytopathology 38:* 918.

GUARANA: (*Paullinia cupana* HBK)

Anthracnose

The anthracnose disease of guarana was collected in the state of Amazonas (Brazil) in 1946 and described 15 years later but is not known to occur elsewhere. By 1959, it was estimated that anthracnose was responsible for an average 80 percent reduction in yield per plant.

Young leaves are especially susceptible to infection that results in circular to elliptical necrotic lesions which vary in size, often with yellow halos, and which may coalesce to involve considerable leaf area. Symptoms on affected leaves are more pronounced on the upper leaf surfaces, and severely diseased leaves may roll and fall prematurely. Defoliation may be sufficient to retard plant growth and induce growth of axillary buds reminiscent of "superbrotamento" symptoms (see Cassava diseases).

The causal fungus has been identified as *Colletotrichum guaranicola* Albuquerque, which produces acervuli 75–83 μm in diameter and simple, hyaline conidiophores 4–6 × 14–20 μm with hyaline, oblong, unicellular conidia 4-6 × 12–20 μm. Disease development is enhanced by rainfall and periods of high humidity.

Application of copper fungicides has been recommended for control. Differences in susceptibility of varietal forms of the host species have not been thoroughly investigated.

1. Albuquerque, F. D. 1960. Antracnose do guaraná. *Bol. Tec. Inst. Agron. Norte 40.* 33 p.

2. Castro, A. M. G., R. M. Moura and R. C. Mafra. 1975. Preliminary reports on the phytosanitary control of guarana (*Paullinia cupana* var. *sorbilis*) in the State of Amazonas. *Fitopatologia 10:* 49.

K

KAVA (AWA, YANGGONA, HOI): *Piper methysticum* Forst.

Wilt

The cause of wilt in Fiji has not been established.

1. Gatty, R. 1956. Kava—Polynesian beverage shrub. *Econ. Bot. 10:* 241–249.
2. Parham, E. V. 1935. Wilt disease of "yangona." *Agr. J. (Fiji) 8:* 28.

KOLA: *Cola nitida* (Vent.) Schott & End.

Nut mold [*Schizophyllum commune* Fr.; *Fusarium solani* (Mart.) App. & Wr.; *Gliocladium roseum* (Link.) Bain.; *Penicillium* sp.]

1. Heim, R. and C. Moreau. 1956. Une moisissure des noix de kola. *Rev. Mycol. Suppl. Colon. (Paris) 20:* 37–38.

Boron deficiency

Dieback of shoot tips and profuse axillary bud growth impart a bunchy habit to affected plants. Leaves often are shorter and wider as well as thicker and often twisted. Flowers, nearly all pistillate on deficient trees, are noticeably larger but set no fruit. Deficiency symptoms develop only when boron content is less than 15 ppm/g leaf tissue.

1. Jacob, V. J., N. E. Egbe and G. E. Okoloko. 1972. Abnormalities in vegetative and floral organs of *Cola nitida* caused by boron deficiency. *Can. J. Bot. 50:* 381–383.

L

LENTIL: *Lens esculenta* Moench.

Downy mildew (*Peronospora lentis* Gäumann)

Indefinite, chlorotic spots on the upper leaf surfaces (of lower leaves) that appear powdery gray below and extend to cover the entire lower leaf surfaces were first noted in Germany in 1923 and India (1968). Affected leaves become completely chlorotic and fall. Conidiophores are aseptate, 185–526 μm in length, dichotomously branched 4–6 times with subglobose conidia 12–25 $\times$ 13–30 μm borne singly on sterigmata to 25 μm in length. Globose oospores 20–42 μm in diameter are formed in leaf tissues and have a thick, rough epispore. Conidia germinate by formation of a germ tube.

1. Beniwal, S. P. S. and S. K. Srivastava. 1968. Downy mildew of lentil (*Lens esculenta*). *Plant Dis. Reptr. 52:* 817–818.

Fusarium wilt

A disease of lentil seedlings in Russia in 1931 was reported to be caused by *Fusarium* sp. and wilts of lentil in Uruguay (1941) and India (1934, 1941) were attributed to one or more species of the same genus. No causal agent was determined for wilt of lentil in Hungary (1937). Incidence greater than 60 percent in experimental plantings in 1948 near New Delhi prompted more thorough examination of the disease, which presumably remains restricted to the above countries.

Lower leaves on an affected plant become curled and wilted, and these symptoms progress upwards until the entire plant is involved. The top of the plant eventually bends over, and the plant succumbs. Roots are discolored and the root system poorly developed.

The fungus that causes wilt is *Fusarium orthoceras* var. *lentis* Vasudeva & Srinivasan (*F. oxysporum* Schlecht. f. "*Redolens*" Snyd. & Hans.) and produces abundant 0–5 septate microconidia 2–4.8 × 6.1–66.3 μm. Macroconidia are produced infrequently, and no pionnotes, sporodochia or sclerotia are formed.

Cardinal temperatures for growth of the fungus *in vitro* are 9, about 28 and 35 C, and the highest number of diseased plants was obtained at 25 percent soil moisture in controlled tests.

Control procedures have not been reported, but crop rotation should be beneficial. Development of resistant varieties probably offers the best prospect for continued production where wilt is a serious problem.

1. Vasudeva, R. S. and K. V. Srinivasan. 1952. Studies on the wilt disease of lentil (*Lens esculenta* Moench.). *Indian Phytopathol. 5:* 23–32.

2. Wilson, V. E. and J. Brandsberg. 1965. Fungi isolated from diseased lentil seedlings in 1963–64. *Plant Dis. Reptr. 49:* 660–662.

OTHER DISEASES

Alternaria leaf blight (*Alternaria tenuis*)

1. Sen Gupta, P. K. and C. R. Das. 1964. *Alternaria* leaf blight of lentil. *Current Sci. 33:* 562–563.

Botrytis seedling blight (*Botrytis cinerea*)

1. Wilson, V. E. and J. Brandsberg. 1965. Fungi isolated from diseased lentil seedlings in 1963–64. *Plant Dis. Reptr. 49:* 660–662.

Root rot (*Rhizoctonia* sp.)

1. Kannaiyan, J. and Y. L. Nene. 1973. A new root rot disease of lentil. *Current Sci. 42:* 257.

Rust (*Uromyces viciae-fabae*) (see Broadbean—Rust)

1. Laundon, G. F. and J. M. Waterston. 1965. *Uromyces viciae-fabae. Commonwealth Mycol. Inst. Descriptions Pathogenic Fungi Bacteria 60.* 2 p.

2. Prasada, R. and U. N. Varma. 1948. Studies on lentil rust, *Uromyces fabae* (Pers.) de Bary in India. *Indian Phytopathol. 1:* 142–146.

Sclerotinia blight (*Sclerotinia sclerotiorum*)

1. Kannaiyan, J. and Y. L. Nene. 1973. *Sclerotinia* blight of lentil. *Current Sci. 42:* 32.

Virus diseases (Alfalfa mosaic; Bean yellow mosaic; Cucumber mosaic; Pea leaf roll)

All 4 viruses have been identified from naturally infected lentil plants in Iran.

1. Bos, L. 1970. Bean yellow mosaic virus. *Commonwealth Mycol. Inst. Descriptions Plant Viruses 40.* 4 p.

2. Bos, L. and E. M. J. Jaspars. 1971. Alfalfa mosaic virus. *Commonwealth Mycol. Inst. Descriptions Plant Viruses 46.* 4 p.

3. Gibbs, A. J. and B. D. Harrison. 1970. Cucumber mosaic virus. *Commonwealth Mycol. Inst. Descriptions Plant Viruses 1.* 4 p.

4. Kaiser, W. J. 1973. Etiology and biology of viruses affecting lentil (*Lens esculenta* Moench.) in Iran. *Phytopathol. Mediter. 12:* 7–14.

MUNG BEAN: *Phaseolus aureus* L.

Bacterial leaf spot and Halo blight (*Xanthomonas phaseoli/Pseudomonas phaseolicola*)

1. Hayward, A. C. and J. M. Waterston. 1965. *Pseudomonas phaseolicola. Commonwealth Mycol. Inst. Descriptions Pathogenic Fungi Bacteria 45.* 2 p.

2. Hayward, A. C. and J. M. Waterston. 1965. *Xanthomonas phaseoli. Commonwealth Mycol. Inst. Descriptions Pathogenic Fungi Bacteria 48.* 2 p.

3. Patel, P. N. and J. K. Jindal. 1972. Bacterial leaf spot and halo blight diseases of mung bean (*Phaseolus aureus*) and other legumes in India. *Indian Phytopathol. 25:* 517–525.

4. Schmitthenner, A. F., J. A. J. Hoitink and M. E. Kroetz. 1971. Halo blight of mung bean incited by a new strain of *Pseudomonas phaseolicola. Phytopathology 61:* 909.

Thanatephorus cucumeris (= *Rhizoctonia solani*)

1. Kaiser, W. J., D. Danesh, M. Okhovat and H. Mossahebi. 1968. Diseases of pulse crops (edible legumes) in Iran. *Plant Dis. Reptr. 52:* 687–691.

2. Lai, M., A. R. Weinhold and J. G. Hancock. 1968. Permeability changes in *Phaseolus aureus* associated with infection by *Rhizoctonia solani. Phytopathology 58:* 240–245.

Cercospora leaf spot (*Cercospora canescens*)

1. Mew, I. C., T. C. Wang and T. W. Mew. 1975. Inoculum production and evaluation of mung bean varieties for resistance to *Cercospora canescens. Plant Dis. Reptr. 59:* 397–401.

2. Mulder, J. L. and P. Holliday. 1969. *Cercospora canescens. Commonwealth Mycol. Inst. Descriptions Pathogenic Fungi Bacteria 462.* 2 p.

Bean common mosaic

1. Bos, L. 1971. Bean common mosaic virus. *Commonwealth Mycol. Inst. Descriptions Plant Viruses 73.* 4 p.

2. Kaiser, W. J. and G. H. Mossahebi. 1974. Natural infection of mungbean by bean common mosaic virus. *Phytopathology 64:* 1209–1214.

3. Nene, Y. L. 1973. Viral diseases of some warm weather pulse crops in India. *Plant Dis. Reptr. 57:* 463–467.

Leaf crinkle

1. Nene, Y. L. 1972. A survey of viral diseases of pulse crops in Uttar Pradesh. *G. B. Pant Univ. Coll. Agr. Expt. Sta. Res. Bull. 4.* 191 p.

2. Williams, F. J., J. S. Grewal and K. S. Amin. 1968. Serious and new diseases of pulse crops in India in 1966. *Plant Dis. Reptr. 52:* 300–304.

Yellow mosaic

1. Nariani, T. K. 1960. Yellow mosaic of mung (*Phaseolus aureus* L.). *Indian Phytopathol. 13:* 24–29.

2. Nene, Y. L. 1972. A survey of viral diseases of pulse crops in Uttar Pradesh. *G. B. Pant Univ. Coll. Agr. Expt. Sta. Res. Bull. 4.* 191 p.

3. Nene, Y. L., J. S. Naresh and N. G. Nair. 1971. Additional hosts of mung bean yellow mosaic virus. *Indian Phytopathol. 24:* 415–417.

N

NUTMEG: *Myristica fragrans* Houll

Fruit rot (*Diplodia natalensis* P. Evans)

A dark, water-soaked rot of the pericarp that extended to the mace and even involved the seed was reported from India in 1954. The black pycnidia measured 340 × 163 μm, and the elliptical, hyaline, unicellular and 1-septate pycnidiospores were 11-18 × 20–30 μm.

1. Ramakrishnan, T. S. and A. P. Sarojini Damodaran. 1954. Fruit rot of nutmeg. *Indian Phytopathol. 7:* 7–15.

O

OPIUM POPPY: *Papaver somniferum* L.

Bacterial blight (*Erwinia aroidea*)

Bacteria are seed-borne and cause foliar infection via stomatal penetration that results in dark spots on aerial plant parts and finally a wet blight. Crop rotation, sanitation and application of insecticides (insects also transmit the bacterium) have been recommended for control.

1. Chattopadhyay, S. B. 1967. *Diseases of plants yielding drugs, dyes and spices*. Indian Council of Research, New Delhi. 100 p.

2. Elliott, C. 1951. *Manual of bacterial plant pathogens*. Chronica Botanica Co., Waltham, Mass. 186 p.

Downy mildew

Mildew on opium poppy has occurred in India since 1918, where losses have been estimated to reach 30 percent.

Symptoms include chlorotic leaf spots, wherein the upper leaf surface of affected tissues bulges upward and gradually becomes necrotic. Such spots may coalesce, and the causal fungus sporulates on the under leaf side of the spots. Plants affected in this manner usually are not killed, despite the presence of numerous necrotic leaf spots. Other plants may sustain more severe disease development with persistent sporulation from the under surfaces of leaf lesions, and the infection may involve stems, buds and capsules. Plants that develop such extensive symptoms are killed, often within a month after the first appearance of symptoms.

The disease is caused by *Peronospora arborescens* (Berk.) de Bary, which produces stout, erect conidiophores branched 5–10 times that emerge from stomata and bear single oval to globose, hyaline conidia 15–26 × 18–20 μm at the ends of branches. Pale oospores 30–54 μm in diameter develop in all infected aerial plant parts.

Conidia do not remain viable unless kept at low temperatures and will germinate (by formation of a germ tube) in 2 hours after exposure to free water. Optimum temperature for conidial germination is 15–20 C. Symptoms may appear in 3–5 days on inoculated leaves, and sporulation can develop in 12–24 hours on leaves placed in a moist chamber. Oospores in the soil, which apparently provide the only means for the fungus to over-season, do not cause infection during the season when they are formed but can cause infection of seedlings the following year and may remain viable in the soil for 3 years.

A crop rotation schedule of 3 years or more will provide protection against infection from oospores in the soil, and a soil drench of Dithane Z-78 (0.4 percent) also will serve to reduce oospore infection. Foliar sprays of Manzate or Dithane (M-22 or Z-78) will help to reduce severity of mildew. Debris, particularly from crops with infected plants, should be collected and destroyed.

1. Kothari, K. L. and N. Prasad. 1970. Downy mildew of opium poppy in Rajasthan. *Indian Phytopathol. 23:* 674–688.

2. Kothari, K. L. and A. C. Verma. 1967. Control of downy mildew of opium poppy by root application of fungicides. *Plant Dis. Reptr. 51:* 686–687.

3. Yossifovitch, M. 1929. *Peronospora arborescens* (Berk.) de Bary, parasite tres important de *Papaver somniferum* en Yugoslavie. *Rev. Pathol. Veg. Entomol. Agr. 16:* 235–270.

Poppy fire [*Pleospora papaveracea* (de N.) Wint.]

1. Ballarin, C. 1950. Untersuchungen über *Helminthosporium papaveris. Phytopathol. Z. 16:* 399–442.

2. Darpoux, H. 1945. Contribution a l'étude des maladies des plantes oléagineuses en France. *Ann. Epiphyt., n.s. 11:* 71–103.

3. Reinmuth, E. 1948. Die helminthosporiose des ölmohns. *Z. Pflanzenkrankh. Pflanzenschutz 55:* 138–141.

Powdery mildew (*Erysiphe polygoni*)

1. Kothari, K. L. and A. C. Verma. 1972. Germination of conidia of poppy mildew (*Erysiphe polygoni* DC.) *Mycopathol. Mycol. Appl. 47:* 253–260.

2. Zaman, M. B. and A. S. Khan. 1969. Studies on some diseases of *Papaver somniferum* Linn. *Pakistan J. Forest. 19:* 185–193.

Leaf smut (*Entyloma fuscum* Schroet.)

1. Savile, D. B. O. 1946. *Entyloma fuscum* and related smuts attacking *Papaveraceae. Can. J. Res., Sect. C, 24:* 109–114.

Alternaria leaf spot [*Alternaria brassicae* (Berk.) Sacc.]

1. Ellis, M. B. 1968. *Alternaria brassicae. Commonwealth Mycol. Inst. Descriptions Pathogenic Fungi Bacteria 162.* 2 p.

2. Schmitt, C. G. and B. Lipscomb. 1975. Pathogens of selected members of the

Papaveraceae—an annotated bibliography. *Agr. Res. Serv., U. S. Dept. Agr. ARS-NE-62.* 186 p.

Bean yellow mosaic

1. Bos, L. 1970. Bean yellow mosaic virus. *Commonwealth Mycol. Inst. Descriptions Plant Viruses 40.* 4 p.

2. Kovachevsky, J. C. 1968. Das bohnengelbmosaik-virus in Bulgarien. *Phytopathol. Z. 61:* 41–48.

Cabbage black ring spot

1. Hamlyn, B. M. G. 1953. Quantitative studies on the transmission of cabbage black ring spot virus by *Myzus persicae* (Sulz.). *Ann. Appl. Biol. 40:* 393–402.

(Beet) Mosaic

1. Schmelzer, K. and W. Rondomanski. 1958. Zur atiologe des mohnmosaiks. *Phytopathol. Z. 33:* 426–429.

Sugar beet yellows

1. Bennett, C. W. 1960. Sugar beet yellows disease in the United States. *U. S. Dept. Agr. Tech. Bull. 1218.* 63 p.

2. Cleij, G. 1961. Beet yellows in poppy. *Euphytica 10:* 225–228.

3. Russell, G. E. 1965. The host range of some English isolates of beet yellowing viruses. *Ann. Appl. Biol. 55:* 245–252.

Yellowing (virus) disease

1. Schmitt, C. G. and B. Lipscomb. 1975. Pathogens of selected members of the *Papaveraceae*—an annotated bibliography. *Agr. Res. Serv., U. S. Dept. Agr. ARS-NE-62.* p. 186.

P

PIGEON PEA: *Cajanus cajan* (L.) Millsp.

Bacterial leaf spot and stem canker

Symptoms consisting of light brown, round, water-soaked leaf lesions (0.5 mm) that enlarged to 1 mm and developed halos as they turned dark brown were described from India (1950). Spots sometimes coalesced to form lesions 2 mm in diameter, and leaf infections often extended to large veins. Severely affected leaves turned chlorotic and dropped.

Stem lesions developed into elongated (2.5 × 12.5 mm) dark brown cankers that resulted in peeling of the bark when numerous. Bacterial exudate was present on leaf and stem lesions.

The disease was caused by *Xanthomonas cajanii* Kulkarni et al., characterized by gram negative rods 0.9–1.4 × 1.3–2.2 μm, motile by single polar flagella. Disease development was most serious under conditions of 80–90 percent relative humidity and a temperature range of 24–31 C on young plants grown in the shade.

Control procedures were not determined, but a planting schedule and procedure to avoid development of succulent plants during periods of high humidity and temperature (in the shade of other plants) should prove beneficial.

1. Kulkarni, Y. S., M. K. Patel and S. G. Abhyankar. 1950. A new bacterial leaf spot and stem canker of pigeon pea. *Current Sci. 19:* 384.

2. Kulkarni, Y. S., M. K. Patel and S. G. Abhyankar. 1952. A new bacterial leaf-spot and stem canker of pigeon pea. *Indian Phytopathol. 5:* 21–22.

Fusarium wilt

Wilt of pigeon pea has been recognized in India since 1910 and remains one of the most important diseases of that crop there. The disease has been a limiting factor for production in some districts, i.e. Bihar, Uttar Pradesh and Bombay, where near total losses in some varieties have been sustained.

Primary leaves on infected plants developed epinasty 10 days after infection in controlled inoculations. Progressive intercostal chlorosis appeared in the primary and first and second trifoliate leaves during the next 2 weeks, followed quickly by loss of turgor and wilting. Affected plants were

noticeably stunted, particularly the roots, which exhibited external browning, and size of leaves was reduced. Infected plants were killed.

Wilt is caused by *Fusarium udum* Butl. [*F. lateritium* f. *cajana* (Padw.) Gord.].

Susceptible varieties were found to be systemically invaded by the fungus, while only the root system of a resistant variety (N. P. 15) was invaded. However, soil amendments with 100 ppm (or more) manganese afforded protection against infection of susceptible varieties.

1. Archibald, J. F. 1961. Transmission of gall-diseases of cacao, mango, and pigeon pea. *Nature 190:* 284.

2. Butler, E. J. 1910. The wilt disease of pigeon-pea and the parasitism of *Neocosmospora vasinfecta* Smith. *Mem. Dept. Agr. India (Bot. Ser.) II, 9:* 1–64.

3. Chadha, K. C. and S. P. Raychaudhuri. 1966. Interaction between sterility virus and *Fusarium udum* Butl. in pigeon-pea. *Indian J. Agr. Sci. 36:* 133–139.

4. Chakravarti, B. P. 1952. Wilt of arhar (*Cajanus cajan*) in Bihar. *Bihar Agr. Coll. Mag. 3:* 9–10.

5. Singh, G. P. and A. Husain. 1964. Presence of fusaric acid in wilt-affected pigeon-pea plants. *Current Sci. 33:* 287.

6. Subramanian, S. 1963. *Fusarium* wilt of pigeon pea. I. Symptomatology and infection studies; II. Changes in host metabolism; III. Manganese nutrition and disease resistance. *Proc. Indian Acad. Sci., Sect. B., 57:* 134–148; 178–194; 259–274.

7. Vasudeva, R. S., A. C. Jain and K. G. Nema. 1952. Investigations of the inhibitory action of *Bacillus subtilis* on *Fusarium udum* Butl., the fungus causing wilt of pigeon-pea [*Cajanus cajan* (L.) Millsp.]. *Ann. Appl. Biol. 39:* 229–238.

Brown leaf spot [*Leptosphaerulina briosiana* (Poll.) Graham & Luttrell]

Reported only from India.

1. Graham, J. H. and E. S. Luttrell. 1961. Species of *Leptosphaerulina* on forage plants. *Phytopathology 51:* 680–693.

2. Pavgi, M. S. and R. A. Singh. 1965. Some parasitic fungi on pigeon pea from India. *Mycopathol. Mycol. Appl. 27:* 97–106.

Cercoseptoria leaf spot (*Cercoseptoria cajanicola* Pavgi & Singh)

Fasciculate, simple conidiophores 2–2.6 × 24–40 μm (1–3 septate) with hyaline, filiform, multiseptate conidia 1.3–2.2 × 41–156 μm occurred in red-brown leaf lesions 0.5–2.5 mm in India.

1. Pavgi, M. S. and R. A. Singh. 1965. Some parasitic fungi on pigeon pea from India. *Mycopathol. Mycol. Appl. 27:* 97–106.

Cercospora leaf spots (*Cercospora cajani* P. Hennings; *C. indica* Singh; *C. instabilis* Rangel)

Conidia of all 3 species, which have been reported from India, are subhyaline, slightly curved and measure 3–7 × 10–50 μm (1–9 septate), 3.3–5.3 × 14–105 μm (0–7 septate) and 2.6–4 × 26.4–145 μm (3–11 septate), respectively.

1. Chupp, C. 1953. *A monograph of the fungus genus* Cercospora. Ithaca, New York. 667 p.

2. Pavgi, M. S. and R. A. Singh. 1965. Some parasitic fungi on pigeon pea from India. *Mycopathol. Mycol. Appl. 27:* 97–106.

3. Singh, U. B. 1934. Studies on *Cercospora indica* n. sp. parasitic on *Cajanus indicus* Spreng. *Indian J. Agr. Sci. 4:* 343–360.

Collar rot (*Pythium aphanidermatum*)

1. Malik, R. P. 1945. Collar rot of pigeon-pea caused by *Pythium aphanidermatum* (Edson) Fitz. *Indian J. Agr. Sci. 15:* 92–93.

2. Waterhouse, G. M. and J. M. Waterston. 1964. *Pythium aphanidermatum. Commonwealth Mycol. Inst. Descriptions Pathogenic Fungi Bacteria 36.* 2 p.

Phoma canker (*Phoma* sp.)

Blotchy, irregular swollen cankers 2–3 cm in length that enlarged longitudinally and split open to expose rough stem tissues were observed in Puerto Rico in 1960.

1. Alvarez Garcia, L. A. 1960. *Phoma* canker of pigeon peas in Puerto Rico. *J. Agr. Univ. Puerto Rico 44:* 28–30.

Phyllosticta leaf spot (*Phyllosticta cajani* Sydow)

Dark, ostiolate pycnidia 40 × 155 μm in which were borne hyaline, ovate conidia 1.3–2.6 × 2.6–5.3 μm were found in zonate leaf lesions in India.

1. Pavgi, M. S. and R. A. Singh. 1965. Some parasitic fungi on pigeon pea from India. *Mycopathol. Mycol. Appl. 27:* 97–106.

2. Sydow, H. P. and E. J. Butler, 1916. Fungi Indiae Orientalis. *Ann. Mycol. 14:* 177–220.

Powdery mildew (*Leveillula taurica*)

1. Mukerji, K. G. 1968. *Leveillula taurica. Commonwealth Mycol. Inst. Descriptions Pathogenic Fungi Bacteria 182.* 2 p.

2. Williams, F. J., J. S. Grewal and K. S. Amin. 1968. Serious and new diseases of pulse crops in India in 1966. *Plant Dis. Reptr. 52:* 300–304.

Pyrenochaeta leaf spot (*Pyrenochaeta cajani* Singh & Pavgi)

1. Pavgi, M. S. and R. A. Singh. 1965. Some parasitic fungi on pigeon pea from India. *Mycopathol. Mycol. Appl. 27:* 97–106.

2. Sydow, H. P. and E. J. Butler. 1916. Fungi Indiae Orientalis. *Ann. Mycol. 14:* 177–220.

Stem cankers [*Diplodia cajani* Raychaudhury; *Dothiorella cajani* (Syd. & Butl.) Petrak & Syd.; *Macrophoma cajani* Syd. & Butl.; *M. cajanicola* Pavgi & Singh]

1. Pavgi, M. S. and R. A. Singh. 1965. Some parasitic fungi on pigeon pea from India. *Mycopathol. Mycol. Appl. 27:* 97–106.

2. Raychaudhury, S. P. 1942. A disease of pigeon pea [*Cajanus cajan* (L.) Millsp.] caused by *Diplodia cajani* spec. nov. *Indian J. Agr. Sci. 12:* 837–847.

Stem blight (*Phytophthora* sp.)

1. Williams, F. J., J. S. Grewal and K. S. Amin. 1968. Serious and new diseases of pulse crops in India in 1966. *Plant Dis. Reptr. 52:* 300–304.

2. Williams, F. J., K. S. Amin and B. Baldev. *Phytophthora* stem blight of *Cajanus cajan*. *Phytopathology 65:* 1029–1030.

Sterility mosaic

A disease of pigeon pea characterized by drastic reduction in yield from affected plants was reported from India in 1931 and found also in Burma the same year. The disease spreads readily through field plantings, and sizeable numbers of plants often become infected, resulting in considerable loss of potential harvest, but a quantitative assessment of yield reduction has not been reported.

Young plants that become infected develop pale green foliage on which can be seen a dark green mottle. The mosaic leaf symptom becomes less noticeable as plants mature, but reduction in number and size of leaflets and fewer or complete absence of flowering branches and flowers render affected plants quite conspicuous. Diseased plants generally are smaller than healthy plants.

Sterility mosaic disease is caused by a virus which is transmitted by an eriophyid mite, *Aceria cajani* Channabasavanna, as well as by grafting, and infrequently by sap inoculation, but is not seed-borne. The virus is not known to infect any plant other than pigeon pea, which becomes systemically infected, although the mosaic symptoms typically are less obvious on mature plants. Production of some fruits and seeds on infected plants serves to emphasize that the name "sterility disease" applies primarily to reduction in yield and not to processes of reproduction.

Several unsuccessful attempts have been made to find resistant varieties, so control presently consists of modified horticultural practices and mite control. New plantings of pigeon pea should be isolated from older plantings, particularly if the disease is present in the area. Application of miticide may help to reduce spread of the disease.

1. Capoor, S. P. 1952. Observations on the sterility disease of pigeon pea. *Indian J. Agr. Sci. 22:* 271–274.

2. Kandaswamy, T. K. and K. Ramakrishnan. 1960. An epiphytotic of pigeon pea sterility mosaic at Coimbatore. *Madras Agr. J. 47:* 440–441.

3. Narayanasamy, P. and K. Ramakrishnan. 1965. Studies on the sterility mosaic disease of pigeon pea. I. Transmission of the disease; II. Carbohydrate metabolism of infected plants. *Proc. Indian Acad. Sci., Sect. B, 62:* 73–86; 130–139.

4. Prasad, S. N. 1968. Studies on sterility disease of "rahar" (*Cajanus cajan*). *Allahabad Farmer 39:* 235–237.

5. Seth, M. L. 1962. Transmission of pigeon pea sterility by an eriophyid mite. *Indian Phytophatol. 15:* 225–227.

Proliferation diseases

Symptoms reminiscent of "witches' broom" diseases of other crops have been observed on pigeon pea plants in Dominican Republic and Puerto Rico. Mycoplasma-like organisms and rhabdovirus-like particles were observed in phloem and phloem parenchyma cells, respectively, of affected plants.

1. Hirumi, H., K. Maramorosch and E. Hichez. 1973. Rhabdovirus and mycoplasmalike organism: natural dual infection of *Cajanus cajan*. *Phytopathology 63:* 202.

2. Maramorosch, K., H. Hirumi, M. Kimura, J. Bird and N. G. Vakili. 1974. Diseases of pigeon pea in the Caribbean area: an electron microscopy study. *FAO Plant Protect. Bull. 22:* 32–36.

3. Maramorosch, K., H. Hirumi, M. Kimura, J. Bird and N. G. Vakili. 1974. Pigeon pea witches' broom disease. *Phytopathology 64:* 582–583.

4. Vakili, N. G. and K. Maramorosch. 1974. "Witches'-broom" disease caused by mycoplasmalike organisms on pigeonpeas (*Cajanus cajan*) in Puerto Rico. *Plant Dis. Reptr. 58:* 96.

Yellow mosaic

This whitefly-transmitted disease of pigeon pea in Puerto Rico is presumed identical to *Rhynchosia* mosaic disease on *Rhynchosia minima* (L.) DC.

1. Bird, J., J. E. Perez, R. Alconero, N. G. Vakili and P. L. Melendez. 1972. A whitefly-transmitted golden yellow mosaic virus of *Phaseolus lunatus* in Puerto Rico. *J. Agr. Univ. Puerto Rico. 56:* 64–74.

2. Bird, J. and J. Sanchez. 1971. Whitefly-transmitted viruses in Puerto Rico. *J. Agr. Univ. Puerto Rico 55:* 461–467.

3. Maramorosch, K., H. Hirumi, M. Kimura, J. Bird and N. G. Vakili. 1974. Diseases of pigeon pea in the Caribbean area: an electron microscopy study. *FAO Plant Protect Bull. 22:* 32–36.

Mineral deficiency symptoms

The youngest leaves on plants deficient in *nitrogen* remain light green and later become chlorotic. Older leaves also turn light green in color and sometimes fall prematurely. Growth of affected plants is progressively reduced.

Plants deficient in *phosphorus* are stunted (shortened internodes), and leaves remain small and colored deep green but fall early.

Margins of the oldest leaves beyond the ends of veins first become yellow on plants lacking sufficient *potassium*. These chlorotic areas extend along the leaf margins to coalesce, followed by progressive necrosis of the leaf apices. Necrotic tissue remains separated from green tissue by a yellow band, and leaves fall when about half of the lamina becomes necrotic.

Calcium deficiency results in plants assuming a nearly prostrate habit followed by chlorotic spotting of the laminae, which later become necrotic.

Upward rolling of oldest leaves, followed by interveinal chlorosis, bronzing and sometimes necrosis, is typical on plants deficient in *magnesium*.

Lack of sufficient *iron* causes the interveinal tissues of the youngest leaves to become chlorotic, sometimes with small, necrotic spots, while the veins remain dark green.

1. Nichols, R. 1964. Studies on the major-element deficiencies of the pigeon pea (*Cajanus cajan*) in sand culture. I. Foliar symptoms of the major-element deficiencies. *Plant Soil 21:* 377–387.

POTATO: *Solanum tuberosum* L.

Bacterial wilt (Brown rot)

Within a few years of the first report of bacterial wilt of potato in the United States in 1890, similar diseases of tobacco, tomato, eggplant (*Solanum melongena* L.), peanut and banana were reported from various areas throughout the world. There is some indication that the disease of tobacco had been recognized more than 200 years earlier in Japan. Erratic occurrence of the disease on different crops in various geographic locations near the turn of the century gave little indication of the importance bacterial wilt of potato and related diseases of other crops would assume shortly thereafter. Economic losses from bacterial wilt typically have varied from insignificant to complete, sufficient to cause abandonment of potato (or other crop) culture and even social dislocation of farmers. Thus, bacterial wilt is one of the most important and widely distributed diseases of potato in warm climates.

The lower leaves on affected plants sometimes become slightly yellowed about the time first wilting becomes apparent. The disease often develops unilaterally in affected plants with symptoms confined to a single stalk. Chlorosis and necrosis of affected leaves quickly result. Rapid progression of symptoms that soon become systemic sometimes causes leaves on all stems of a diseased plant to wilt and die simultaneously with little change in color (Fig. 74). Dark brown discoloration of vascular tissue in stems and roots extends into the tubers, which often show no external evidence of infection. Slight lateral pressure to an infected tuber cut in half will cause creamy droplets of bacterial ooze to form in the area of the vascular ring. Exudation

Fig. 74. Responses of four potato cultivars to bacterial wilt infection. Yema de Huevo and 1395 are cultivars of *Solanum phureja*. (Courtesy of H. D. Thurston, Cornell University, Ithaca.)

through the "eyes" of severely affected tubers causes soil to adhere to the tuber surfaces. A diagnostic aid to verify bacterial wilt infection consists of placing a section of stem from an infected plant in a glass of water, and within minutes fine, white strands of bacteria will stream from the cut end.

Bacterial wilt is caused by *Pseudomonas solanacearum*, which is a short, gram negative, rod-shaped bacterium, 0.5–0.7 × 0.8–1.5 μm, usually with single, polar flagella. Colonies *in vitro* are first round to irregular, smooth and opalescent white but soon turn gray-white with brown pigmentation that diffuses into the agar in the center of colonies. On medium B of King, colonies are not fluorescent; cardinal temperatures for growth *in vitro* are about 10, 33 and 40 C. Differences in pathogenicity have been demonstrated for colony types distinguishable on a medium that contains 2,3,5-triphenyltetrazolium chloride (0.005 percent), i.e. fluidal white colonies with pink centers are strongly pathogenic, and butyrous, red colonies are avirulent. Three pathotypes of the bacterium have been distinguished on the basis of host ranges, including responses of tobacco leaves infiltrated with bacteria. Race 1 of *P. solanacearum* is pathogenic to tobacco, tomato, some other solanaceous plants and some diploid bananas, while race 2 is pathogenic only to triploid bananas and race 3 is pathogenic to potato and tomato but is only weakly pathogenic to other solanaceous plants. Subgroups have been distinguished within each race of the bacterium.

Infection typically occurs through root wounds, and vascular discoloration and wilting result soon thereafter, although length of the incubation period is dependent on age of the plant at infection, ambient temperature, and other factors. Optimum temperature range for disease development is 25–35 C, but serious losses have been sustained at cooler temperatures, i.e. elevations above 2000 m. High soil moisture also has been found to enhance development of the disease.

The source of primary inoculum for many serious outbreaks of the disease on land not previously cultivated has not been determined. Neither has there been satisfactory explanation for progressive decline in severity of the disease on some soils planted with susceptible crops in successive seasons. The bacterium can survive in soil for a few months, but rapid loss of viability of the bacterium in infected plant debris may result from desiccation. Dissemination of the bacterium through infected tubers or plants has been demonstrated.

Strict attention should be given to selection of seed tubers from plants free of the disease and planting of potatoes in rotation with crop plants not susceptible to bacterial wilt. Care should be taken to avoid transport of plants or soil from infested fields to other areas, and potatoes should be planted only in soils that are well-drained. A breeding program is in progress to transfer to potato resistance to bacterial wilt from *Solanum phureja* Juz. & Buk.

1. Buddenhagen, I. 1964. Biological and physiological aspects of bacterial wilt caused by *Pseudomonas solanacearum*. *Annu. Rev. Phytopathol. 2:* 203–230.

2. Buddenhagen, I. W., L. Sequeira and A. Kelman. 1962. Designation of races of *Pseudomonas solanacearum*. *Phytopathology 52:* 727.

3. French, E. R. and L. Sequeira. 1970. Strains of *Pseudomonas solanacearum* from Central and South America: a comparative study. *Phytopathology 60:* 506–512.

4. Harrison, D. E. 1961. Bacterial wilt of potatoes. I. Field symptoms of the disease and studies on the causal organism, *Pseudomonas solanacearum* variety *asiaticum*. *Australian J. Agr. Res. 12:* 854–871.

5. Hayward, A. C. and J. M. Waterston. 1964. *Pseudomonas solanacearum*. *Commonwealth Mycol. Inst. Descriptions Pathogenic Fungi Bacteria 15.* 2 p.

6. Kelman, A. 1953. The bacterial wilt caused by *Pseudomonas solanacearum*. *N. C. Agr. Expt. Sta. Tech. Bull. 99.* 194 p.

7. Kelman, A. 1954. The relationship of pathogenicity in *Pseudomonas solanacearum* to colony appearance on a tetrazolium medium. *Phytopathology 44:* 693–695.

8. Lozano, J. C. and L. Sequeira. 1970. Differentiation of races of *Pseudomonas solanacearum* by a leaf infiltration technique. *Phytopathology 60:* 833–838.

9. Nielsen, L. W. and F. L. Haynes, Jr. 1960. Resistance in *Solanum tuberosum* to *Pseudomonas solanacearum*. *Am. Potato J. 37:* 260–267.

10. Ribeiro, R. L. D., O. Kimura, F. Akiba, S. Sudo, C. F. Robbs and L. A. D. Castro. 1973. Obtenção de clones de batatinha *(Solanum phureja* × *Solanum tuberosum)* altamente resistentes a isolados nacionais de *Pseudomonas solanacearum*. *Fitopatologia 8:* 18.

11. Robinson, R. A. 1968. The concept of vertical and horizontal resistance as illustrated by bacterial wilt of potatoes. *Commonwealth Mycol. Inst. Phytopathol. Paper 10.* 37 p.

12. Rowe, P. R. and L. Sequeira. 1970. Inheritance of resistance to *Pseudomonas solanacearum* in *Solanum phureja*. *Phytopathology 60:* 1499–1501.

13. Sequeira, L. and P. R. Rowe. 1969. Selection and utilization of *Solanum phureja* clones with high resistance to different strains of *Pseudomonas solanacearum*. *Am. Potato J. 46:* 451–462.

14. Thurston, H. D. 1963. Bacterial wilt of potatoes in Colombia. *Am. Potato J. 40:* 381–390.

15. Thurston, H. D. 1965. Testing the Colombian potato collection for resistance to *Pseudomonas solanacearum*. *Am. Potato J. 42:* 298.

16. Thurston, H. D. and J. C. Lozano. 1968. Resistance to bacterial wilt of potatoes in Colombian clones of *Solanum phureja*. *Am. Potato J. 45:* 51–55.

Blackleg (Pierna negra)

Blackleg was first described from Holland in 1902, but the disease is known worldwide now and can cause serious losses estimated to be in excess of 50 percent in some cases.

Foliage symptoms of diseased plants become apparent soon after infection as chlorosis of the leaflets, which exhibit reduced turgor and upward curling of the margins. Leaf petioles are more erect and impart a distinctive habit to affected plants, which also are recognizable by progressive yellowing and wilting. Dark brown, internal discoloration, first apparent in the lower stem, usually is confined to the pith in early stages of disease development but soon extends into the vascular elements, which may be darkened some distance in advance of pith and external stem symptoms. The lower stem often remains distinctly wet in appearance, and diseased plants are easily pulled from the soil. Although only a single stem of a plant may show symptoms, infection frequently spreads through the stolons into tubers. Symptoms in tubers range from simple browning of the vascular bundles to extensive rotting of internal tissues, which may remain relatively unaffected by change in color until the tuber is cut, after which the cut surfaces quickly darken (Fig. 75).

The causal bacterium has been identified as *Erwinia atroseptica* (van Hall) Jennison, a gram negative rod 0.7 × 1–2 μm with 1–6 peritrichous flagella. Cardinal temperatures for growth *in vitro* are 3, 27 and 35 C. Similarity to *E. carotovora* in many physiological characters and in capability for rotting potato tubers has prompted extensive taxonomic debate about identity of the 2 organisms, but only *E. atroseptica* has been shown to cause infection in potato plants.

Inoculum originates mainly from contaminated seed and the seed-cutting process serves admirably to disseminate further seed-borne bacteria. Infection, which usually is dependent on wounding as a predisposing factor,

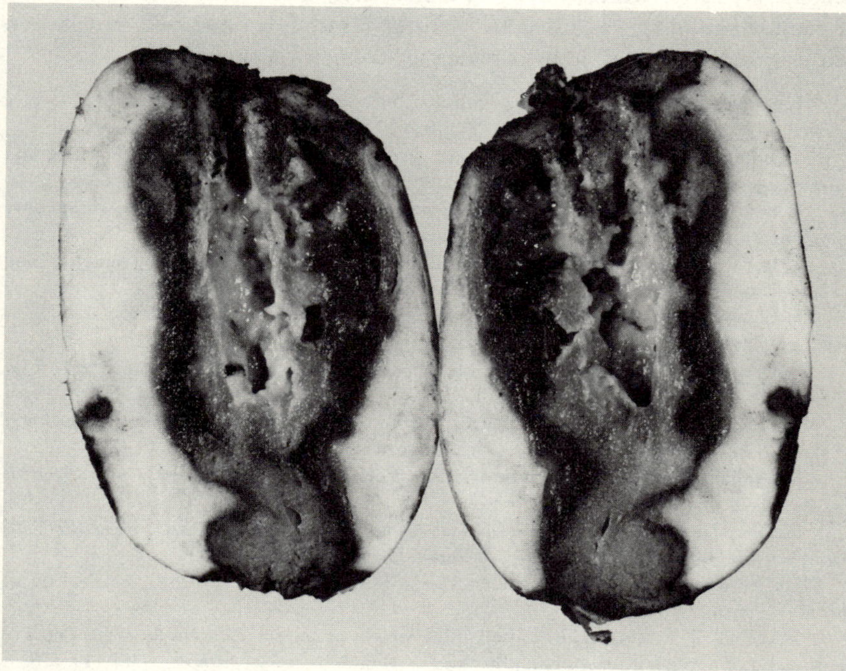

Fig. 75. Tuber symptoms of blackleg. (Courtesy of H. D. Thurston, Cornell University, Ithaca.)

does not often result from contamination of the cut surface of seed pieces, because the bacteria can be effectively excluded by formation of wound cork. In one of the classical demonstrations of the role insects can assume in plant pathological problems, it was revealed that the seed-corn maggot (fly) [*Hylemya cilicrura* (Rondani)] was an efficient vector of the blackleg bacterium. Inoculum on larvae, which are externally contaminated by crawling over their infested egg cases after hatching, is effectively spread over the cut surface of the seedpieces by the feeding of the maggot, which then permits the maggot to become internally contaminated. The larvae, which continue to feed throughout the entire seed-piece, ultimately inoculate (by wounding) the associated plant and cause blackleg symptoms to develop. The adult fly that develops after pupation remains internally contaminated from feeding as a larva and eventually oviposits eggs in moist soil near plant stems or seed-pieces. The eggs are externally contaminated with the blackleg bacterium during the process of oviposition. The potato maggot (*H. trichodactyla* Rondani) later was found to be associated with blackleg in like manner.

It has been demonstrated that blackleg bacteria can overwinter in soil and that low oxygen supply in soil predisposes seed-pieces to infection. Improper curing of cut seed prior to planting, wherein wound cork formation is inhibited, can result in blackleg infection. Seed either should be planted immediately after cutting or stored temporarily under conditions favorable for the formation of wound cork. Planting on well-drained soils in rotation with crops not susceptible to blackleg is strongly recommended.

1. Bonde, R. 1939. Comparative studies of the bacteria associated with potato blackleg and seedpiece decay. *Phytopathology 29:* 831–853.

2. Bonde, R. 1939. The role of insects in the dissemination of potato blackleg and seed-piece decay. *J. Agr. Res. 59:* 889–918.

3. Bonde, R. 1950. Factors affecting potato blackleg and seed-piece decay. *Maine Agr. Expt. Sta. Bull. 482.* 31 p.

4. Burkholder, W. H. and W. L. Smith, Jr. 1949. *Erwinia atroseptica* (van Hall) and *Erwinia carotovora* (Jones) Holland. *Phytopathology 39:* 887–897.

5. Ciampi, L. R. 1972. Enfermedades bacterianas de la papa en Chile. *Fitopatologia 7:* 5–14.

6. Dowson, W. J. 1940. Identity of the bacterium causing potato blackleg. *Nature 145:* 263.

7. Fredericks, A. L. and H. N. Metcalf. 1970. Potato blackleg disease. *Am. Potato J. 47:* 337–343.

8. Graham, D. C. and P. C. Harper. 1967. Potato blackleg and tuber soft rot. *Scottish Agr. 46:* 68–74.

9. Hidalgo, O. A., L. W. Nielsen and E. R. French. 1969. Pierna negra (*Erwinia atroseptica*) de la papa en el Peru. *Rev. Fitopatol. 4:* 15–18.

10. Jennison, H. M. 1923. Potato blackleg, with special reference to the etiological agent. *Ann. Missouri Bot. Garden 10:* 1–72.

11. Leach, J. G. 1926. The relation of the seed-corn maggot (*Phorbia fusciceps* Zett.) to the spread and development of potato blackleg in Minnesota. *Phytopathology 16:* 149–176.

12. Leach, J. G. 1927. The nature of seed-piece transmission of potato blackleg. *Phytopathology 17:* 155–160.

13. Leach, J. G. 1930. Potato blackleg: the survival of the pathogen in the soil and some factors influencing infection. *Phytopathology 20:* 215–228.

14. Lund, B. M. and G. M. Wyatt. 1972. The effect of oxygen and carbon dioxide concentrations on bacterial soft rot of potatoes. I. King Edward potatoes inoculated with *Erwinia carotovora* var. *atroseptica*. *Potato Res. 15:* 174–179.

15. Morse, W. J. 1917. Studies upon the blackleg disease of the potato with special reference to the relationship of the causal organism. *J. Agr. Res. 8:* 79–126.

16. Perombolen, M. C. M. 1972. The extent and survival of contamination of potato stocks in Scotland by *Erwinia carotovora* var. *carotovora* and *E. carotovora* var. *atroseptica*. *Ann. Appl. Biol. 71:* 111–117.

17. Perombolen, M. C. M. 1972. A reliable and rapid method for detecting contamination of potato tubers by *Erwinia carotovora*. *Plant Dis. Reptr. 56:* 552–554.

18. Perombolen, M. C. M. 1973. Sites of contamination and numbers of *Erwinia carotovora* present in stored seed potato stocks in Scotland. *Ann. Appl. Biol. 74:* 59–65.

19. Ramsey, G. B. 1919. Studies on the viability of the potato blackleg organism. *Phytopathology 9:* 285–288.

20. Smith, W. L., Jr. 1950. Pathogenic differences manifested by *Erwinia atroseptica* and *Erwinia carotovora*. *Phytopathology 40:* 1011–1017.

Ring rot

The ring rot disease of potato was described from Germany in 1913 but was found within a few years in Canada and the United States. Serious losses were incurred soon after the disease was discovered, and the extremely infectious nature of the pathogen prompted enforcement of strict control measures. Incidence of ring rot has been significantly reduced where the disease is not tolerated in seed-tuber production.

Plants infected at an early stage may exhibit general chlorosis and wilting, which increase in intensity until the plants are killed, but ring rot symptoms typically occur after mid-season. Leaflets develop a progressive interveinal chlorosis that imparts a partial "herring-bone" effect. This symptom is accompanied by a characteristic necrosis and downward cupping of margins with the extreme edges of leaflets tightly rolled upward. Systemic wilt and subsequent death of diseased plants precede general decline of foliage on healthy plants as they mature.

Light brown vascular discoloration of underground stems may not extend into tubers but often can be traced upward to near soil level. Extrusion of whitish strands of bacteria from the vascular tissues of underground stems upon application of gentle external pressure is a diagnostic symptom of ring rot infection. Tuber infection may become apparent only after harvest (in storage), first as yellowish discoloration of vascular tissue that darkens and becomes more extensive with age (Fig. 76). Infected tubers typically develop

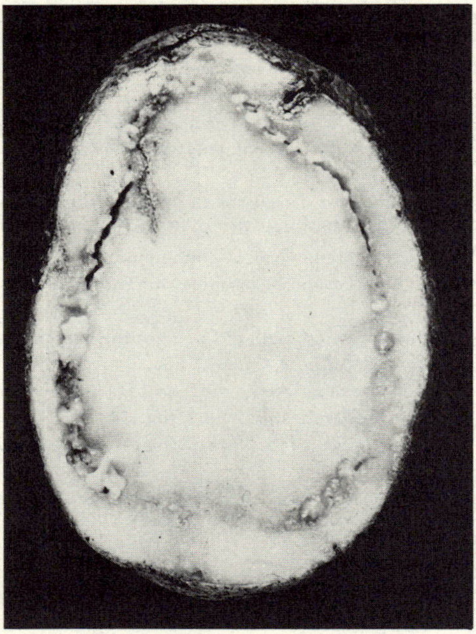

Fig. 76. Symptoms of ring rot confined to the vascular tissues in an infected tuber. (Courtesy of E. K. Wade, University of Wisconsin, Madison.)

external cracks during storage. External pressure applied to the edge of the cut tubers near areas of discolored vascular and associated tissues causes extrusion of characteristic creamy yellow droplets of bacterial ooze. Further rotting of tubers in storage usually follows.

The ring rot disease is caused by *Corynebacterium sepedonicum* (Spieck. & Kotth.) Skaptason & Burkholder, characterized by non-motile cells 0.4–0.6 × 0.8–1.2 μm that are gram positive and usually rod-shaped but may assume pleomorphic forms. Optimum temperature for growth *in vitro* ranges from 18–21 C, with maximum temperature for growth about 30 C.

Dissemination of the bacterium occurs largely through use of infected tubers, and, although there is no indication that the organism can survive in soil, there is evidence that inoculum can be dispersed in irrigation water. Infection occurs through wounds, including the exposed surfaces of tubers cut for seed, but there is no foliar invasion. Systemic progression of the infection is confined to the vascular tissue at first but later extends into the xylem parenchyma, and visible cavities can develop in tubers. Development of foliar symptoms may be retarded at temperatures below 20 C but progress rapidly at about 25 C.

The causal bacterium remains viable between seasons on machinery, bags and bins, as well as in debris, and volunteer plants from infected tubers left in the field also can supply primary inoculum. Transmission of the disease by insects has been demonstrated, but the most important source of inoculum usually is infected tubers used for seed. The bacterium can be spread by the cutting knife from one infected tuber to the seed cut from several tubers in succession thereafter.

Control has been achieved by strict exclusion of the disease from seed-tuber production fields and storage verified by repeated visual inspection and laboratory certification of plants and tubers. Equipment for cutting seed should be subjected to continuous disinfection (e.g. boiling water bath) and all machinery (5.25 percent sodium hypochlorite diluted 1:9 with water), as well as storage space (0.8 kg copper sulfate in 40 ℓ water), treated before use. (Quaternary ammonium compounds, i.e. Hyamine 2389, Roccal or Germ-I-Tol, can be used instead, but storage facilities must be rinsed afterwards with potable water.) Complete decontamination of all premises and equipment and disposal of all seed stocks on hand is recommended following identification of the disease in the field or in storage. Subsequently, only healthy seed stocks should be permitted on premises. Resistant cultivars (Saranac, Teton) have been developed but should be used only with the knowledge that such types produce a crop despite being infected and could subject other simultaneous or subsequent plantings (of susceptible cultivars) to infection.

1. Ark, P. A. 1946. Some laboratory and field data on ring-rot of potatoes in California. *Am. Potato J. 23:* 170–181.

2. Bonde, R. 1939. Bacterial wilt and soft rot of the potato. *Am. Potato J. 16:* 109–114.

3. Bonde, R. 1942. Ring rot in volunteer plants. *Am. Potato J. 19:* 131–133.

4. Bonde, R. and M. Covell. 1950. Effect of host variety and other factors on pathogenicity of potato ring-rot bacteria. *Phytopathology 40:* 161–172.

5. Bonde, R., F. J. Stevenson, C. J. Clark and R. V. Akeley. 1942. Resistance of certain potato varieties and seedling progenies to ring rot. *Phytopathology 32:* 813–819.

6. Burkholder, W. H. 1938. The occurrence in the United States of the tuber ring rot and wilt of the potato [*Phytomonas sepedonica* (Spieckermann u. Kotthoff) Bergey et al.]. *Am. Potato J. 15:* 243–245.

7. Burkholder, W. H. 1942. Diagnosis of the bacterial ring rot of the potato. *Am. Potato J. 19:* 208-212.

8. Duncan, J. and H. Genereux. 1960. La transmission par les insectes de *Corynebacterium sepedonicum* (Spieck. & Kott.) Skaptason et Burkholder. *Can. J. Plant Sci. 40:* 110–116.

9. Eddins, A. H. 1939. Some characteristics of bacterial ring rot of potatoes. *Am. Potato J. 16:* 309–322.

10. Hayward, A. C. and J. M. Waterston. 1964. *Corynebacterium sepedonicum. Commonwealth Mycol. Inst. Descriptions Pathogenic Fungi Bacteria 14.* 1 p.

11. Katznelson, H. and M. D. Sutton. 1956. Laboratory detection of *Corynebacterium sepedonicum,* causal agent of bacterial ring rot of potatoes. *Can. J. Bot. 34:* 48–53.

12. Knorr, L. C. 1944. Efficacy of the rotary knife in the control of potato ring rot. *Am. Potato J. 21:* 250–261.

13. Knorr, L. C. 1948. Suscept range of the potato ring rot bacterium. *Am. Potato J. 25:* 361–371.

14. Kreutzer, W. A., G. H. Lane and J. L. Paschal. 1946. Comparative effectiveness of certain knife disinfectants and the use of the double-edged knife for the control of ring rot of potatoes. *Am. Potato J. 23:* 291–299.

15. Logsdon, C. E. 1967. Effect of soil temperature on potato ring rot (*Corynebacterium sepedonicum*). *Am. Potato J. 44:* 281–286.

16. MacLachlan, D. S., H. A. Monro, H. M. Racicot and J. E. King. 1953. Fumigation of used bags with toxic gases for the control of bacterial ring rot of potato. *Can. J. Agr. Sci. 33:* 132–140.

17. Patel, M. K., N. B. Kulkarni and Y. S. Kulkarni. 1952. Ring disease of potato in India. *Current Sci. 21:* 47–48.

18. Raeder, J. M. 1949. Ring rot of potatoes. *Am. Potato J. 26:* 126–131; 203–207.

19. Riedl, W. A., F. J. Stevenson and R. Bonde. 1946. The Teton potato, a new variety resistant to ring rot. *Am. Potato J. 23:* 379–389.

20. Sherf, A. F. 1944. Infection experiments with potato ring rot and the effect of soil temperature on the disease. *Am. Potato J. 21:* 27–29.

21. Skaptason, J. B. 1943. Studies on the bacterial ring-rot disease of potatoes. *N. Y. (Cornell) Agr. Expt. Sta. Mem. 250.* 30 p.

22. Skaptason, J. B. and W. H. Burkholder. 1942. Classification and nomenclature of the pathogen causing bacterial ring rot of potatoes. *Phytopathology 32:* 439–441.

23. Sniesko, S. F. and R. Bonde. 1943. Studies on the morphology, physiology, serology, longevity, and pathogenicity of *Corynebacterium sepedonicum. Phytopathology 33:* 1032–1044.

24. Spieckermann, A. and P. Kotthoff. 1914. Untersuchungen über die kartoffelpflanze und ihre krankheiten. I. Die bakterienringfaule der kartoffelpflanze. *Landw. Jahrb. 46:* 659–732.

25. Starr, G. H. 1944. Hot water for the control of potato ring rot bacteria on the cutting knife. *Am. Potato J. 21:* 161–163.

26. Starr, G. H. 1944. The longevity of *Corynebacterium sepedonicum* on potato bags when placed under different environmental conditions. *Am. Potato J. 21:* 177–179.

27. Starr, G. H. 1951. Some factors influencing infection by *Corynebacterium sepedonicum* in potato plants. *Am. Potato J. 28:* 551–558.

28. Stevenson, F. J. and J. R. Livermore. 1949. The Saranac potato: a new variety promising in Australia. *Am. Potato J. 26:* 45–46.

Common scab

Recognition of the common scab disease preceded by several years determination of the causal agent in 1890 in the United States. The disease probably occurs wherever potatoes are grown, although more attention apparently has been given to its occurrence in temperate climates. Losses that result from common scab consist largely of disfiguration of tuber surfaces, which can result in considerable waste in peeling of tubers for culinary use.

Symptoms, which are confined primarily to tubers, assume different forms that range from a superficial russeting to slightly raised, somewhat roughened, corky areas, often with irregular cracks, to shallow, concave lesions or deep, crater-like dry pits that may be 3 mm in depth (Fig. 77). Areas affected by each of these symptom types are mostly circular to irregular in outline but typically include dry, corky, uneven centers. Coloration of

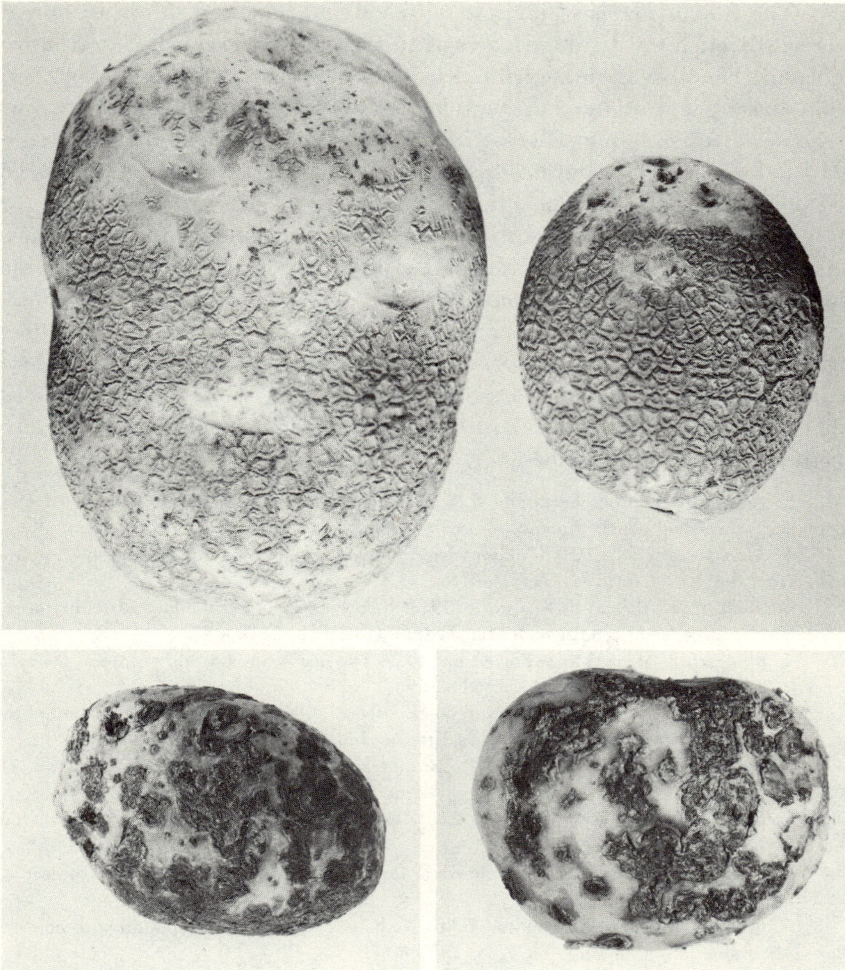

Fig. 77. Three manifestations of common scab on potato tubers. (Courtesy of E. K. Wade, University of Wisconsin, Madison.)

more extremely damaged tissues is usually dark brown, while more superficial lesions are often lighter in color but darker than healthy tuber areas. Small, inconspicuous, superficial lesions may develop on roots and stolons.

Common scab is caused by *Streptomyces scabies* (Thaxt.) Waks. & Henrici which produces only mycelium and approximately cylindrical, hyaline spores (0.6–0.7 × 1–2 μm) *in vitro*. The fungus grows in culture over a temperature range of about 5–40 C (optimum 25–30 C) with optimum temperature for spore germination near the upper limit.

The fungus persists indefinitely in soil and has been shown to survive passage through the digestive tracts of animals and remain viable in manure. Inoculum can also originate from diseased seed tubers. Infection may result from direct penetration or through lenticels or wounds, but tubers are most susceptible when young. Disease incidence occurs throughout the range of pH 5.2–8 soil reaction but has been found to be less severe with increase in soil moisture under experimental conditions.

Several varieties, e.g. Antigo, Cherokee, Menominee and Seneca, have been reported resistant to scab, and other russeted varieties, e.g. Russet Burbank, are not usually seriously affected. Tubers selected for seed should be free of scab. The disease usually is not serious on soils of pH 5.2 or less, although the use of sulfur as a soil amendment to achieve this reaction level may be quite expensive as well as less than a reliable control measure for all soil types. Rotation of potatoes with acid-soil tolerant crops or green manure crops has been recommended.

1. Adams, A. P., N. Sandar and D. C. Nelson. 1970. Some properties of soils affecting russet scab and silver scurf of potatoes. *Am. Potato J. 47:* 49–57.

2. Afanasiev, M. M. 1937. Comparative physiology of *Actinomyces* in relation to potato scab. *Nebraska Agr. Expt. Sta. Res. Bull. 92.* 63 p.

3. Blodgett, F. M. and E. K. Cowan. 1935. Relative effects of calcium and acidity of the soil on the occurrence of potato scab. *Am. Potato J. 12:* 265–274.

4. Blodgett, F. M. and F. J. Stevenson. 1946. The new scab-resistant potatoes, Ontario, Seneca, and Cayuga. *Am. Potato J. 23:* 315–329.

5. Cooper, D. C., G. W. Stokes and G. H. Rieman. 1954. Periderm development of the potato tuber and its relationship to scab resistance. *Am. Potato J. 31:* 58–66.

6. Darling, H. M. 1937. A study of scab resistance in the potato. *J. Agr. Res. 54:* 305–317.

7. Duff, G. H. and C. G. Welch. 1927. Sulphur as a control agent for common scab of potato. *Phytopathology 17:* 297–314.

8. Fellows, H. 1926. Relation of growth in the potato tuber to the potato-scab disease. *J. Agr. Res. 32:* 757–781.

9. Fink, H. C. 1956. Vapam and P. C. N. B. soil treatments for potato scab control. *Plant Dis. Reptr. 40:* 190–192.

10. Goss, R. W. 1937. The influence of various soil factors upon potato scab caused by *Actinomyces scabies. Nebraska Agr. Expt. Sta. Res. Bull. 93.* 40 p.

11. Harrison, M. D. 1962. Potato russet scab, its cause and factors affecting its development. *Am. Potato J. 39:* 368–387.

12. Hooker, W. J. and G. C. Kent. 1950. Sulfur and certain soil amendments for potato scab control in the peat soils of northern Iowa. *Am. Potato J. 27:* 343–365.

13. Hooker, W. J. and O. T. Page. 1960. Relation of potato tuber growth and skin maturity to infection by common scab, *Streptomyces scabies. Am. Potato J. 37:* 414–423.

14. Jones, L. R., H. H. McKinney and H. Fellows. 1922. The influence of soil temperature on potato scab. *Wisconsin Agr. Expt. Sta. Res. Bull. 53.* 35 p.

15. Lapwood, D. H. 1966. The effects of soil moisture at the time potato tubers are forming on the incidence of common scab (*Streptomyces scabies*). *Ann. Appl. Biol. 58:* 447–456.

16. Lapwood, D. H. 1973. Irrigation as a practical means to control potato common scab (*Streptomyces scabies*); final experiment and conclusions. *Plant Pathol. 22:* 35–41.

17. Lapwood, D. H., L. W. Wellings and W. R. Rosser. 1970. Control of common scab of potatoes by irrigation. *Ann. Appl. Biol. 66:* 397–405.

18. Larson, R. H., A. R. Albert and J. C. Walker. 1938. Soil reaction in relation to potato scab. *Am. Potato J. 15:* 325–330.

19. Martin, W. H. 1923. Influence of soil moisture and acidity on potato scab. *Soil Sci. 16:* 69–73.

20. Millard, W. A. 1922/1923. Common scab of potatoes. *Ann. Appl. Biol. 9:* 156–164; *10:* 70–88.

21. Millard, W. A. and S. Burr. 1926. A study of twenty-four strains of *Actinomyces* and their relation to types of common scab of potato. *Ann. Appl. Biol. 13:* 580–644.

22. Nonnecke, I. L., W. E. Torfason and L. C. Young. 1966. Chinook: a new high quality potato variety resistant to common scab. *Am. Potato J. 43:* 1–5.

23. Nugent, T. J. 1956. Soil treatments with PCNB (Terraclor) for control of potato scab. *Plant Dis. Reptr. 40:* 428.

24. O'Keefe, B. R. and H. O. Werner. 1965. Platte and Hi-Plains, two new scab resistant processing potato varieties. *Am. Potato J. 42:* 361–368.

25. Peterson, C. E., N. K. Ellis, R. V. Akeley and F. J. Stevenson. 1954. Cherokee: a new medium-maturing potato variety resistant to common scab, late blight, mild mosaic, and net necrosis. *Am. Potato J. 31:* 53–58.

26. Potter, H. S., W. J. Hooker, W. Cargo and G. T. Stachwick. 1959. Pentachloronitrobenzene and urea-formaldehyde for potato scab control in Michigan. *Plant Dis. Reptr. 43:* 633–637.

27. Rieman, G. H. and D. H. Young. 1955. Antigo: a new white, medium-maturing potato variety resistant to common scab. *Am. Potato J. 32:* 407–410.

28. Sanford, G. B. 1926. Some factors affecting the pathogenicity of *Actinomyces scabies*. *Phytopathology 16:* 525–547.

29. Schaal, L. A. 1944. Variation and physiologic specialization in the common scab fungus (*Actinomyces scabies*). *J. Agr. Res. 69:* 169–186.

30. Shapovalov, M. 1915. Effect of temperature on germination and growth of the common potato-scab organism. *J. Agr. Res. 4:* 129–134.

31. Waksman, S. A. 1922. The influence of soil reaction upon growth of *Actinomyces* causing potato scab. *Soil Sci. 14:* 61–79.

32. Walker, J. C., R. H. Larson and A. R. Albert. 1938. Studies of resistance to potato scab in Wisconsin. *Am. Potato J. 15:* 246–252.

33. Weinhold, A. R., J. W. Oswald, T. Bowman, J. Bishop and D. Wright. 1964. Influence of green manures and crop rotation on common scab of potato. *Am. Potato J. 41:* 265–273.

34. Wheeler, E. J., F. J. Stevenson and H. C. Moore. 1944. The Menominee potato: a new variety resistant to common scab and late blight. *Am. Potato J. 21:* 305–311.

Powdery scab (Roña)

The powdery scab disease of potato was discovered in Germany in 1841 and described from there a year later. Within the next 50 years, the disease was found widespread throughout Europe and in several South American countries, including Colombia, Ecuador and Peru. The disease has been found subsequently throughout the world, i.e. Asia, Africa, Australia and New

Zealand, in addition to North America. Damage caused by the disease consists largely of disfiguration of tuber surfaces which can be extensive under favorable conditions. Confusion with symptoms caused by common scab disease has made accurate assessment of losses difficult.

Affected tubers develop small, slightly raised, brownish areas, which may enlarge quickly to nearly 5 mm in diameter. Coloration of tissues under the epidermis of the tuber changes to light brown, as the consistency of these tissues also is altered. A powdery mass of spores is exposed when the epidermis ruptures but characteristically remains attached at the margins of lesions and imparts a ragged appearance. Infection is confined to underground portions of plants.

Powdery scab disease is caused by *Spongospora subterranea* (Wallr.) Lagerheim, which penetrates the tuber as a plasmodium that induces some hypertrophy and additional periderm formation. Repeated invasion of new periderm with hypertrophy precedes cleavage of the multinucleate plasmodium into intracellular sporangia which germinate by liberation of zoospores that form plasmodia to repeat the infection process. Resting spores develop from multinucleate plasmodia and form into balls or clusters of unicellular, uninucleate spores, each of which germinates as a spherical swarmspore (or zoospore) 2.5–3.5 μm in diameter with 2 unequal flagella. Sori on tuber surfaces that contain dusty masses of resting spores are foci for initiation of dry rot.

It has been established that the causal fungus can persist in soil and that environmental conditions which include high soil moisture and temperature below 15 C are required for development of powdery scab on potato tubers. Further, the fungus is pathogenic on several solanaceous plants (not including petunia) and causes distinct galls to form on the roots. (Galls on roots of potato caused by the powdery scab fungus are similar in appearance to nodules caused by root knot nematodes.) In addition to being an important pathogen, the powdery scab fungus is the only known vector of the mop-top virus (see Potato—Mop-top).

Little attention has been directed to control, since the disease presumably occurs only under conditions of cool temperature and high soil moisture. However, only healthy tubers should be used for planting material, and neither potatoes nor other susceptible solanaceous crops should be planted in successive seasons on land where powdery scab has appeared.

1. Douchez, Y. 1942. Sur les galles des racines de pomme de terre provoquées par le *Spongospora subterranea* (Wallr.) T. Johnson. *Ann. Inst. Pasteur (Paris) 68:* 351–353.

2. Güssow, H. T. 1913. Powdery scab of potatoes [*Spongospora subterranea* (Wallr.) Johns.]. *Phytopathology 3:* 18–19.

3. Hims, M. J. and T. F. Preece. 1975. *Spongospora subterranea* f. sp. *subterranea. Commonwealth Mycol. Inst. Descriptions Pathogenic Fungi Bacteria* 477. 2 p.

4. Kunkel, L. O. 1915. A contribution to the life history of *Spongospora subterranea. J. Agr. Res. 4:* 265–278.

5. Ledingham, G. A. 1935. Occurrence of zoosporangia in *Spongospora subterranea* (Wallroth) Lagerheim. *Nature 135:* 394.

6. Letham, D. B. 1972. Powdery scab of potatoes (*Spongospora subterranea*). *Agr. Gaz. New S. Wales 83:* 358–359.

7. Manzer, F. E., R. V. Akeley and D. Merriam. 1964. Resistance to powdery scab in *Solanum tuberosum. Am. Potato J. 41:* 374–376.

8. Melhus, I. E. 1914. Powdery scab (*Spongospora subterranea*) of potatoes. *U. S. Dept. Agr. Bull. 82.* 16 p.

9. Melhus, I. E., J. Rosenbaum and E. S. Schultz. 1916. *Spongospora subterranea* and *Phoma tuberosa* on the Irish potato. *J. Agr. Res. 7:* 213–254.

10. Osborn, T. G. B. 1911. *Spongospora subterranea* (Wallroth) Johnson. *Ann. Bot. (London) 25:* 327–341.

11. Ramsey, G. B. 1918. Influence of moisture and temperature upon infection by *Spongospora subterranea. Phytopathology 8:* 29–31.

12. Von Lagerheim, G. 1892. Remarks on the fungus of a potato scab (*Spongospora solani* Brunch.). *J. Mycol. 7:* 103–104.

13. Weber, G. F. 1923. Potato diseases and insects. *Univ. Florida Agr. Expt. Sta. Bull. 169:* 103–163.

14. Wild, N. 1929. Untersuchungen über den pulverschorf der kartoffelknollen [*Spongospora subterranea* (Wallr.) Johnson]. *Phytopathol. Z. 1:* 367–452.

Wart

Wart disease of potatoes was described from Hungary in 1895 and subsequently found in many other countries in western Europe. The disease was found later in Africa, India, the United States, Mexico, Russia, Canada (including Newfoundland and Nova Scotia), Peru, and Brazil and, recently, in New Zealand. It was considered such a serious threat to potato production in Europe nearly 60 years after its discovery (1953) that the European Plant Protection Organization requested all member nations to submit annual reports concerning occurrence of the disease. Early application of quarantine restrictions and implementation of additional control procedures have served to keep the disease restricted in distribution.

Infection of all underground plant parts, except roots, can occur, but tuber symptoms are most striking. Protuberances, or "warts," roughly spherical but variable in size and shape and which have a soft rather than firm consistency, develop on tubers. Warts consist of hyperplastic and disfigured tissue that bears some resemblance to irregular clumps of cauliflower (*Brassica oleracea* var. *botrytis* L.) and are nearly the same color as healthy tubers at first but may develop some green color if near the soil surface. Warts darken with age, and disintegration may commence before harvest of tubers begins.

Synchytrium endobioticum (Schilb.) Perc. (*Chytridiales*), the fungus that causes wart disease, persists in soil or debris as resting sporangia that germinate by internal production of uniflagellate zoospores. After liberation and direct penetration of host tissue, the zoospore soon gives rise to a uninucleate, intracellular thallus which, as it grows in size, induces host hypertrophy. Progression through the prosorus stage to development of a vesicle, in which the nucleus divides repeatedly, precedes delimitation in each vesicle of 4–9 multinucleate sporangia, i.e. a sorus. The nuclei in each sporangium continue to divide prior to liberation of sporangia and subsequent release therefrom of uniflagellate zoospores. Reinfection of the host results in further host tissue hypertrophy and proliferation, but some individual zoospores may fuse with another to form a zygote. Following penetration of host tissue as before, the zygote infection, by hyperplastic host response with little hypertrophy, is carried more deeply into the tuber, where multinucleate resting sporangia are formed.

Resting sporangia liberated into soil may persist for more than a year, but viability can last for 5 years in dried wart material (debris). Germination of sporangia and zoospore activity is dependent on soil saturation, but infection was found, experimentally, to occur from 12–24 C. Several physiological races, or pathotypes, of the fungus have been identified using cultivars of potato with definite response to local fungus isolates.

Control has been accomplished by application of quarantine restrictions to contain infestations within defined geographic areas wherein only cultivation of resistant potato varieties has been permitted. Some countries do not permit importation of potatoes from countries where wart occurs.

1. Bojnansky, V. 1968. The effect of soil type on the development and severity of potato wart disease. *European Potato J. 11:* 100–110.

2. Braun, H. 1942. Biologische spezialisierung bei *Synchytrium endobioticum* (Schilb.) Perc. *Z. Pflanzenkrankh. 52:* 481–486.

3. Cartwright, K. 1926. On the nature of the resistance of the potato to wart disease. *Ann. Bot. 40:* 391–395.

4. Curtis, K. M. 1921. The life-history and cytology of *Synchytrium endobioticum* (Schilb.) Perc., the cause of wart disease in potato. *Phil. Trans. Roy. Soc., Ser. B., 210:* 409–478.

5. Dingley, J. 1970. An outbreak of potato wart disease in New Zealand. *Commonwealth Phytopathol. News, September, 1970.* 2 p.

6. Glynne, M. D. 1925. Infection experiments with wart disease of potatoes, *Synchytrium endobioticum* (Schilb.) Perc. *Ann. Appl. Biol. 12:* 34–60.

7. Glynne, M. D. 1926. The viability of the winter sporangium of *Synchytrium endobioticum* (Schilb.) Perc., the organism causing wart disease in potato. *Ann. Appl. Biol. 13:* 19–36.

8. Glynne, M. D. 1926. Wart disease of potatoes: the development of *Synchytrium endobioticum* (Schilb.) Perc. in "immune" varieties. *Ann. App. Biol. 13:* 358–359.

9. Glynne, M. D. 1934. Infectivity of summer sporangia of potato wart disease in incipient infections on varieties immune in the field. *Nature 134:* 253.

10. Gough, G. C. 1920. Wart disease of potatoes. A study of its history, distribution, and the discovery of immunity. *J. Roy. Hort. Soc. 45:* 301–312.

11. Hampson, M. C. 1975. Induction of secondary fluorescence in the resting sporangium of the potato wart disease fungus, *Synchytrium endobioticum* (European race 2). *Phytopathology 65:* 374–379.

12. Köhler, E. 1931. Über die verschiedenen typen der krebsresistenz und krebsempfänglichkeit bei den kartoffelsorten. *Züchter 3:* 249–252.

13. Mygind, H. 1969. Methods for the detection of resting sporangia of potato wart [*Synchytrium endobioticum* (Schilb.) Perc.] in infested soil. *Acta Agr. Scand. 4:* 317–343.

14. Nelson, G. A. and O. A. Olsen. 1967. Staining reactions of resting sporangia of *Synchytrium endobioticum* with a tetrazolium compound. *Phytopathology 57:* 965–968.

15. Niederhauser, J. S. 1953. *Synchytrium endobioticum* on wild potatoes in Mexico. *Phytopathology 43:* 480.

16. Noble, M. and M. D. Glynne. 1970. Wart disease of potatoes. *FAO Plant Protect. Bull. 18:* 125–135.

17. Olsen, O. A. 1961. Potato wart investigations in Newfoundland. *Can. Plant Dis. Surv. 41:* 147–155.

18. Orton, C. R. and F. D. Kern. 1919. The potato wart disease. A new and serious disease recently discovered in Pennsylvania. *Penn. State Coll. Bull. 156.* 16 p.

19. Phadtare, S. G. and K. P. Sharma. 1971. Additional hosts of *Synchytrium endobioticum. Indian Phytopathol. 24:* 389–392.

20. Roach, W. A. and M. D. Glynne. 1928. The toxicity of certain sulphur compounds to *Synchytrium endobioticum,* the fungus causing wart disease of potatoes. *Ann. Appl. Biol. 15:* 168–190.

Late blight (Tizon tardio, Rancha)

Late blight of potato is one of the plant diseases directly involved historically in the recognition that microorganisms can cause disease in plants. The outbreak of late blight in Ireland in 1845 dramatically emphasized the limited diet of the poor that consisted almost entirely of potatoes. Loss of the potato crop, which resulted in widespread famine, occurred before scientists fully understood about fungi and related such organisms to plant diseases. Despite microscopic illustration of a fungus found consistently associated with the "potato murrain", acceptance of the fungus as the cause of this disease was not immediate. It is a bit ironic that a clergyman, the Reverend Berkeley, was instrumental in proving that late blight was caused by a fungus and did not result from the supernatural.

Although the causative agent was first recognized in 1845 and a description published a year later, the late blight disease of potato had been observed in Canada a year earlier, and there is some evidence of much earlier recognition of the disease in South America. The disease now is worldwide in occurrence and, even though annual losses have been reduced significantly, late blight remains a continuing threat to successful production because of the potential damage that it can cause.

Foliage symptoms appear on leaflets, petioles or stems as dark colored areas variable in size with indefinite margins and often with a brown or purplish cast. Leaf lesions frequently are surrounded by zones of light green tissue and, under favorable conditions, expand rapidly to cause extensive blight symptoms. If favorable conditions persist, foliage is quickly and com-

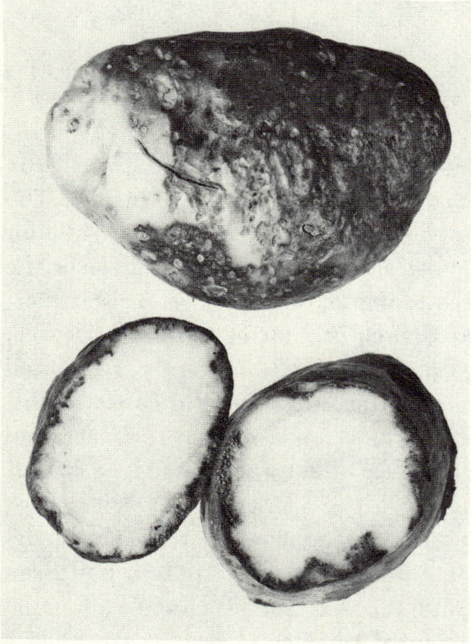

Fig. 78. Late blight infection (top) in the field and (bottom) symptoms on tubers. (Courtesy of H. D. Thurston, Cornell University, Ithaca.)

pletely destroyed (Fig. 78). A characteristic odor can be noted in affected fields where the disease is developing rapidly. Sporulation of the fungus commonly develops on the undersurfaces of leaf lesions at the margins of the discolored tissues during periods of high humidity. When unfavorable weather conditions intervene, lesion development is interrupted, no sporulation occurs, and affected leaf tissues assume a dark brown, dry appearance.

Disease development begins again promptly with the recurrence of cool, wet weather.

Tuber infection may occur in the field, and symptoms appear as brown or purple, external areas of discoloration associated with an internal brown, dry rot that usually penetrates no deeper than 10–15 mm into the tuber (Fig. 78). Subsequent contamination of diseased tissues by secondary microorganisms often results in a slimy, wet rot of affected tubers after harvest.

Sporangia of *Phytophthora infestans* (Mont.) DBy are oval, hyaline, measure 12–23 × 21–38 μm and germinate directly by a germ tube, which may result in formation of a secondary sporangium, but more often they germinate by production of approximately 8 biciliate zoospores. Zoospores germinate, after a short period of activity, by production of coenocytic mycelium. Oogonia vary from 31–50 μm in diameter, and oospores (24–56 μm in diameter) germinate by production of 1–4 germ tubes on which are borne terminal sporangia that may germinate either directly or indirectly as described above for sporangia produced asexually.

The first indication of variation in pathogenic capability of different isolates of the fungus resulted in designation of a "potato race" infectious only to potato and a "tomato race" pathogenic to both potato and tomato. More than a quarter-century later, it was determined that 4 resistance (R) factors derived from *Solanum demissum* Lindl. were dominant and inherited independently. Isolates of the fungus were characterized by pathogenicity on potato cultivars with these dominant (or recessive, i.e. "r") resistance factors and combinations of them until 5 additional (R) resistance factors, also from *S. demissum*, independent and dominant, were discovered in the next 10 years. Although similar resistance factors have been identified in other species of *Solanum* (*S. stoloniferum* Schlecht.), identity with factors from *S. demissum* remains to be established. It has been suggested that the above designations of resistance factors (identified by response to distinctive biotypes of the fungus) be reserved for heritable characters derived from *S. demissum* only and that other identification schemes be devised for heritable characters from other *Solanum* species. The pattern of pathogenicity exhibited by fungus isolates from tomato when inoculated to potato cultivars is quite erratic and bears no relation to the identification scheme for *S. demissum* characters based on responses to fungus isolates from potato. Variation in virulence or pathogenic aggressiveness among isolates of the same pathotype has been demonstrated, which adds further etiological complication, and 2 mating types (A^1 and A^2) have been identified and are influential in formation of the perfect or sexual stage of the fungus. Both mating types occur in Mexico, but only one has been found in the United States, Canada, western Europe and the West Indies.

Environmental conditions, specifically temperature and humidity, are very important factors that influence development of late blight in the field.

Infected tubers left in refuse or volunteer plants from such tubers can supply inoculum as sporangia, which are produced only at a relative humidity of 91 percent or higher. Sporangia develop over a range of 3–26 C with maximum production at 18–22 C, but sporangia produced at 15 C produce zoospores more quickly (1–2 hours). Optimum temperature for zoospore production from sporangia is 12 C, while sporangial germination by production of germ tubes is more common at 25 C. Oospore formation *in vivo* occurs readily at about 20 C, and oospore germination has been observed from 12–25 C.

Potato plants of all ages are susceptible to infection, and sporangia are readily disseminated by wind currents or splashing rain. Continuous susceptibility of potato plants, combined with capability of the causal fungus to propagate quite rapidly and cause epiphytotic losses, has prompted repeated attempts to forecast late blight occurrence in localized areas. Attempts to predict natural development of late blight have been based on correlation of earlier occurrences of late blight with recorded combinations of environmental conditions, including length of nighttime dew periods, night temperature, daytime cloud cover and rainfall. Such information has been useful in prescribing routines for more economical application of protective fungicides. Tuber infection has been reduced by killing or mechanical removal of foliage a few days before harvest.

All volunteer potato plants from previous crops should be destroyed before planting only healthy seed tubers, and potato fields should not be situated near tomato fields, particularly if blight is present. Several varieties with resistance to late blight have been developed, including Calrose, Empire, Menominee, Kennebec and Sebago, but disease control with these and other varieties can be greatly improved by application of dithiocarbamate or copper-containing fungicides (Dithane M-45, Manzate 200 etc.). Killing or removal of all foliage a few days before harvest is recommended. Infected tubers should be discarded at harvest and strict sanitation procedures observed for disposal of storage refuse.

1. Beaumont, A. 1947. The dependence on the weather of the dates of outbreak of potato blight epidemics. *Trans. Brit. Mycol. Soc. 31:* 45–53.

2. Black, W. 1952. Inheritance of resistance to blight (*Phytophthora infestans*) in potatoes: interrelationships of genes and strains. *Proc. Roy. Soc. Edinburgh, B, 64:* 312–352.

3. Black, W. 1952. A genetical basis for classification of strains of *Phytophthora infestans. Proc. Roy. Soc. Edinburgh, B, 65:* 36–51.

4. Black, W. and M. E. Gallegly, 1957. Screening of *Solanum* species for resistance to physiologic races of *Phytophthora infestans. Am. Potato J. 34:* 273–281.

5. Black, W., C. Mastenbroek, W. R. Mills and L. C. Peterson. 1953. A proposal for an international nomenclature of races of *Phytophthora infestans* and of genes controlling immunity in *Solanum demissum* derivatives. *Euphytica 2:* 173–179.

6. Bonde, R. and E. S. Schultz. 1945. The control of potato late blight tuber rot. *Am. Potato J. 22:* 163–167.

7. Clark, C. F. 1946. The Calrose potato: a new variety possessing resistance to late blight. *Am. Potato J. 23:* 343–347.

8. Clinton, G. P. 1910. Oospores of potato blight, *Phytophthora infestans*. *Conn. Agr. Expt. Sta. Bien. Rep. 1909–1910:* 753–774.

9. Cook, H. T. 1949. Forecasting late blight epiphytotics of potatoes and tomatoes. *J. Agr. Res. 78:* 545–563.

10. Crosier, W. 1934. Studies in the biology of *Phytophthora infestans* (Mont.) De Bary. *N. Y. (Cornell) Agr. Expt. Sta. Mem. 155.* 40 p.

11. De Bary, A. 1876. Researches into the nature of the potato-fungus, *Phytophthora infestans*. *J. Roy. Agr. Soc. England, Ser. 2, 12:* 239–269.

12. De Bruyn, H. L. G. 1926. The overwintering of *Phytophthora infestans* (Mont.) De By. *Phytopathology 16:* 121–140.

13. Dionne, L. A., K. M. Graham and W. J. Hodgson. 1960. The inheritance of resistance to higher races of *Phytophthora infestans*. *Am. Potato J. 37:* 346.

14. Galindo, J. and M. E. Gallegly. 1960. The nature of sexuality in *Phytophthora infestans*. *Phytopathology 50:* 123–128.

15. Gallegly, M. E. 1968. Genetics of pathogenicity of *Phytophthora infestans*. *Annu. Rev. Phytopathol. 6:* 375–396.

16. Gallegly, M. E. and J. Galindo. 1958. Mating types and oospores of *Phytophthora infestans* in nature in Mexico. *Phytopathology 48:* 274–277.

17. Giddings, N. J. and A. Berg. 1919. A comparison of the late blights of tomato and potato. A preliminary report. *Phytopathology 9:* 209–210.

18. Guzmán-N., J. 1964. Nature of partial resistance of certain clones of three *Solanum* species to *Phytophthora infestans*. *Phytopathology 54:* 1398–1404.

19. Hirst, J. M. and O. J. Stedman. 1960. The epidemiology of *Phytophthora infestans*. II. The source of inoculum. *Ann. Appl. Biol. 48:* 489–517.

20. Jeffrey, S. I. B., J. L. Jinks and M. Grindle. 1962. Intraracial variation in *Phytophothora infestans* and field resistance to potato blight. *Genetica 32:* 323–338.

21. Jinks, J. L. and M. Grindle. 1963. Changes induced by training in *Phytophthora infestans*. *Heredity 18:* 245–264.

22. Malcolmson, J. F. and W. Black. 1966. New R genes in *Solanum demissum* Lindl. and their complementary races of *Phytophthora infestans* (Mont.) de Bary. *Euphytica 15:* 199–203.

23. Mastenbroek, C. 1952. Investigations into the inheritance of the immunity from *Phytophthora infestans* deB. of *Solanum demissum* Lindl. *Euphytica 1:* 187–198.

24. Melhus, I. E. 1915. Hibernation of *Phytophthora infestans* of the Irish potato. *J. Agr. Res. 5:* 71–102.

25. Mills, W. R. 1940. *Phytophthora infestans* on tomato. *Phytopathology 30:* 830–839.

26. Murphy, P. A. 1927. The production of the resting-spores of *Phytophthora infestans* on potato tubers. *Proc. Roy. Dublin Sci. Soc., n.s. 18:* 407–412.

27. Niederhauser, J. S., J. Cervantes and L. Servín. 1954. Late blight in Mexico and its implications. *Phytopathology 44:* 406–408.

28. Niederhauser, J. S. and W. R. Mills. 1953. Resistance of *Solanum* species to *Phytophthora infestans* in Mexico. *Phytopathology 43:* 456–457.

29. Pristou, R. and M. E. Gallegly. 1954. Leaf penetration by *Phytophthora infestans*. *Phytopathology 44:* 81–86.

30. Pristou, R. and M. E. Gallegly. 1956. Differential reactions of potato hosts to foreign and domestic potato physiologic races of *Phytophthora infestans*. *Am. Potato J. 10:* 287–295.

31. Reddick, D. and W. Mills. 1938. Building up virulence in *Phytophthora infestans*. *Am. Potato J. 15:* 29–34.

32. Reddick, D. and L. C. Peterson. 1945. Empire. A blight resistant variety. *Am. Potato J. 22:* 357–362.

33. Reddick, D. and L. C. Peterson. 1950. Additional blight-resistant varieties. *Am. Potato J. 27:* 1–10.

34. Romero, S. and D. C. Erwin. 1967. Genetic recombination in germinated oospores of *Phytophthora infestans*. *Nature 215:* 1393–1394.

35. Romero, S. and D. C. Erwin. 1969. Variation in pathogenicity among single-oospore cultures of *Phytophthora infestans*. *Phytopathology 59:* 1310–1317.

36. Romero, S. and M. E. Gallegly. 1963. Oogonium germination in *Phytophthora infestans*. *Phytopathology 53:* 899–903.

37. Smoot, J. J., F. J. Gough, H. A. Lamey, J. J. Eichenmuller and M. E. Gallegly. 1958. Production and germination of oospores of *Phytophthora infestans*. *Phytopathology 48:* 165–171.

38. Thurston, H. D., L. E. Heidrick and J. Guzmán-Naranjo. 1962. Partial resistance to *Phytophthora infestans* (Mont.) de Bary within the Colección Central Colombiana. *Am. Potato J. 39:* 63–69.

39. Van Der Plank, J. E. 1966. Horizontal (polygenic) and vertical (oligogenic) resistance against blight. *Am. Potato J. 43:* 43–52.

40. Ward, H. M. 1887. Illustrations of the structures and life history of *Phytophthora infestans*, the fungus causing the potato disease. *Quart. J. Microscop. Soc. 27:* 413–425.

Pink rot

The first report and description of the causal agent of pink rot disease of potato came from Ireland in 1913, although the disease had been observed 4 years earlier. Pink rot was found throughout the world in the next 35 years, but losses have been localized and irregular in occurrence.

Affected plants in late season exhibit a progressive wilt characterized by yellowing of lower leaves that later dry and fall. Dark colored, water-soaked stem lesions develop below soil level and may extend upwards beyond lower leaves. A wet, basal, stem rot with water-soaking of the pith and brownish discoloration of vascular tissues typically develops thereafter in combination with production of aerial tubers in leaf axils.

Tubers usually are infected via the stolon by a rot that rapidly progresses throughout as the surface of the infected area turns dark brown. The margin between healthy and diseased tissues is evident as a dark band, while lenticels and eyes often assume purple or black coloration. The skin is easily removed by rubbing the surface of diseased portions of tubers to expose slightly darkened tissues that shrink little, although they may become somewhat spongy. Pressure exerted on diseased tubers causes droplets of exudate to form at eyes and lenticels. Upon cutting infected tubers, healthy tissues cannot be differentiated from diseased tissues immediately, but, upon being exposed to air, the diseased tissues turn first pink and later brown to black in minutes.

Sporangia of *Phytophthora erythroseptica* Pethyb. measure 26 × 43 μm with oogonia 30–35 μm in diameter and oospores that nearly fill the oogonia. Optimum and maximum temperatures for growth *in vitro* have been determined to be approximately 24–28 C and about 34 C, respectively.

Pink rot occurs over the range of temperatures from 10–34 C but develops most rapidly at temperatures optimum for growth in culture, i.e. about 25 C. Another tuber rot with similar symptoms, but caused by *P. drechsleri* Tucker, is favored by warmer temperatures near 30 C. Excessive soil moisture enhances disease development caused by both organisms.

Roguing of infected plants and rotation of potato with other crop plants have been recommended for control, but cultivation of plants and storage of tubers under conditions that exclude excessive moisture are essential.

1. Cairns, H. and A. E. Muskett. 1933. Pink rot of the potato. *Ann. Appl. Biol. 20:* 381–403.

2. Cairns, H. and A. E. Muskett. 1939. *Phytophthora erythroseptica* Pethybr. in relation to its environment. *Ann. Appl. Biol. 26:* 470–480.

3. Fernández-Northcote, E. N., E. Huapaya R. and H. Torres M. 1972. La podredumbre rosada de la papa; búsqueda de fuentes de resistencia en tubérculos. *Fitopatologia 6:* 18–27.

4. French, E. R. et al. 1972. Enfermedades de la papa en el Perú. *Estac. Expt. Agr. La Molina Bol. Tec. 77.* 36 p.

5. Goss, R. W. 1949. Pink rot of potatoes caused by *Phytophthora erythroseptica* Pethyb. *Nebraska Agr. Expt. Sta. Res. Bull. 160.* 27 p.

6. Jones, W. 1945. Pink rot disease of potatoes in British Columbia. *Sci. Agr. 25:* 597–600.

7. McKay, R., J. B. Loughnane and T. Kavanagh. 1957. Haustoria of *Phytophthora erythroseptica* in potato tubers affected with pink rot. *Trans. Brit. Mycol. Soc. 40:* 407–408.

8. Pethybridge, G. H. 1913. On the rotting of potato tubers by a new species of *Phytophthora* having a method of sexual reproduction hitherto undescribed. *Sci. Proc. Roy. Dublin Soc. 13:* 529–565.

9. Pethybridge, G. H. 1914. Further observations on *Phytophthora erythroseptica* Pethyb. and on the disease produced by it in the potato plant. *Sci. Proc. Roy. Dublin Soc. 14:* 179–198.

10. Vargas, L. A. and L. W. Nielsen. 1972. *Phytophthora erythroseptica* in Peru: its identification and pathogenesis. *Am. Potato J. 49:* 309–320.

Powdery mildew(s)

Powdery mildew infections of potato that consist of scant mycelial growth and white powdery sporulation on both leaf surfaces and result in necrotic, brown, irregular lesions have been recorded from Peru, Israel and the United States, but identity of the causal fungus has not been established with certainty in all instances. Cleistothecia (90–135 μm in diameter and mostly globose) of *Erysiphe cichoracearum* DC are characterized by numerous, simple, flexuous external appendages and contain asci 25–50 × 60–90 μm, each with 2 ascospores 12–18 × 20–30 μm. Ellipsoid conidia 14–26 × 25–45 μm have been observed in the United States. Infections elsewhere have been attributed to the imperfect (*Oidiopsis*) stage of *Leveillula taurica* (cylindrical conidia 14–20 × 29–95 μm). Foliar applications of Karathane or materials of similar chemical nature should provide effective control.

1. Bazán de Segura, C. 1975. *Enfermedades de cultivos frutícolas y hortícolas.* Lima, Peru. 276 p.

2. Chorin, M. 1946. The powdery mildews of potatoes in Palestine. *Palestine J. Bot. Hort. Sci., Ser. R, 5:* 259–261.

3. French, E. R. et al. 1972. Enfermedades de la papa en el Perú. *Estac. Expt. Agr. La Molina Bol. Tec.* 77. 36 p.

4. Kapoor, J. N. 1967. *Erysiphe cichoracearum. Commonwealth Mycol. Inst. Descriptions Pathogenic Fungi Bacteria 152.* 2 p.

5. Kunkel, L. O. 1936. Powdery mildew of potato in New Jersey. *Phytopathology 26:* 392–393.

6. Menzies, J. D. 1950. *Erysiphe cichoracearum* DC as a parasite of potatoes. *Plant Dis. Reptr. 34:* 140–141.

7. Mukerji, K. G. 1968. *Leveillula taurica. Commonwealth Mycol. Inst. Descriptions Pathogenic Fungi Bacteria 182.* 2 p.

8. Thomas, D. G. 1946. Powdery mildew of potato. *Nature 158:* 417–418.

9. Valleau, W. D. 1941. Powdery mildew of potato in Kentucky. *Phytopathology 31:* 357–358.

Sclerotinia disease (*Sclerotiniose*)

Potato is infrequently attacked by *Sclerotinia sclerotiorum* (see Bean—*Sclerotinia* wilt), and the first report of such infection came from Ireland in 1910. Subsequent incidence of the disease has been reported from many parts of the world, and losses have ranged from negligible to 70 percent in some fields in Florida (United States) in 1934.

The stem is typically invaded at the soil line and results in a water-soaked lesion that extends linearly as the epidermis is darkened and killed with subsequent external development of white mycelium. Upward progress of the infection causes wilting of the plant prior to complete collapse and death. Black sclerotia of irregular shape and to 2.5 cm in length develop in and on diseased plant parts, but tubers are not usually infected. The sclerotia in soil and plant debris serve to perpetuate the fungus from one season to the next and germinate under cool, humid conditions to produce brown, stipitate apothecia to 10 mm in diameter. The ascospores, which are wind-borne from apothecia, can also cause infection of potato foliage.

No completely satisfactory control procedure has been devised, but sanitation practices to eliminate diseased plant debris and rotation with crop plants not susceptible to infection, e.g. *Gramineae*, have been recommended.

1. Eddins, A. H. 1937. *Sclerotinia* rot of Irish potatoes. *Phytopathology 27:* 100–103.

2. French, E. R. et al. 1972. Enfermedades de la papa en el Perú. *Estac. Expt. Agr. La Molina Bol. Tec.* 77. 36 p.

3. Henson, L. and W. D. Valleau. 1940. The production of apothecia of *Sclerotinia sclerotiorum* and *S. trifoliorum* in culture. *Phytopathology 30:* 869–873.

4. Moore, W. D. 1950. Flooding as a means of destroying the sclerotia of *Sclerotinia sclerotiorum. Phytopathology 40:* 920–927.

5. Moore, W. D., R. A. Conover and D. L. Stoddard. 1949. The sclerotiniose disease of vegetable crops in Florida. *Florida Agr. Expt. Sta. Bull. 457.* 20 p.

6. Niederhauser, J. S. 1946. *Sclerotinia sclerotiorum* on potatoes in Long Island. *Plant Dis. Reptr. 30:* 270.

7. Ramsey, G. B. 1941. *Botrytis* and *Sclerotinia* as potato tuber pathogens. *Phytopathology 31:* 439–448.

Rusts

Two rust diseases have been found on potato at elevations of 2500 m or above in Central and South America. *Puccinia pittieriana* P. Hennings (2-celled teliospores 17–25 × 23–38 μm, rounded on both ends, and hyaline, ovoid basidiospores 8–10 × 11–13 μm), a microcyclic rust fungus, was first reported from Costa Rica in 1903–1904 and subsequently from Mexico, Colombia, Ecuador and Peru. Occurrence of the disease is restricted to elevations of 2750–3350 m, where widely differing crop losses have been reported.

Teliospores germinate over a range of temperatures from 4–20 C with optimum germination at 8–12 C and first production of (4) basidiospores (from each teliospore) after 3 hours. Basidiospores may be air-borne, and the fungus can persist on potato in continuous cultivation or on other solanaceous plants including tomato, but the infections cause no deformation of plant parts (Fig. 79). Differences in susceptibility of potato cultivars have been noted, but maneb has been used successfully for rust control at higher altitudes in Colombia.

A second rust fungus on potato, *Aecidium cantensis* Arth., has been known to occur in Peru and Honduras since 1931 at altitudes of 2500–2750 m. Only the aecial stage of the fungus has been observed. It occurs on all

Fig. 79. Rust lesions (*Puccinia pittieriana*) on the underside of potato leaflets. (Courtesy of H. D. Thurston, Cornell University, Ithaca.)

foliar parts of the plant and causes orange-colored lesions on the undersides of leaves and elongated orange lesions on stems and petioles with considerable deformation of affected plant parts. The fungus is also infectious on tomato and ullucu (*Ullucus tuberosus* Caldas).

1. Abbott, E. V. 1929. A new disease of potato. *Vida Agr. 6:* 181.

2. Abbott, E. V. 1931. Further notes on plant diseases in Peru. *Phytopathology 21:* 1061–1071.

3. Arthur, J. C. 1920. Two destructive rusts ready to invade the United States. *Science 51:* 246–247.

4. Arthur, J. C. 1921. Origin of potato rust. *Science 53:* 228–229.

5. Arthur, J. C. 1929. Las royas de los vegetales (*Uredinales*) del Peru. *Bol. Estac. Expt. Agr. Soc. Nac. Agr. (Lima, Peru) 2.* 14 p.

6. Bazán de Segura, C. 1975. *Enfermedades de cultivos frutícolas y hortícolas.* Lima, Peru. 276 p.

7. Buritica, P. and J. Orjuela. 1968. Estudios fisiologicos de *Puccinia pittieriana* Henn., causante de la roya de la papa (*Solanum tuberosum* L.). *Fitotecnic Latinoamericana 5:* 81–88.

8. Castaño, J. J. 1952. Roya de la papa. *Agr. Trop. (Colombia) 9:* 47–48.

9. Diaz, J. R. and J. Echeverria. 1963. Chemical control of *Puccinia pittieriana* on potato in Ecuador. *Plant Dis. Reptr. 47:* 800–81.

10. Diaz, J. R. and J. Echeverria. 1963. Frecuencia de la roya de la papa, *Puccinia pittieriana* y su control quimico. *Cienc. Natur. 6:* 12–18.

11. Laundon, G. F. and A. F. Rainbow. 1971. *Puccinia pittieriana. Commonwealth Mycol. Inst. Descriptions Pathogenic Fungi Bacteria 286.* 2 p.

12. Reddick, D. 1932. Some diseases of wild potatoes in Mexico. *Phytopathology 22:* 609–612.

13. Thurston, H. D. 1973. Threatening plant diseases. *Annu. Rev. Phytopathol. 11:* 27–52.

Smut (Gangrena, Buba)

There is mild confusion about when smut was recognized as a distinct disease of potato, but it was described in 1943 from observations made in the Andean mountains of Venezuela 4 years earlier. It was suggested at that time that the disease may have been present there for many years before being distinguished from powdery scab, which also was prevalent. Since it was originally described, smut has been reported from Colombia, Ecuador, Peru, Mexico and Panama, as well as Venezuela. Recent estimates of losses have ranged to 80 percent, while in some localities, viz. the Cañete Valley south of Lima, cultivation of some varieties is effectively prohibited.

Affected tubers are misshapen and develop irregular protuberances often referred to as "warty swellings." Cuts across these diseased tissues reveal dark spots (sori) scattered throughout the cortex to a depth of a cm or more, while lighter colored spots occur throughout the entire tuber (Fig. 80). Frequently, some tubers from an infected plant may remain unaffected, but diseased tubers are soon reduced to a dry, brown, powdery mass. Symptoms are confined to tubers.

Fig. 80. Tuber symptoms of potato smut infection. (Courtesy of E. R. French, International Potato Center, Lima.)

Sori of *Thecaphora solani* Barrus are oval to irregular in cross-section, range to a mm in length and may coalesce. Sporeballs produced in sori are mostly subspherical, 12–35 × 12–48 μm and composed of 2–8 adherent, subspherical spores 8–18 × 8–20 μm with a thick, contiguous, noticeably verrucose wall. Spore germination has not been observed.

Little epidemiological information has been established, but infection usually originates from infested soil, which commonly results from infected tubers being left in the field. The disease was first found only at elevations of 2500 m or more, but, since 1954, smut has become increasingly widespread at elevations near sea level in Peru.

Crop rotation to avoid planting potato after potato is strongly recommended. Two cultivars, Mariva and Sipeña, have been effectively resistant in preliminary trials in Peru.

1. Barrus, M. F. 1944. A *Thecaphora* smut on potatoes. *Phytopathology 34:* 712–714.

2. Barrus, M. F. and A. S. Muller. 1943. An Andean disease of potato tubers. *Phytopathology 33:* 1086–1089.

3. Bazán de Segura, C. 1960. The gangrena disease of potato in Peru. *Plant. Dis. Reptr. 44:* 257.

4. Bazán de Segura, C. and A. Q. Diaz. 1955. The Andean disease of potato tubers in the Peruvian coast area. *Plant Dis. Reptr. 39:* 477.

5. Zachmann, R. and D. Baumann. 1975. *Thecaphora solani* on potatoes in Peru: present distribution and varietal resistance. *Plant Dis. Reptr. 59:* 928–931.

Rhizoctonia disease (Black scurf)

The disease of potato now commonly referred to simply as "*Rhizoctonia*" was originally described from Germany in 1858 but now occurs wherever

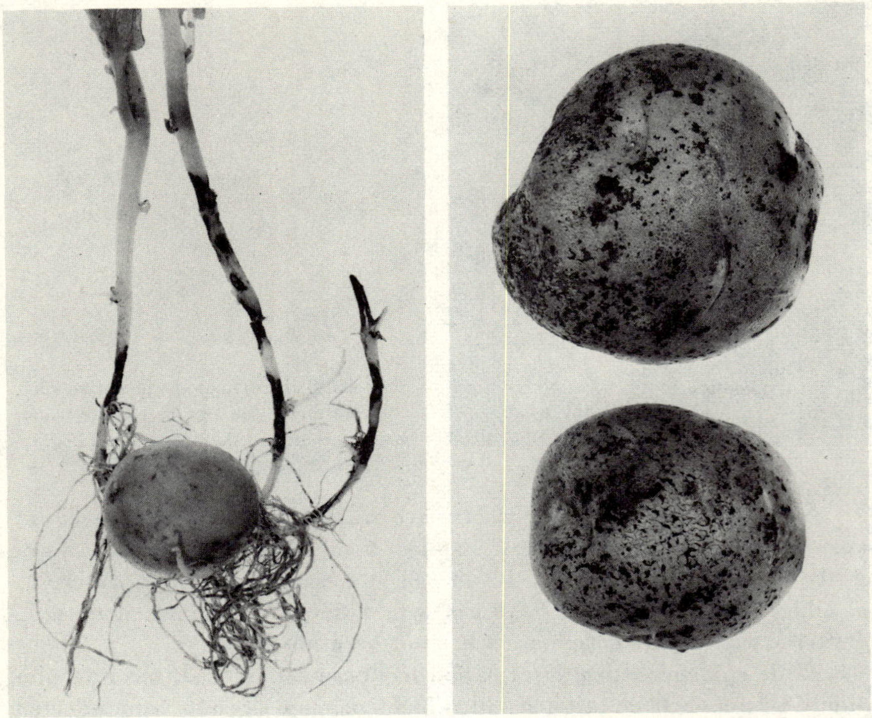

Fig. 81. *Rhizoctonia* lesions on underground potato stems (left) and (right) sclerotia on tuber surfaces ("dirt that won't wash off"). (Courtesy of H. D. Thurston, Cornell University, Ithaca.)

potatoes are grown. Missing hills are the most troublesome aspect of this disease, but incidence usually does not approach serious proportions.

Growing tips of sprouts from seed-pieces are attacked and killed, often before emerging above soil level, by a rapidly developing black, cortical rot. Repeated occurrences of similar infection and killing of sprouts from buds that germinate subsequent to killing of the previous sprouts may result in missing hills. Stems become more resistant to infection with aging, and later infections cause more restricted dark lesions and cankers but do not kill the stems (Fig. 81). Above-ground foliage may be noticeably stunted, sometimes with purple discoloration.

Flattened, black sclerotia develop closely appressed to the surfaces of tubers (black scurf symptoms are sometimes called "dirt that won't wash off") but are associated with completely superficial infections (Fig. 81).

The common name of this disease is derived from the generic name of the asexual stage of the causal fungus *Thanatephorus cucumeris* (Frank) Donk (= *Rhizoctonia solani*), which produces fragile, extramatrical growth with clavate basidia 8–11 × 12–18 μm and ellipsoid, flattened basidiospores 4–7 × 7–13 μm. Cardinal temperatures for growth of the imperfect form of the fungus *in vitro* are approximately 8, 25–30 and 35 C. Sclerotia germinate throughout most of this temperature range, but optimum temperature for stem lesion formation is near 20 C. The fungus can persist between seasons both in soil as well as sclerotia on tubers. Considerable variation among isolates has been noted *in vitro*, and the fungus does infect several other host plants, but no identifiable pathotypes have been designated.

Rotation of potato with other crop plants, even including susceptible species, has provided some reduction in disease incidence. Shallow planting of potato seed-pieces serves to reduce probability of early infection, the most susceptible period, after which soil may be pushed around stems by cultivating after they become older and more resistant. Only tubers free of infection should be used for seed.

1. Afanasiev, M. M. and H. E. Morris. 1948. Time of infection and accumulative effect of *Rhizoctonia* on successive crops of potatoes. *Am. Potato J. 25:* 17–23.

2. Blair, I. D. 1943. Behaviour of the fungus *Rhizoctonia solani* Kühn in the soil. *Ann. Appl. Biol. 30:* 118–127.

3. Edgington, L. V. and L. V. Busch. 1967. Control of *Rhizoctonia (solani)* stem canker in potato. *Can. Plant Dis. Surv. 47:* 28–29.

4. Elmer. O. H. 1942. Effect of environment on the prevalence of soil-borne *Rhizoctonia. Phytopathology 32:* 972–977.

5. El Zarka, A. M. 1965. Studies on *Rhizoctonia solani* Kühn, the cause of the black scurf disease of potato. *Wageningen Landbouwhogeschol Lab. Fytopathol. Mededel. 217.* 73 p.

6. Graham, D. C. 1964. The use of organo-tin compounds as potato tuber disinfectants, particularly against *Rhizoctonia solani. European Potato J. 7:* 33–44.

7. Hide, G. A. 1972. Effects of black scurf (*Rhizoctonia solani*) on potatoes. *Ann. Appl. Biol. 74:* 139–148.

8. LeClerg, E. L. 1941. Pathogenicity studies with isolates of *Rhizoctonia solani* obtained from potato and sugar beet. *Phytopathology 31:* 49–61.

9. Matsumoto, T. 1921. Studies in the physiology of the fungi. XII. Physiological specialization in *Rhizoctonia solani* Kühn. *Ann. Missouri Bot. Gard. 8:* 1–62.

10. Mordue, J. E. M. 1974. *Thanatephorus cucumeris. Commonwealth Mycol. Inst. Descriptions Pathogenic Fungi Bacteria 406.* 2 p.

11. Papavizas, G. C., C. B. Davey and R. S. Woodard. 1962. Comparative effectiveness of some organic amendments and fungicides in reducing activity and survival of *Rhizoctonia solani* in soil. *Can. J. Microbiol. 8:* 847–853.

12. Person, L. H. 1945. Pathogenicity of isolates of *Rhizoctonia solani* from potatoes. *Phytopathology 35:* 132–134.

13. Ramsey, G. B. 1917. A form of potato disease produced by *Rhizoctonia. J. Agr. Res. 9:* 421–426.

14. Richards, B. L. 1921. Pathogenicity of *Corticium vagum* on the potato as affected by soil temperature. *J. Agr. Res. 21:* 459–482.

15. Richards, B. L. 1923. Further studies on the pathogenicity of *Corticium vagum* on the potato as affected by soil temperature. *J. Agr. Res. 23:* 761–770.

16. Richter, H. and R. Schneider. 1953. Untersuchungen zur morphologischen und biologischen differenzierung von *Rhizoctonia solani* K. *Phytopathol. Z. 20:* 167–226.

17. Sanford, G. B. 1952. Persistence of *Rhizoctonia solani* Kühn in soil. *Can. J. Bot. 30:* 652–664.

18. Singh, M. 1965. Studies on four isolates of *Rhizoctonia solani* Kuehn, the causal agent of black scurf disease of potato in the Punjab. I. Symptoms, strains, and pathogenicity. *Indian Phytopathol. 17:* 23–26.

19. Singh, R. S. 1972. Studies on control of black scurf disease of potato (*Rhizoctonia solani*). *Indian Phytopathol. 25:* 343–349.

20. Small, T. 1945. Black scurf and stem canker of potato (*Corticium solani* Bourd. & Galz.). Field studies on the use of clean and contaminated seed potatoes and on the contamination of crop tubers. *Ann. Appl. Biol. 32:* 206–209.

21. Storey, I. F. 1941. A comparative study of strains of *Rhizoctonia solani* (Kühn) with special reference to their parasitism. *Ann. Appl. Biol. 28:* 219–228.

22. Thirumalachar, M. J. 1953. *Rhizoctonia solani* infections of potato tubers in India. *Phytopathology 43:* 645–647.

Southern blight

The fungus responsible for southern blight disease was first reported as a pathogen of tomato in Florida (United States) in 1892, but it was not named until 1911, and it was not identified as a pathogen of potato until 1915 (in India). Occurrence of the fungus in warmer climates throughout the world has been established, and there have been infrequent outbreaks in cooler, temperate climates. Losses of other crops caused by the southern blight fungus range from negligible to complete, but this disease is not usually of major consequence on potatoes.

Infection typically occurs in the portion of the stem just below soil level and prompts an exaggerated upright plant habit, followed by wilting and yellowing of leaves. A water-soaked lesion on the stem at the infection site rapidly enlarges until the stem is girdled, and white external mycelium develops on the stem and in the soil nearby. Numerous light to dark tan, nearly spherical sclerotia soon develop within the mycelial growth and often aspressed to the stem surface. Plants are killed quickly, and a band of mycelium and sclerotia with soil intermixed adheres to plants when pulled, but stems often are destroyed so completely that only a few vascular strands remain from underground stems. Infection may extend through the roots to cause small, circular, sunken tuber lesions 2–3 mm that enlarge to form a firm, wet, transparent rot. This rot precedes formation of a soft, white, rubbery rotting with formation of sclerotia in the resulting cavities caused by shrinking of affected tissues.

Round, necrotic, light brown leaf spots to 1.5 cm, often with narrow borders and visible external mycelium followed by formation of single sclerotia in the centers, have been observed on several hosts susceptible to the causal fungus.

Sclerotium rolfsii, the imperfect stage of *Botryobasidium rolfsii*, produces hyaline hyphae with clamp connections and large cells that measure 2–9 × 150–250 μm, while smaller cells measure about 2 μm in diameter. Nearly spherical sclerotia in various shades of brown measure 0.5–0.8 mm in diameter. The hymenium of the perfect stage develops as dense white mycelial masses in which develop obovoid basidia 4–5 × 7–9 μm. Each basidium bears an elliptical to pyriform, hyaline basidiospore 3.5–5 × 6–7 μm on each of 2–4 sterigmata 2.5–4 μm in length. Isolates of the fungus have been found to differ morphologically, but growth *in vitro* occurs from 10–40 C with 25–35 C optimum.

Disease development usually is enhanced by high soil moisture and temperatures that favor growth in culture, but serious losses have occurred in dry weather. Sclerotia, which are the primary means for over-seasoning, remain viable when dry for as long as 5 years but lose viability in weeks under moist conditions.

Control can be accomplished by soil fumigation with chloropicrin or methyl bromide or treatment of soil with pentachloronitrobenzene (rate depends on soil type and method of application).

1. Aycock, R. 1966. Stem rot and other diseases caused by *Sclerotium rolfsii*. *N. C. Agr. Expt. Sta. Tech. Bull. 174.* 202 p.

2. Edson, H. A. and M. Shapovalov. 1923. Parsitism of *Sclerotium rolfsii* on Irish potatoes. *J. Agr. Res. 23:* 41–46.

3. Higgins, B. B. 1927. Physiology and parasitism of *Sclerotium rolfsii* Sacc. *Phytopathology 17:* 417–448.

4. Kulkarni, N. B. and L. Ahmed. 1965. Studies on the basidial formation by *Sclerotium rolfsii* Sacc. V. Basidial stage of *S. rolfsii* isolate from potato on a new medium containing organic nitrogen compound. *Sydowia 19:* 162–164.

5. Kulkarni, N. B. and L. Ahmed. 1966. Studies on the basidial formation by *Sclerotium rolfsii* Sacc. III. Perfect stage of *S. rolfsii* isolates from potato, colocasia and groundnut on a new medium. *Mycopathol. Mycol. Appl. 28:* 107–109.

6. Mordue, J. E. M. 1974. *Corticium rolfsii. Commonwealth Mycol. Inst. Descriptions Pathogenic Fungi Bacteria 410.* 2 p.

7. Muller, A. S. 1940. La marchitez y podredumbre de papas causada por *Sclerotium rolfsii* Sacc. *Agr. Venezolano 4:* 19–20.

8. Weber, G. F. 1943. Southern blight, *Corticium rolfsii,* of potato tubers. *Phytopathology 33:* 615–617.

Black dot (Anthracnose, Dartrose, Punteado negro)

A number of similar, if not identical, potato disease syndromes were reported from throughout the world beginning about 1825, but description of the causal agent for black dot dates from 1850. The disease is usually of minor economic import.

Affected plants usually appear after midseason and develop yellowing at the tips of leaflets that gradually extends to involve entire leaflets, which also may become rolled. Chlorosis of the foliage becomes general, and diseased plants wilt and dry 2 weeks or more before neighboring healthy plants. Earlier infection may result in some dwarfing.

Stem lesions that develop from near the soil-line to the seed-piece consist mostly of cortical infections, which cause the affected tissues to be easily removed upon drying to reveal purple discoloration in the vascular cylinder. Sclerotia develop in vascular tissues as well as on cortical remnants (Fig. 82). Stolons and roots are affected in like manner, although lesions on stolons frequently involve all tissues. Fragments of dried stolons frequently remain attached at harvest to tubers which may show no signs of infection other than reduction in size or may have varying numbers of sclerotia on the surfaces.

Conidiophores of *Colletotrichum coccodes* (Wallr.) Hughes are cylindrical, 10–30 μm in length and range to 3-septate and produce hyaline conidia (pink *en masse*) 3–8 × 18–22 μm. Tapered setae are dark and measure 5–7 × 80–350 μm, while sclerotia range from 100–500 μm in diameter. The fungus is a weak pathogen that typically attacks plants in a debilitated condition or under environmental conditions unfavorable for potato culture. Sclerotia may remain viable in debris or soil for more than a year.

Field sanitation and use of healthy seed combined with crop rotation (excluding tomato) and other recommended cultivation practices are the only possibilities for control at present.

1. Cheal, W. F. 1928. The black dot fungus of potatoes. *Gard. Chron. 84:* 508–509.

2. Dickson, B. T. 1926. The "black dot" disease of potato. *Phytopathology 16:* 23–40.

3. French, E. R. et al. 1972. Enfermedades de la papa en el Perú. *Estac. Expt. Agr. La Molina Bol. Tec.* 77. 36 p.

4. Griffiths, D. A. and W. P. Campbell. 1972. Fine structure of conidial development in *Colletotrichum atramentarium. Trans. Brit. Mycol. Soc. 59:* 483–489.

5. Mordue, J. E. M. 1967. *Colletotrichum coccodes. Commonwealth Mycol. Inst. Descriptions Pathogenic Fungi Bacteria 131.* 2 p.

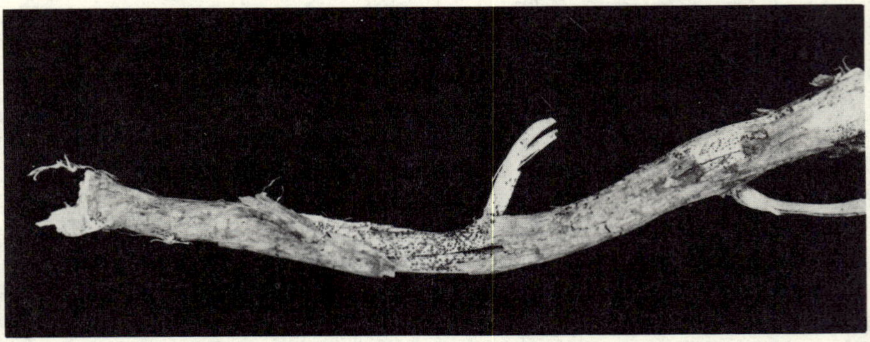

Fig. 82. Sporulation of the black dot (anthracnose) fungus on old potato stem. (Courtesy of E. R. French, International Potato Center, Lima.)

6. Stevenson, W. R., R. J. Green and G. B. Bergeson. 1976. Occurrence and control of potato black dot root rot in Indiana. *Plant Dis. Reptr. 60:* 248–251.

7. Thirumalachar, M. J. 1967. Pathogenicity of *Colletotrichum atramentarium* on some potato varieties. *Am. Potato J. 44:* 241–244.

Charcoal rot

The fungus that causes charcoal rot of potato, *Macrophomina phaseolina*, is also infectious to numerous other crop plants (see Bean—Charcoal rot) and is widespread in tropical and subtropical environments, although serious losses occur occasionally in temperate climates. Persistence of the fungus in plant debris and soil predisposes seed-pieces to infection, which progresses into the lower stem and tubers. A dark, cortical rot of the lower stem, roots and stolons is typical, with abundant production of small, black sclerotia. Tuber infection, often localized near the stolon end, also is largely superficial even though blackening may extend inward for some distance. High temperatures and excessive soil moisture increase incidence of charcoal rot. No completely satisfactory control has been devised, and only adherence to recommended cultural practices can be advised.

1. Ashby, S. C. 1927. *Macrophomina phaseoli* (Maubl.) comb. nov. the pycnidial stage of *Rhizoctonia bataticola* (Taub.) Butl. *Trans. Brit. Mycol. Soc. 12:* 141–147.

2. Boewe, G. H. 1942. Charcoal rot on potatoes in Illinois. *Plant Dis. Reptr. 26:* 142–143.

3. Dutt, B. L. and S. Raj. 1970. Potato charcoal rot and its control in Punjab. *Indian Farming 20:* 31–32.

4. Holliday, P. and E. Punithalingam. 1970. *Macrophomina phaseolina. Commonwealth Mycol. Inst. Descriptions Pathogenic Fungi Bacteria 275.* 2 p.

5. Luttrell, E. 1946. Pycnidial strain of *Macrophomina phaseoli. Phytopathology 36:* 978–980.

6. Vinot, M. and P. Bernaux. 1948. La pourriture charbonneuse de la pomme de terre dans la region méditerranéen. *Ann. Epiphyt. Phytogenitique 14:* 91–102.

7. Watson, R. D. 1944. Charcoal rot of Irish potatoes. *Phytopathology 34:* 433–435.

Early blight

Description of the causal agent of early blight appeared (1882) before the foliar disease was distinguished in the United States in 1891. Tuber symptoms were not recognized until 1925. The disease occurs wherever potatoes are grown and can cause serious defoliation as well as loss of tubers.

Foliar symptoms develop first on lower leaves as dark colored spots that are circular or angular in shape and soon become necrotic. Lesions range to 4 mm in diameter and are encircled by narrow, chlorotic halos but characteristically include concentric ridges in a "target board" pattern (Fig.

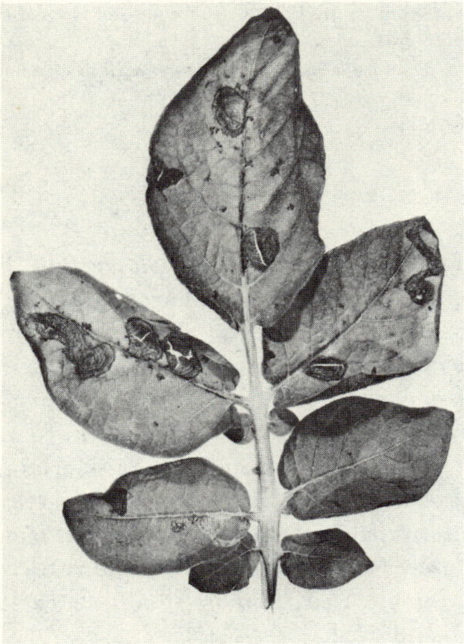

Fig. 83. Early blight symptoms on potato leaflets. Note concentric zonation ("target board appearance") in lesions. (Courtesy of H. D. Thurston, Cornell University, Ithaca.)

83). Infection progresses upward to involve the entire plant, and leaves seriously affected turn senescent and drop. Defoliation can be severe. Lesions do develop infrequently on stems.

Tuber infections appear as dark, circular to irregular lesions, sometimes slightly raised, to 2 cm in diameter, with margins well demarcated from healthy tissue. Penetration is limited to about 5 mm, but the diseased tissue assumes the brown, corky texture of a dry rot making the margins of lesions easily distinguished.

Conidia of *Alternaria solani* (Ell. & Martin) L. R. Jones & Grout measure 12–20 × 120–296 μm, are dark, muriform, beaked and may be borne in chains. The fungus is rather variable in culture and cardinal temperatures for growth *in vitro* are about 2, 27, and 45 C.

The conidia are easily detached and disseminated by wind and germinate quickly in free water. Penetration of leaf surfaces occurs directly, and sporulation can occur in the resulting lesions within 3 days. The fungus can persist in debris for more than a year and is also infectious on a number of other crop plants, including tomato.

Field sanitation and rotation with crops not susceptible to infection will reduce the probability of serious loss. Foliar applications of fungicides, including Dithane M-45, Maneb (80 percent), Manzate 200 or Polyram (80 percent), have been recommended for control.

1. Bonde, R. 1929. Physiological strains of *Alternaria solani*. *Phytopathology 19:* 533–548.

2. Braun, H. 1935. *Alternaria solani* als parasit der kartoffelknolle. *Nachrichtenbl. Deut. Pflanzenschutzdienst 15:* 10–111.

3. Folsom, D. and R. Bonde. 1925. *Alternaria solani* as a cause of tuber rot in potatoes. *Phytopathology 15:* 282–286.

4. Harrison, M. D., C. H. Livingston and N. Oshima. 1965. Epidemiology of potato early blight in Colorado. I. Initial infection, disease development and the influence of environmental factors. *Am. Potato J. 42:* 279–291.

5. Harrison, M. D., C. H. Livingston and N. Oshima. 1965. Control of potato early blight in Colorado. I. Fungicial spray schedules in relation to the epidemiology of the disease. *Am. Potato J. 42:* 319–327.

6. Henning, R. G. and L. J. Alexander. 1952. Evidences of existence of physiologic races of *Alternaria solani* (Ell. and Mart.) Jones and Grout. *Phytopathology 42:* 467.

7. Jones, L. R. 1892. A new potato disease. *Vermont Agr. Expt. Sta. Ann. Rep. 5:* 131–132.

8. Jones, L. R. 1893. The new potato disease or early blight. *Vermont Agr. Expt. Sta. Ann. Rep. 6:* 66–70.

9. Jones, L. R. 1896. Various forms of potato blight and their causes; studies upon *Macrosporium solani* E. & M. *Vermont Agr. Expt. Sta. Ann. Rep. 9:* 72–88.

10. Klaus, H. 1940. Untersuchungen über *Alternaria solani* Jones et Grout, inbesondere über seine pathogenitat an kartoffelknollen in abhängigkeit von den ausserfaktoren. *Phytopathol. Z. 13:* 126–195.

11. Mathur, R. S., B. K. Singh and M. S. Nagarkoti. 1971. Control of early blight of potato with fungicides. *Indian Phytopathol. 24:* 58–62.

12. Rotem, J. and S. Feldman. 1965. The relation between the ratio of yield to foliage and the incidence of early blight in potato and tomato. *Israeli J. Agr. Res. 15:* 115–122.

13. Stall, R. E. 1957. Numbers of nuclei in cells of vegetative structures of *Alternaria solani.* Phytopathology 47: 33–34.

14. Whetzel, H. H. 1923. The *Alternaria* blight of potatoes in Bermuda. *Phytopathology 13:* 100–103.

Kasahui [*Ulocladium consortiale* (Thüm.) Simmons]

Small, dark leaf spots more conspicuous on the lower surfaces of leaflets have been observed in the higher elevations of Peru. Foliar application of fungicides for late blight have been effective for control of kasahui.

1. Bazán de Segura, C. 1975. *Enfermedades de cultivos fruticolas y hortícolas.* Lima, Peru. 276 p.

2. French, E. R. et al. 1972. Enfermedades de la papa en el Perú. *Estac. Expt. Agr. La Molina Bol. Tec. 77.* 36 p.

Leptosphaerulina leaf spot (*Leptosphaerulina* sp.)

1. Vessey, J. C. 1975. Some potato diseases of the Peruvian humid tropics. *Plant Dis. Reptr. 59:* 1004–1007.

Septoria leaf spot (Mancha anular)

A leaf spot disease of potato caused by a "*Septoria* sp." was described from the Andes of Venezuela in 1968 after having been observed 2 years earlier in Central America. The disease is not a major problem in potato production except under specific environmental conditions, when some yield loss can result from excessive defoliation.

Symptoms are restricted to foliage where mostly circular lesions with dark margins and light centers contain dark, concentric circles. Dark colored pycnidia frequently develop in leaf lesions that may reach 4.5 mm in diameter, while lesions on petioles and stems are elongated. Severe infection typically results in extensive defoliation.

The disease is caused by *Septoria lycopersici* Speg. that produces dark pycnidia 100–150 × 90–230 μm. Pycnidiospores are hyaline, filiform and measure 2 × 52–95 μm with 2–6 septations. Maximum increase in colony diameter of cultures *in vitro* occurs at 20 C, while the fungus grows within a temperature range of about 5–28 C. The pathotype of the fungus that causes typical symptoms on tomato causes only restricted, non-sporulating lesions on potato, which has prompted designation of the potato pathogen as *S. lycopersici* strain A (for Andean). The pathotype from potato also causes typical symptoms on tomato.

Leaf spot on potato occurs only under cool, humid weather conditions like those found above 2000 m elevation. Sporulation occurred only in laboratory cultures exposed to light and not in cultures placed in the dark. Pycnidiospores are extruded from pycnidia in cirri and dispersed by splashing rain prior to germination and indirect infection.

Control can be accomplished by planting to avoid cool temperature conditions accompanied by high humidity as fog, mist or rain. Foliar application of copper or dithiocarbamate fungicides should be beneficial. Some difference in susceptibility of varieties has been noted.

1. Ciccarone, A. T. and G. Malaguti. 1966. Prime osservazioni su una septoriosi della patata (*Solanum tuberosum* L.) osservata nell' America centrale. *Israel J. Bot. 15:* 138–143.

2. Ciccarone, A. T. and G. Malaguti. 1968. Una interesante septoriosis de la papa en los Andes Venezolanos. *Fitotec. Latinoamer.* 5: 51–54.

3. Endrinal, D. M. and M. S. Celino. 1940. *Septoria* leaf spot of tomato. *Phillipine Agr. 29:* 593–610.

4. Harris, H. A. 1935. Morphologic studies of *Septoria lycopersici*. *Phytopathology 25:* 790–799.

5. Jimenez, A. T. and E. R. French. 1972. Mancha anular foliar (*Septoria licopersici* subgroup A) de la papa. *Rev. Fitopatol. 5:* 15–20.

6. MacNeill, B. H. 1950. Studies in *Septoria lycopersici* Speg. *Can. J. Res. 45:* 515–516.

7. Sutton, B. C. and J. M. Waterston. 1966. *Septoria lycopersici* Speg. *Commonwealth Mycol. Inst. Descriptions Pathogenic Fungi Bacteria 89.* 2 p.

8. Torres, H., E. R. French and L. W. Nielsen. 1970. Potato diseases in Peru, 1965–68. *Plant Dis. Reptr. 54:* 315–318.

Silver scurf

A disease of new potatoes in Vienna (Austria) in 1871 was identified as silver scurf, but, while widespread in Europe, it was not found in the United States until nearly 40 years later, at which time it also was reported from Australia. Damage is confined originally to disfigurement of tubers, but subsequent deterioration in storage can result in loss of salable produce.

Tuber infection results in superficial lesions often circular, variable in diameter, and localized, but sometimes affected areas coalesce to cover a sizeable portion of the tuber (Fig. 84). Affected tissues are brownish in color first but may appear glazed or silvery when moist. Coloration in red-skinned varieties may completely disappear. Lesions darken with age, and the surfaces may slough, causing tubers so affected to shrivel and wrinkle from excessive dehydration. Infection is confined to the tubers, which are susceptible in all stages of development.

Silver scurf is caused by *Spondylocladium atrovirens* Harz., which produces dark, multiseptate conidiophores 7–8 × 150–375 μm that bear multiseptate (2–9) conidia variable in size (6–12 × 22–56 μm, rounded at the base and pointed at the apex) in indistinct whorls. Germination of conidia occurs at 98 percent relative humidity or above, while cardinal temperatures

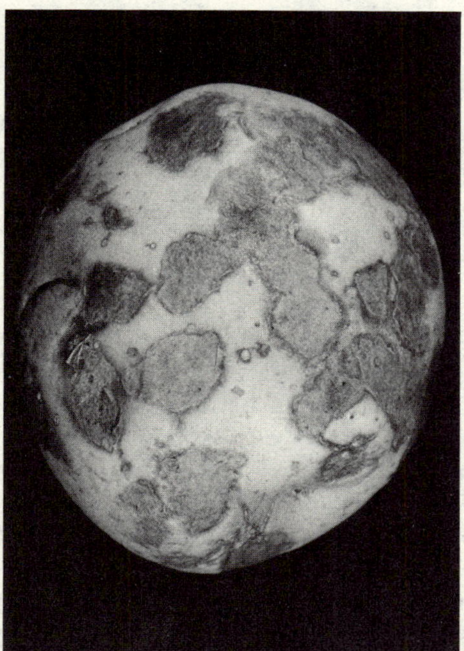

Fig. 84. Silver scurf lesions on potato tuber. (Courtesy of H. D. Thurston, Cornell University, Ithaca.)

for growth *in vitro* are about 6, 24 and 28 C. However, conidia and some mycelium can survive temperatures below freezing.

Infection of tubers in the field results from contact of developing tubers with the diseased seed-piece, but infection can occur in storage as well under favorable conditions of moisture and temperature. Dissemination of the disease occurs largely through infected tubers, although the fungus is capable of surviving adverse environmental conditions.

Varietal differences in susceptibility have been noted but offer no promise for control at present. Healthy tubers to be used for seed the next season should be selected at harvest and stored apart from other stocks at temperature and humidity levels as low as is consistent with maintenance of good quality.

1. Burke, O. D. 1938. The silver scurf disease of potatoes. *N. Y. (Cornell) Agr. Expt. Sta. Bull. 692.* 30 p.

2. Kramer, M. 1942. A "sarna prateada" da batatinha. *Biológico 8:* 83–86.

3. Melhus, I. E. 1913. Silver scurf, a disease of the potato. *U. S. Dept. Agr. Bur. Plant Ind. Circ. 127:* 15–24.

4. Santerre, J. 1961. La tache argentée dans la province du Québec. *Can. Plant Dis. Surv. 41:* 321–324.

5. Santerre, J. 1966. Absence apparente de l'organisme de la tache argentée des pommes de terre, *Helminthosporium atrovirens*, dans les sols nouvellement défrichés. *Can. J. Plant Sci. 46:* 647–652.

6. Santerre, J. 1967. Evaluation de l'importance relative du sol et de la semence comme source d'inoculum de la tache argentée des pommes de terre au moyen d'essais de répression. *Can. J. Plant Sci. 47:* 695–702.

7. Santerre, J. 1969. La semence infectée source d'inoculum de la tache argentée dans les sols affectés à la culture des pommes de terre. *Can. J. Plant Sci. 49:* 83–86.

8. Santerre, J. 1972. Nouvelles etudes sur la transmission de la tache argentée chez la pomme de terre par des semences contaminees. *Can. J. Plant Sci. 52:* 625–632.

9. Schultz, E. S. 1916. Silver-scurf of the Irish potato caused by *Spondylocladium atrovirens*. *J. Agr. Res. 6:* 339–350.

10. Taubenhaus, J. J. 1916. A contribution to our knowledge of silver scurf of the white potato. *N. Y. Bot. Gard. Mem. 6:* 549–560.

Skin spot

The skin spot disease was first named in Ireland in 1915, although the condition may have been recognized a few years earlier. It is not usually considered a major disease of potato, and incidence in tubers harvested for table use is not considered of serious consequence. The disease has been responsible, however, for reductions in stand to 30 percent after planting infected tubers.

Infections on the tuber surface may result either in small raised protuberances or, in other varieties, small, dark, sunken, circular spots with slightly raised centers. Neither symptom is more than 1 mm in diameter or 2 mm in depth, but several lesioned areas may coalesce to affect a sizeable portion of the tuber surface. The underground parts of potato plants, including stem, stolons and adventitious roots also may be infected and result in small, discrete, light brown spots that may coalesce to form large, brown lesions. Small, transverse cracks in the affected areas are followed by development of deep, wide, longitudinal cracks, but infection does not extend beyond the cortex. There are no foliar symptoms.

Skin spot is caused by *Oospora pustulans* Owen & Wakefield that produces septate hyphae 5–7 μm in diameter and hyaline, cylindrical to oblong, uniseptate conidia 2–2.5 × 6–12 μm in chains on erect conidiophores. Tomato also is susceptible to infection.

Inoculum for current season disease development in the field originates largely from infections on seed tubers. The humid conditions of soil after planting enhance sporulation and infection of underground plant parts, but tubers, which may be infected through lenticels and "eye" buds, are not infected until near maturity and harvest. Incidence of tuber infection has been correlated with high rainfall, when tubers begin to mature followed by cool temperatures during harvest. Infections in seed tubers may not become apparent until during storage.

No completely satisfactory control has been devised, although tuber treatment immediately after harvest was better than treatment just prior to planting in controlled experiments. Only healthy seed with no evidence of skin spot infection should be planted in well-drained soils.

1. Allen, J. D. 1957. The development of potato skin-spot disease. *Ann. Appl. Biol. 45:* 293–298.

2. Boyd, A. E. W. 1957. Field experiments on potato skin-spot disease caused by *Oospora pustulans* Owen & Wakef. *Ann. Appl. Biol. 45:* 283–292.

3. Boyd, A. E. W. and J. H. Lennard. 1961. Some effects of potato skin spot (*Oospora pustulans*) in Scotland. *European Potato J. 4:* 361–377.

4. Boyd, A. E. W. and J. H. Lennard. 1962. Seasonal fluctuation in potato skin spot. *Plant Pathol. 11:* 161–166.

5. Edie, H. H. and A. E. W. Boyd. 1966. The effect of delay in treatment on the control of potato skin spot (*Oospora pustulans*). *European Potato J. 9:* 216–225.

6. Greeves, T. H. and A. E. Muskett. 1939. Skin spot (*Oospora pustulans* Owens & Wakef.) of the potato, and its control by tuber disinfection. *Ann. Appl. Biol. 26:* 481–496.

7. Hide, G. A., J. M. Hirst and E. J. Mundy. The phenology of skin spot (*Oospora pustulans* Owen & Wakef.) and other fungal diseases of potato tubers. *Ann. Appl. Biol. 64:* 265–279.

8. Hide, G. A., J. M. Hirst and O. J. Stedman. 1973. Effects of skin spot (*Oospora pustulans*) of potatoes. *Ann. Appl. Biol. 73:* 151–162.

9. Hirst, J. M. and G. A. Salt. 1959. *Oospora pustulans* Owen & Wakefield as a parasite of potato root systems. *Trans. Brit. Mycol. Soc. 42:* 59–66.

10. Nagdy, G. A. and A. E. W. Boyd. 1965. Susceptibility of potato varieties to skin spot (*Oospora pustulans*) in relation to the structure of the skin and eye. *European Potato J. 8:* 200–214.

11. Owen, M. N. 1919. The skin spot disease of potato tubers. *Kew Bull. 8:* 289–301.

Verticillium wilt (Early maturity disease)

A wilt disease of potato was reported from Germany in 1879 and the causal organism identified and described. The disease became widespread throughout that country and northern United States in the first 15 years of the twentieth century but assumed epiphytotic proportions in North America about 35 years later when losses of 25–50 percent were often reported. *Verticillium* wilt has been reported more recently from New Zealand, Australia and Peru but probably occurs wherever potatoes are grown.

Lower leaves on affected plants are the first to wilt, usually after midseason, but this condition gradually progresses upward to involve all the

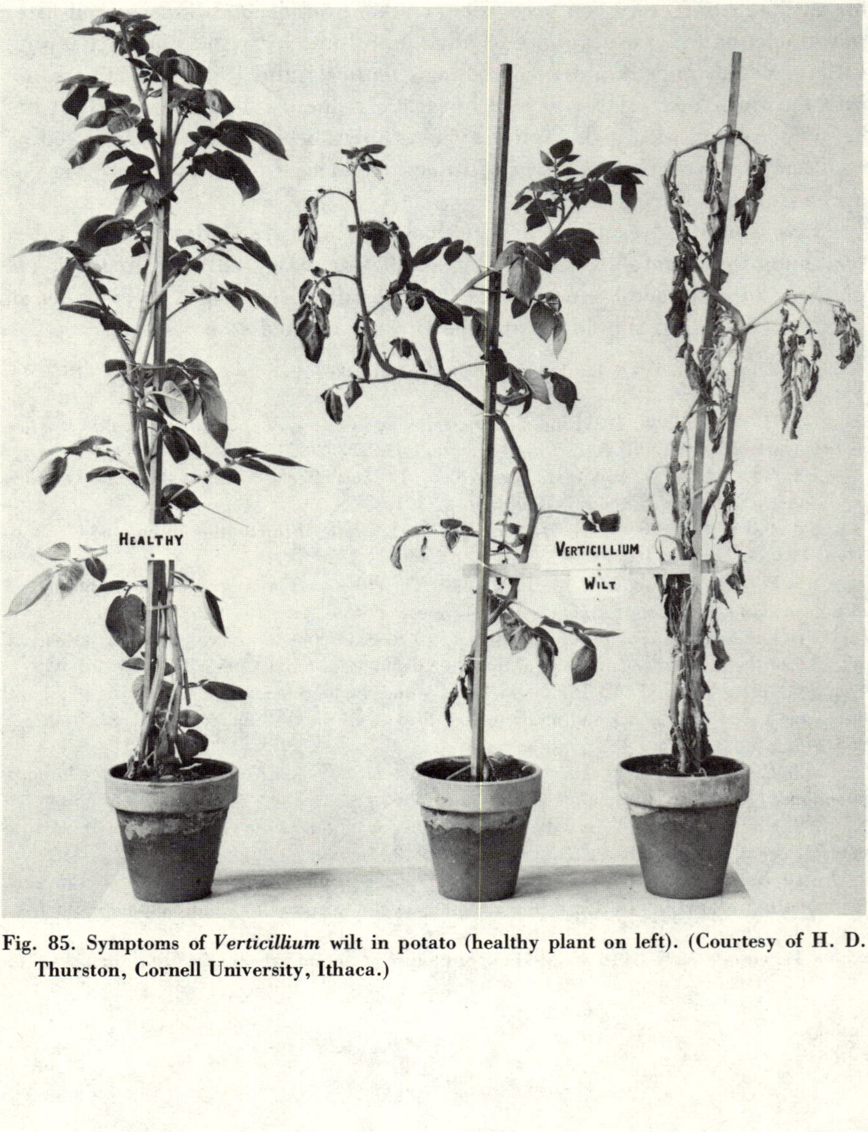

Fig. 85. Symptoms of *Verticillium* wilt in potato (healthy plant on left). (Courtesy of H. D. Thurston, Cornell University, Ithaca.)

leaves on the stalk. The coloring of affected leaflets fades through general chlorosis to necrosis, and death of the stalk results (Fig. 85). Frequently, only a single stalk is affected with no obvious wilting of leaflets before they become generally clorotic, while in other cases wilting may appear suddenly, and the stalk or the whole plant is killed quickly.

Red-brown vascular discoloration in the lower portions of affected stems is a characteristic symptom that often extends through the stolons into the vascular ring of developing tubers. However, vascular discoloration in tubers is not a diagnostic symptom of *Verticillium* wilt infection.

Verticillium albo-atrum, the causal fungus, is characterized by verticillate conidiophores (2–4 phialides 1–3 × 20–30 μm at each node) and unicellular, subcylindrical, hyaline conidia 2–4 × 4–11 μm. Morphological variants have been distinguished by type of mycelium produced *in vitro* and presence or absence of pseudosclerotia. Production of microsclerotia by anastomosis of conidia has been reported.

Persistence in soil for an indefinite time results from extended viability of sclerotial forms and has been estimated at more than 10 years. Infection occurs on the underground stem or roots, and severity of subsequent disease development may vary according to predominant biotype and host cultivar, as well as ambient temperature.

Attempts to control *Verticillium* wilt of potato by a single foil fumigation with 1,3-dichloropropene alone or in combination with chloropicrin served to reduce, but not eliminate, tuber infection and produce some increase in yield. However, annual soil fumigation for 5 successive years reduced incidence of diseased plants and increased yields. Preharvest burning of vines to destroy microsclerotia in debris and reduce soil inoculum was required for 2 years in succession to increase yields, although it did not reduce soil inoculum levels. Crop rotation has not been completely successful as a control measure because the fungus is pathogenic on a wide range of plants. Differences in susceptibility of potato cultivars to biotypes of the fungus have been demonstrated and offer some possibility for increased yields.

1. Ayers, G. W. and R. R. Hurst. 1952. Studies on *Verticillium* wilt of potato. *Am. Potato J. 29:* 201–205.

2. Biehn, W. L. 1970. Control of *Verticillium* wilt of potato by soil treatment with benomyl. *Plant Dis. Reptr. 54:* 171–173.

3. Biehn, W. L. 1970. Evaluation of seed treatments for control of seedborne *Verticillium* wilt of potato. *Plant Dis. Reptr. 54:* 254–255.

4. Biehn, W. L., A. Hawkins and P. M. Miller. 1971. Effect of 1,3-dichloropropene and a combination of 1,3-dichloropropene + chloropicrin on infection of potatoes by *Verticillium albo-atrum. Plant Dis. Reptr. 55:* 968–971.

5. Boyd, O. C. 1950. *Verticillium* wilt of potato. *Plant Dis. Reptr. 34:* 339–340.

6. Busch, L. V. 1966. Susceptibility of potato varieties to Ontario isolates of *Verticillium albo-atrum. Am. Potato J. 43:* 439–442.

7. Busch, L. V. and L. V. Edgington. 1967. Correlation of photoperiod with tuberization and susceptibility of potato to *Verticillium albo-atrum*. *Can. J. Bot. 45:* 691–693.

8. Chamberlain, E. E. and R. M. Brien. 1933. *Verticillium* wilt of potatoes and tomatoes in New Zealand. *New Zeal. J. Sci. Technol. 14:* 366–377.

9. Easton, G. D., M. E. Nagle and D. L. Bailey. 1972. Effect of annual soil fumigation and preharvest vine burning on *Verticillium* wilt of potato. *Phytopathology 62:* 520–524.

10. Folsom, D. 1957. *Verticillium* wilt of potatoes in relation to fungicides added to the fertilizer. *Am. Potato J. 34:* 1–5.

11. Green, R. J., Jr. 1969. Survival and inoculum potential of conidia and microsclerotia of *Verticillium albo-atrum* in soil. *Phytopathology 59:* 874–876.

12. Harrison, D. E. 1967. *Verticillium* wilt of potato. *J. Agr. (Melbourne) 65:* 113–122.

13. Hawksworth, D. L. and P. W. Talboys. 1970. *Verticillium albo-atrum. Commonwealth Mycol. Inst. Descriptions Pathogenic Fungi Bacteria 255.* 2 p.

14. Hoyman, W. G. and E. Dingman. 1967. Temik, a systemic insecticide effective in delaying *Verticillium* wilt of potato. *Am. Potato J. 44:* 3–8.

15. Isaac, I. 1949. A comparative study of pathogenic isolates of *Verticillium. Trans. Brit. Mycol. Soc. 32:* 137–157.

16. Isaac, I. 1953. A further comparative study of pathogenic isolates of *Verticillium. Trans. Brit. Mycol. Soc. 36:* 180–195.

17. Ludbrook, W. L. 1933. Pathogenicity and environal studies on *Verticillium* hadromycosis. *Phytopathology 23:* 117–154.

18. McKay, M. B. 1926. Potato wilt and its control. *Oregon Agr. Expt. Sta. Bull. 221.* 23 p.

19. McKeen, C. D. 1943. A study of some factors affecting the pathogenicity of *Verticillium albo-atrum* R. & B. *Can. J. Res., Sect. C, 21:* 95–117.

20. Miller, P. M., L. V. Edgington and A. Hawkins. 1967. Effects of soil fumigation on *Verticillium* wilt, nematodes and other diseases of potato roots and tubers. *Am. Potato J. 44:* 316–323.

21. Neilson, L. W. 1948. *Verticillium* wilt of potatoes in Idaho. *Idaho Agr. Expt. Sta. Res. Bull. 13.* 23 p.

22. Pethybridge, G. H. 1916. The *Verticillium* disease of the potato. *Proc. Roy. Dublin Soc. 15:* 63–92.

23. Powelson, R. L. 1973. Efficacy of soil fumigants for control of *Verticillium* wilt of potatoes. *Am. Potato J. 50:* 162–167.

24. Reinke, J. and G. Berthold. 1879. Die zersetzung der kartoffel durch pilze. *Untersuch. Bot. Lab. Univ. Gottingen 1:* 1–100.

25. Robinson, D. B., R. H. Larson and J. C. Walker. 1957. *Verticillium* wilt of potato. *Univ. Wisconsin Res. Bull. 202.* 48 p.

26. Rudolph, B. A. 1931. *Verticillium* hadromycosis. *Hilgardia 5:* 197–353.

27. Schreiber, L. R. and R. J. Green, Jr. 1962. Comparative survival of mycelium, conidia, and microsclerotia of *Verticillium albo-atrum* in mineral soil. *Phytopathology 52:* 288–289.

28. Thanassoulopoulos, C. C. and W. J. Hooker. 1968. Factors influencing infection of field grown potato by *Verticillium albo-atrum. Am. Potato J. 45:* 203–216.

29. Tolmsoff, W. J. 1960. Relation of age of potato plants to infection by *Verticillium albo-atrum. Phytopathology 50:* 86.

30. Waggoner, P. E. 1956. Variation in *Verticillium albo-atrum* from potato. *Plant Dis. Reptr. 40:* 429–431.

31. Wilhelm, S. 1954. Aerial microsclerotia of *Verticillium* resulting from conidial anastomosis. *Phytopathology 44:* 609–610.

32. Wilhelm, S. 1955. Longevity of the *Verticillium* wilt fungus in the laboratory and field. *Phytopathology 45:* 180–181.

33. Young, R. A. 1956. Control of the early maturity disease of potatoes by soil treatment with Vapam. *Plant Dis. Reptr. 40:* 781–784.

Virus diseases of potatoes

The "degeneration" of potatoes recognized in Europe in the 18th century was eventually identified and resolved as a complex of virus diseases. Several viruses now known to be infectious to potato may have been included in this complex, while a number of other virus diseases more recently identified were not recognized as components of degeneration. The viruses that continue to cause major diseases of potato today all have particular characteristics repeatedly demonstrated, but the peculiar nature of potato cultivation, i.e. repeated asexual propagation, presents a special problem for virus disease control.

Virus diseases of potato are readily perpetuated in tubers from infected plants, which constitute the most important source of primary inocula. Recognition that improved yields could only be achieved by growing healthy plants, which could be obtained only from healthy seed tubers, prompted development of a simple routine now widely used to produce virus disease-free seed tubers. Further elaboration of this procedure has led to seed production programs widely practiced through which incidence of virus (and fungal and bacterial) diseases can be restricted to a level that does not result in seriously reduced yields.

The routine for production of healthy seed tubers consists essentially of planting selected stocks for seed production, only separate and apart from other potato fields, where insect control can be vigorously pursued and visual inspections made several times during the season. All seed-pieces cut from a single seed tuber are planted in sequence or identifiable position in the field, i.e. as a "tuber unit," and the plants that result therefrom inspected visually on a regular basis. (Seed tubers may be assayed or indexed earlier by growing a single plant from an "eye" in the greenhouse.) If one or more plants from a tuber unit series of plants exhibit disease symptoms, all plants of that unit, including tubers, are rogued immediately. Only tubers from plants in units in which all plants have remained disease-free during the growing season are harvested for later use as "foundation" seed. Greater quantities of seed tubers with restricted levels of (virus) disease incidence can be obtained by planting this foundation seed and routinely roguing diseased plants (and tubers) throughout the season, harvesting only tubers from plant populations with low incidence of specified diseases ("certified seed"). These procedures are presently the only satisfactory means for control of virus diseases of potato, except latent mosaic, which causes no visible symptoms in many cultivars and remains a concomitant character in most cultivars.

Corky ring spot (Internal brown spot, Kringerigheid, Pfropfenbildung, Stem mottle, Tobacco rattle)

Symptoms of the potato disease now called corky ringspot were first observed and described from Germany in 1925. The disease was found in the United States (Florida) 20 years later but apparently has not been identified elsewhere, although the causal agent has been reported from Brazil and Japan as well as several other locations in the United States. Wide seasonal variation in losses to 50 percent were noted originally in Florida, but satisfactory control procedures have been developed, and the disease is no longer a serious threat to potato culture.

Tubers exhibit the only diagnostic field symptoms of corky ringspot, which consist of sometimes irregular, necrotic rings or arcs 1–3 cm in diameter on the surfaces of affected tubers that extend inward, as seen in cut tubers, as necrotic arcs. Rings occur at random over the tuber surface and may coalesce. Enlargement of the tubers during growth causes cracking of necrotic tissues with subsequent suberization and dehydration of affected tuber tissues to form concave corky areas under surface rings or arcs.

Plants grown from infected tubers under controlled conditions exhibited extreme stunting with twisted and elongated leaflets and a yellow mottle more noticeable in the top leaves.

Corky ringspot of potato is caused by the tobacco rattle virus which forms tubular particles comprised of 2 (or 3) lengths, i.e. 190 nm and 45–115 nm. Inactivation of the virus in sap from infected *Nicotiana clevelandii* results upon exposure for 10 minutes to 80–85 C or dilution beyond 10^{-6}. Infectivity *in vitro* is maintained for more than 6 weeks at 20 C. More than 400 species in 50 botanical families, including the *Amaranthaceae*, *Chenopodiaceae*, *Compositae*, *Cucurbitaceae*, *Leguminosae* and *Solanaceae*, are susceptible to infection, but not all are systemically infected. Numerous isolates or strains of the virus have been described, some of which are more easily transmitted mechanically than others, but several species of *Trichodorus* (stubby root nematodes) serve as vectors and may require only an hour (or more) acquisition and transmission feeding intervals. Infectivity was retained for 20 weeks by nematodes (in soil) that were not permitted to feed on plants.

Perpetuation of the disease can be effected both by planting infected tubers and by infectious nematodes. Incidence of corky ringspot is more frequent on light sandy soils, but distribution within a field is often uneven and may reflect nematode populations. Symptom expression is favored by temperatures below 24 C.

Control can be achieved by selection of healthy seed tubers for planting and elimination of *Trichodorus* species of nematodes prior to planting.

1. Cook, H. T. 1949. Corky ringspot found in potatoes from Indiana. *Plant Dis. Reptr. 33:* 95.

2. Eddins, A. H. 1959. Susceptibility of potato varieties and seedling selections to corky ringspot. *Am. Potato J. 36:* 187–190.

3. Eddins, A. H., E. Q. Proctor and E. West. 1946. Corky ringspot of potatoes in Florida. *Am. Potato J. 23:* 330–333.

4. Harrison, B. D. 1970. Tobacco rattle virus. *Commonwealth Mycol. Inst. Descriptions Plant Viruses 12.* 4 p.

5. Kerling, L. C. P. 1930. Microscopic investigation of "pseudonetnecrosis" and "kringerigheid" of the potato. *Phytopathology 20:* 138.

6. Oswald, J. W. and T. Bowman. 1958. Studies on a soil-borne potato virus disease in California. *Phytopathology 48:* 396.

7. Paul, H. L. and O. Bode. 1955. Elektronenmikroskopische untersuchungen über kartoffelviren. II. Vermessung ter teilchen von drei stämmen des rattle-virus. *Phytopathol. Z. 24:* 341–351.

8. Quanjer, H. M. 1943. Bijdrage tot de kennis van de in Nederland voorkomende ziekten van tabak en van de tabaksteelt op kleigrond. *Tijdschr. Plantenziekten 49:* 37–51.

9. Ramsey, G. B. and M. A. Smith. 1947. Corky ringspot of potatoes in Washington. *Plant Dis. Reptr. 31:* 8–9.

10. Rozendaal, A. and J. P. H. van der Want. 1948. Over de identiteit van het ratelvirus van de tabak en het stengelbontvirus van de aardappel. *Tijdschr. Plantenziekten 54:* 113–133.

11. Semancik, J. S. 1966. Purification and properties of two isolates of tobacco rattle virus from pepper in California. *Phytopathology 56:* 1190–1193.

12. Van Der Want, J. P. H. and A. Rozendaal. 1948. Electronenmicroscopisch onderzoek van het virus, dat de ratelziekte van tabak en het stengelbont van de aardappel veroorzaakt. *Tijdschr. Plantenziekten 54:* 134–141.

13. Van Hoof, H. A. 1965. Rattle virus in potato tubers—inoculation and isolation. *Ghent. Landbouwhogeschol Meded. 30:* 1743–1750.

14. Walkinshaw, C. H., G. D. Griffin and R. H. Larson. 1961. *Trichodorus christiei* as a vector of potato corky ringspot (tobacco rattle) virus. *Phytopathology 51:* 806–808.

15. Walkinshaw, C. H. and R. H. Larson. 1959. Corky ringspot of potato, a soil-borne virus disease. *Univ. Wisconsin Res. Bull. 217.* 31 p.

16. Webb, R. E. and E. J. Schultz. 1957. Preliminary studies on corky ringspot of potato. *Am. Potato J. 34:* 193–199.

Latent mosaic (Potato X virus disease)

Presence of a transmissible agent in symptomless potato plants was recognized in 1924 in Ireland. A year later, 3 symptom types were distinguished on tobacco plants inoculated with sap from apparently healthy potatoes, but the latent mosaic disease of potatoes was not identified as a distinct pathological entity until 1931. The latent mosaic disease is widely prevalent throughout potato cultivars grown in all parts of the world. Losses have been estimated to be greater than for all other virus diseases combined because of the widespread occurrence of the disease. Local losses may range from 10–45 percent, depending on severity of the prevalent virus strain(s).

The severity of latent mosaic symptoms in potato varies considerably according to host cultivar involved, predominant virus strain and environmental conditions. Symptoms are confined to the foliage and range from complete masking to mild systemic mottle or, in some cases, severe mottle with rugosity of leaflets and necrosis either as flecking, vein necrosis or interveinal necrosis or top necrosis. Affected plants usually are not killed, but symptoms may extend as necrotic streaks to petioles and stems and result in some defoliation.

Latent mosaic is caused by a virus (potato X virus) with elongated, flexuous particles 13 × 515 nm inactivated in sap from infected tobacco by exposure for 10 minutes to 68–76 C or dilution greater than 10^{-5}–10^{-6}. Infectivity may be retained *in vitro* for several weeks at 20 C and for more than a year in glycerol. The virus is easily transmissible by sap inoculation, but transmission in nature occurs largely by physical contact between plants. A grasshopper [*Melanoplus differentialis* (Thomas)] and the potato wart fungus (*Synchytrium endobioticum*) also have been found to transmit the virus. Transmission can be effected by the cutting knife. Host range of the virus includes several members of the *Solanaceae* (tomato, pepper and several species of *Nicotiana*) and some members of the *Amaranthaceae* (including globe amaranth, a useful local lesion host), *Chenopodiaceae*, *Labiatae* and *Scrophulariaceae*. Strains have been distinguished according to symptoms induced on tobacco, serological relations, responses of potato cultivars and physical properties of the causal virus, but there is no consistent relation between strains identified by these different criteria.

Attempts at control have been directed toward selection and use of seed tubers free of infection, utilization of "field resistant" cultivars that respond with top necrosis when infected (and thereby effectively eliminate diseased plants) and actual immunity to infection. (Some cultivars previously designated as immune have been found more recently to be infected but with low virus titer.) Controlled inoculation of cultivars with mild strain(s) of the virus (to provide immunity against infection with more severe strains) also has been recommended. None of these methods has proved completely satisfactory for control of latent mosaic, which remains widespread in many cultivars grown today.

1. Bagnall, R. H. 1961. Recovery of virus X from leaves of graft inoculated immune potato plants. *Phytopathology 51:* 338–340.

2. Bawden, F. C. and E. M. Crook. 1947. Some properties of potato virus X in leaf extracts made in different ways. *Brit. J. Expt. Pathol. 28:* 403–415.

3. Bawden, F. C., B. Kassanis and F. M. Roberts. 1948. Studies on the importance and control of potato virus X. *Ann. Appl. Biol. 35:* 250–265.

4. Bercks, R. 1970. Potato virus X. *Commonwealth Mycol. Inst. Descriptions Plant Viruses 4.* 4 p.

5. Bradley, R. H. E. 1953. A rapid method of testing plants in the field for potato virus X. *Am. Potato J. 29:* 289–291.

6. Clinch, P. E. M. 1944. Observations on a severe strain of potato virus X. *Sci. Proc. Roy. Dublin Soc. 23:* 273–299.

7. Close, R. 1959. The incidence of virus X in some New Zealand potato varieties. *New Zeal. J. Agr. Res. 2:* 1137–1142.

8. Francki, R. I. B. and G. D. McLean. 1968. Purification of potato virus X and preparation of infectious ribonucleic acid by degradation with lithium chloride. *Australian J. Biol. Sci. 21:* 1311–1318.

9. Hansen, A. J. and R. H. Larson. 1959. A yellow strain of potato virus X. *Am. Potato J. 36:* 98–104.

10. Hooker, W. J., C. E. Peterson and R. G. Timian. 1954. Virus X resistance in potato. *Am. Potato J. 31:* 199–212.

11. Johnson, J. 1925. Transmission of viruses from apparently healthy potatoes. *Wisconsin Agr. Expt. Sta. Res. Bull. 63.* 12 p.

12. Ladeburg, R. C., R. H. Larson and J. C. Walker. 1950. Origin, interrelation and properties of ringspot strains of virus X in American potato varieties. *Wisconsin Agr. Expt. Sta. Res. Bull. 165.* 47 p.

13. Larson, R. H. 1950. The spread of ringspot virus X by cutting knife. *Am. Potato J. 27:* 53–54.

14. Lombard, P. M. 1950. The effect of a mild strain of latent mosaic virus X on yield of some of the new potato varieties in Maine. *Am. Potato J. 27:* 445–449.

15. Loughnane, J. B. and P. A. Murphy. 1938. Dissemination of potato viruses X and F by leaf contact. *Sci. Proc. Roy. Dublin Soc. 22:* 1–15.

16. Mai, W. F. 1947. Virus X in the newer potato varieties and the transmission of this virus by the cutting knife. *Am. Potato J. 24:* 341–351.

17. Matthews, R. E. F. 1949. Studies on potato virus X. I. Types of change in potato virus X infections; II. Criteria of relationships between strains. *Ann. Appl. Biol. 36:* 448–459; 460–474.

18. Roberts, F. M. 1948. Experiments on the spread of potato virus X between plants in contact. *Ann. Appl. Biol. 25:* 266–278.

19. Salaman, R. N. 1933. The potato virus "X": its strains and reactions. *Phil. Trans. Roy. Soc. London 229:* 137–217.

20. Schultz, E. S. and R. Bonde, 1944. The effect of latent mosaic (virus X) on yield of potatoes in Maine. *Am. Potato J. 21:* 278–283.

21. Schultz, E. S., C. F. Clark, F. J. Stevenson and W. P. Raleigh. 1937. Resistance of the potato to latent mosaic. *Am. Potato J. 14:* 124–127.

22. Shepard, J. F. and L. E. Claflin. 1975. Critical analyses of the principles of seed potato certification. *Annu. Rev. Phytopathol. 13:* 271–293.

23. Silberschmidt, K. 1961. The spontaneous occurrence of strains of potato virus X and Y in South America. *Phytopathol. Z. 42:* 175–192.

24. Singh, S. 1972. Spinach (*Spinacia oleracea*) a new test plant for potato virus X. *Indian Phytopathol. 25:* 147–148.

25. Smith, K. M. 1931. On the composite nature of certain potato virus diseases of the mosaic group as revealed by the use of plant indicators and selective methods of transmission. *Proc. Roy. Soc. London, Ser. V, 109:* 251–267.

26. Timian, R. G., W. J. Hooker and C. E. Peterson. 1955. Immunity to virus X in potato: studies of clonal lines. *Phytopathology 45:* 313–319.

27. Van Der Plank, J. E. 1949. Some suggestions on the history of potato virus X. *J. Linn. Soc. London (Bot.) 53:* 251–262.

Leaf roll

Leaf roll symptoms were first described in 1747 from Germany, but reported occurrence in other European countries followed soon thereafter. That leaf roll disease was a distinct entity and only a part of the potato degeneration complex was not recognized until 1905. The infectious nature of the disease was demonstrated in 1916. Leaf roll is one of the major virus diseases of potato and occurs wherever potatoes are grown, but yield losses attributable to leaf roll vary according to variety of potatoes affected.

Evidence of leaf roll infection in potato may be distinguished as primary or early symptoms and secondary symptoms, but simultaneous infection with latent mosaic, which commonly occurs, does not alter symptom expression. The first sign of infection is upward rolling from the margins of apical leaflets, which develops after 3–4 weeks incubation. Affected leaflets may become chlorotic and then tinged with red and purple color, depending on host variety. Apical symptoms may not be clearly evident unless infection occurs early in the growing season.

Secondary symptoms also include upward rolling of leaflets, but at the base of the plant, and the rolling is much more severe and extensive (Fig. 86). Affected leaflets are noticeably stiffened, feel leathery to the touch and often assume some (light yellow) coloration. Necrosis in primary phloem is considered a diagnostic symptom of leaf roll infection but necrosis of phloem tissue in tubers ("net necrosis") is somewhat erratic and less important in diagnosis.

Leaf roll is caused by a virus distinguished by isometric particles 24 nm in diameter that (in sap of *Physalis floridana* Rybd.) are inactivated by exposure to about 70 C for 10 minutes, storage *in vitro* for more than 4 days (2 C) or dilution greater than 10^{-4}. The virus is not transmissible mechanically but is transmitted by several aphid species, of which *Myzus persicae* is the most efficient, in the persistent manner. A number of hosts, mostly in the *Solanaceae* but also including *Amaranthus caudatus* L., *Celosia argentea* and globe amaranth, are susceptible to infection.

Minimum acquisition feeding of 2 hours and inoculation feeding of 30 minutes with a latent period from 10–120 hours are necessary for aphid transmission, but efficiency of transmission by individual insects is increased by extended feeding periods. Aphids permitted to feed on infected plants for 5 days or more may remain infectious as long as they live. Symptom de-

Fig. 86. Tight rolling of lower leaflets characteristic of leaf roll virus infection in potato. (Courtesy of H. D. Thurston, Cornell University, Ithaca.)

velopment is progressively more rapid and severe with increase in temperature from 16 to 28 C, but different strains of the virus have been reported, and not all virus strains respond in like manner to the influence of temperature. Severity of symptoms also was found to be influenced by light, with most severe symptoms produced under highest light intensity within the range of 500–4000 foot candles.

Satisfactory control can only be achieved by use of healthy seed tubers, but application of insecticide to reduce the aphid population will retard current-season spread of the disease.

1. Appel, O. 1915. Leaf roll diseases of the potato. *Phytopathology 5:* 139–148.

2. Artschwager, E. F. 1923. Occurrence and significance of phloem necrosis in the Irish potato. *J. Agr. Res. 24:* 237–245.

3. Atanasoff, D. 1926. Net necrosis of the potato. *Phytopathology 16:* 929–940.

4. Bald, J. G. and E. M. Hutton. 1950. Some effects of leaf-roll virus on the development and growth of the potato plant. *Australian J. Agr. Res. 1:* 3–17.

5. Davidson, T. R. and G. B. Sanford. 1954. Effect of age of potato foliage on expression of leaf-roll symptoms. *Can. J. Bot. 32:* 311–317.

6. Day, M. F. 1955. The mechanism of the transmission of potato leaf roll virus by aphids. *Australian J. Biol. Sci. 8:* 498–513.

7. DeBokx, J. A. 1967. The callose test for the detection of leafroll virus in potato tubers. *European Potato J. 10:* 221–234.

8. Dykstra, T. P. 1933. Weeds as possible carriers of leaf roll and rugose mosaic of potato. *J. Agr. Res. 47:* 17–32.

9. Folsom, D. 1946. Leafroll net necrosis and stem-end browning of potato tubers in relation to temperature and certain other factors. *Phytopathology 36:* 1016–1034.

10. Guthrie, J. W. 1961. Net necrosis associated with primary and secondary infection of leafroll virus in Russet Burbank. *Am. Potato J. 38:* 435–439.

11. Harrison, B. D. 1958. Ability of single aphids to transmit both avirulent and virulent strains of potato leaf roll virus. *Virology 6:* 278–286.

12. Hovey, C. and R. Bonde. 1948. *Physalis angulata* L. a test plant for the potato leaf roll virus. *Am. Potato J. 25:* 52.

13. Hutton, E. M. 1949. The significance of the necrotic phloem reaction in the potato to the leaf-roll virus. *Australian J. Sci. Res., Ser. B, 2:* 249–270.

14. Kassanis, B. 1952. Some factors affecting the transmission of leaf-roll virus by aphids. *Ann. Appl. Biol. 39:* 157–167.

15. Kirkpatrick, H. C. 1948. Indicator plants for studies with the leaf roll virus of potatoes. *Am. Potato J. 25:* 283–290.

16. Kirkpatrick, H. C. and F. Ross. 1952. Aphid-transmission of potato leafroll virus to solanaceous species. *Phytopathology 42:* 540–546.

17. Kojima, M., E. Shikata, M. Sugawara and D. Murayama. 1968. Isolation and electron microscopy of potato leafroll virus from plants. *Virology 35:* 612–615.

18. MacCarthy, H. R. 1954. Aphid transmission of potato leafroll virus. *Phytopathology 44:* 167–174.

19. Mackinnon, J. P. 1969. High levels of resistance to potato leaf roll virus found in greenhouse tests. *Am. Potato J. 46:* 54–56.

20. Natti, J. J., H. C. Kirkpatrick and A. F. Ross. 1953. Host range of potato leafroll virus. *Am. Potato J. 30:* 55–64.

21. Peters, D. 1970. Potato leafroll virus. *Commonwealth Mycol. Inst. Descriptions Plant Viruses 36.* 4 p.

22. Sheffield, F. M. L. 1943. Value of phloem necrosis in the diagnosis of potato leaf-roll. *Ann. Appl. Biol. 30:* 131–136.

23. Sikka, L. C. and J. Munro. 1968. Resistance to the potato leafroll virus in certain *Solanum tuberosum* seedlings. *Indian Phytopathol. 21:* 167–170.

24. Smith, K. M. 1929. Studies on potato virus diseases. V. Insect transmission of potato leaf-roll. *Ann. Appl. Biol. 16:* 209–229.

25. Webb, R. E., R. H. Larson and J. C. Walker. 1952. Relationships of potato leaf roll virus strains. *Univ. Wisconsin Res. Bull. 178.* 39 p.

26. Webb, R. E., E. S. Schultz and R. V. Akeley. 1955. Some variations in symptomatology and transmission of leafroll in potato. *Am. Potato J. 32:* 60–66.

27. Wilson, J. H. 1955. Effects of nutrition and light intensity on symptoms of leaf-roll virus infection in the potato plant. *Ann. Appl. Biol. 43:* 273–287.

Mild mosaic (Crinkle, Potato A virus disease)

Mild mosaic of potato, also sometimes known as "crinkle," was described from Ireland in 1932, but it is now widely distributed and present in most potato growing areas. Yields from infected plants may be seriously reduced.

Only mild mottle symptoms in leaflets also affected with slight rugosity can be seen in varieties that exhibit typical evidence of infection. Some varieties show no symptoms, while others develop top necrosis, but the disease complex identified by the name "crinkle" typically involves double infection with potato A and latent mosaic viruses.

The virus that causes mild mosaic is typified by flexuous, elongated particles 730 nm in length that are inactivated (in sap from infected *Nicandra physalodes*) by exposure for 10 minutes to 44–52 C, dilution greater than 4×10^{-1} or storage *in vitro* (18 C) for more than 18 hours. Transmission by several aphids, including *Myzus persicae*, can be effected following acquisition and transmission feedings of 20 seconds each, but insects often do not remain viruliferous for more than a few minutes after being removed from an infected plant. Low virus titer in some plants can reduce the success of mechanical transmission. Host range of potato A virus is limited to members of the *Solanaceae*.

Insecticide application may reduce aphid population and thereby reduce current-season dissemination, but satisfactory control can be achieved only through use of healthy (certified) seed materials.

1. Allen, T. C. 1965. Field spread of potato virus A inhibited by oil. *Plant Dis. Reptr. 49:* 557.

2. Bagnall, R. H. 1961. Hypersensitivity to viruses A and X in Canadian and American potato varieties. *Am. Potato J. 38:* 192–202.

3. Bartels, R. 1971. Potato virus A. *Commonwealth Mycol. Inst. Descriptions Plant Viruses 54.* 4 p.

4. Bawden, F. C. 1936. The viruses causing top necrosis (acronecrosis) of the potato. *Ann. Appl. Biol. 23:* 487–497.

5. Clinch, P. and J. B. Loughnane. 1933. A study of the crinkle disease of potatoes and of its constituent or associated viruses. *Sci. Proc. Roy. Dublin Soc. 20:* 567–596.

6. Cockerham, G. 1943. The reactions of potato varieties to viruses X, A, B, and C. *Ann. Appl. Biol. 30:* 338–344.

7. Kohler, E. 1952. Beitrage zur kenntnis des A-virus der kartoffel. *Phytopathol. Z. 19:* 295–306.

8. Loughnane, J. B. 1933. Insect transmission of virus A of potatoes. *Nature 131:* 838–839.

9. MacLachlan, D. S., R. H. Larson and J. C. Walker. 1954. Potato virus A. *Am. Potato J. 31:* 67–72.

10. Murphy, P. A. and R. McKay. 1932. The compound nature of crinkle and its production by means of a mixture of viruses. *Sci. Proc. Roy. Dublin Soc. 20:* 227–247.

11. Roland, G. 1956. Sur l'identification du virus A. *Parasitica 12:* 125–128.

12. Salaman, R. N. 1930. Crinkle "A," an infectious disease of the potato. *Proc. Roy. Soc. London, Ser. B, 106:* 50–83.

13. Smith, K. M. 1930. Studies on potato virus diseases. VII. Some experiments with the virus of a potato crinkle with notes on interveinal mosaic. *Ann. Appl. Biol. 17:* 223–240.

14. Webb, R. E. and R. W. Buck. 1955. A diagnostic host for potato virus A. *Am. Potato J. 32:* 248–253.

15. Webb, R. E. and E. S. Schultz. 1961. Resistance of *Solanum* species to potato viruses A, X, and Y. *Am. Potato J. 38:* 137–142.

Mop-top (Corcho)

Mop-top disease of potato was first discovered and transmitted in 1957 in England, but it was not described and identified until 1966. The disease was found soon thereafter in several countries in Europe and in Peru and has recently been found in Bolivia. Estimates of economic losses attributable to mop-top may have been inaccurate because of confusion with similar symptoms caused by tobacco rattle virus, but the disease once was the most commonly found virus disease in the best seed stocks in England.

Foliage symptoms do not develop during the season when tuber infection occurs, but plants grown from infected tubers may develop any of 3 kinds of symptoms. Symptoms described as "mop-top" consist of plants with shortened stem internodes and compacted leaves that contribute to a dwarfed, bunched habit. The most common foliar symptom is yellow "aucuba pattern" blotches or ring and line patterns on lower leaves, which, on some varieties, may develop on middle and upper leaves on stems that are not otherwise affected. A third but inconsistent and often indistinct foliar symptom is appearance of chlorotic "V"'s on upper leaves of stems where aucuba symptoms appeared on lower leaves earlier in the season.

Symptoms in tubers differ in clarity and severity according to variety, as do foliar symptoms, and current season infection is evident as usually necrotic, slightly raised, concentric rings of 1–5 cm that may develop on any part of the tuber. Internal necrosis in the form of arcs or streaks often, but not always, extends inward from external rings. These necrotic tuber symptoms, often referred to as "spraing," are characteristic of primary infection in the cv. Arran Pilot. Tubers from plants grown from infected tubers often are more severely affected with symptoms, which include malformation, some large cracks and a reticulate pattern of small cracks in cultivars affected by mop-top foliage symptoms. Necrotic rings on the surface may encircle the tuber transversely and be associated with internal necrosis. Necrotic flecks may develop in the central part of the tuber near the stolon end.

Particles of the mop-top virus measure 18–20 × 100–150 or 250–300 nm (only the longest are infectious), withstand dilution in tobacco sap from 10^{-2}–10^{-4} and are inactivated by heating for 10 minutes at 75–80 C. Most infectivity is lost after one day storage *in vitro* at 20 C but some infectivity remains for up to 10 weeks. Infectivity is restricted to members of the *Solanaceae* and *Chenopodiaceae* and *Tetragonia expansa*, but many susceptible solanaceous plants do not develop visible symptoms. It has been suggested that the mop-top virus is an aberrant strain of tobacco mosaic virus because of similar particle structure and dimension and remote serological relationship. Although mop-top is mechanically transmissible with some difficulty, the only known vector is the powdery scab fungus *Spongospora subterranea*, which is infectious to several solanaceous plants and *T. expansa*. Internal virus infection of spore balls of the fungus, which are quite resistant to drying, can persist for 2 years or more, while infection of plants presumably is accomplished by zoospores (see Potato—Powdery scab), but the method of fungus infection has not been demonstrated.

Geographic distribution of mop-top is directly dependent on prevalence of powdery scab disease and was found to occur in Scotland more commonly in locations with annual rainfall in excess of 76 cm. Rainfall and soil pH were found to influence spread of mop-top. Bait seedlings of *Nicotiana debneyi* Domin have been used successfully to assay for the presence of infected *S. subterranea* in soil, but treatment of soil with captan prevents infection of bait plants.

Control of mop-top can be achieved by avoiding powdery scab disease, which includes not planting in fields where either disease has occurred previously or following other (susceptible to *S. subterranea*) solanaceous crops. Strict attention should be given to selection of healthy seed tubers.

1. Calvert, E. L. and B. D. Harrison. 1966. Potato mop-top, a soil-borne virus. *Plant Pathol. 15:* 134–139.

2. Cooper, J. I. and B. D. Harrison. 1973. Distribution of potato mop-top virus in Scotland in relation to soil and climate. *Plant Pathol. 22:* 73–78.

3. Harrison, B. D. 1974. Potato mop-top virus. *Commonwealth Mycol. Inst. Descriptions Plant Viruses 138.* 4 p.

4. Harrison, B. D. and R. A. C. Jones. 1970. Host range and some properties of potato mop-top virus. *Ann. Appl. Biol. 65:* 393–402.

5. Harrison, B. D. and R. A. C. Jones. 1971. Effects of light and temperature on symptom development and virus content of tobacco leaves inoculated with potato mop-top virus. *Ann. Appl. Biol. 67:* 377–387.

6. Harrison, B. D. and R. A. C. Jones. 1971. Factors affecting the development of spraing in potato tubers infected with potato mop-top virus. *Ann. Appl. Biol. 68:* 281–289.

7. Hinostroza, A. and E. R. French. 1972. Potato mop-top virus in cork-diseased Peruvian potatoes. *Am. Potato J. 49:* 234–239.

8. Jones, R. A. C. 1975. Presence of potato mop-top virus in Bolivia. *Plant Dis. Reptr. 59:* 193.

9. Jones, R. A. C. and B. D. Harrison. 1969. The behaviour of potato mop-top virus in soil, and evidence for its transmission by *Spongospora subterranea* (Wallr.) Lagerh. *Ann. Appl. Biol. 63:* 1–17.

10. Jones, R. A. C. and B. D. Harrison. 1972. Ecological studies on potato mop-top virus in Scotland. *Ann. Appl. Biol. 71:* 47–57.

11. Kassanis, B., R. D. Woods and R. F. White. 1972. Some properties of potato mop-top virus and its serological relationship to tobacco mosaic virus. *J. Gen. Virol. 14:* 123–132.

12. McKay, J. 1969. Diseases of the potato caused by soil-borne viruses. I. The spraing-mop-top syndrome. *J. Dept. Agr. Repub. Ireland 66:* 40–52.

13. Melhus, I. E., J. Rosenbaum and E. S. Schultz. 1916. *Spongospora subterranea* and *Phoma tuberosa* on the Irish potato. *J. Agr. Res. 7:* 213–254.

14. Salazar, L. F. and R. A. C. Jones. 1975. Some studies on the distribution and incidence of potato mop-top virus in Peru. *Am. Potato J. 52:* 143–150.

15. Torres, H., E. R. French and L. W. Nielsen. 1970. Potato diseases in Peru, 1965–1968. *Plant Dis. Reptr. 54:* 315–318.

16. Van Hoof, H. A. and A. Rozendaal. 1969. Het voorkomen van "potato mop-top virus" in Nederland. *Neth. J. Plant Pathol. 75:* 275.

Rugose mosaic (Potato Y virus disease)

The potato disease commonly called rugose mosaic and the other symptoms described below are caused by the combined infection of plants with potato Y virus and potato X virus (see Potato—Latent mosaic disease), although both causative components of these syndromes are quite capable of infectious activity independently. Potato Y virus was identified originally in England in 1931, but it is now recognized to occur worldwide. It is one of the most important virus diseases of potato and can cause serious reduction in yield.

Symptoms may vary somewhat according to when and how a potato plant becomes infected. Current season infection appears as a distinct mottle, particularly in young leaves, which may precede development of some intercostal necrosis along veins on the undersides of leaflets and symptoms termed "leaf-drop streak." Symptoms in plants that originate from tubers infected with potato Y virus consist of noticeable stunting and shortening of internodes (Fig. 87). Some vein necrosis usually is evident on the undersides of leaflets, which also exhibit mottling and puckering symptoms described as "rugose mosaic." Tuber production and size are reduced irrespective of whether infection originates during the current season or from infected tubers, but yield reduction is generally more severe in the latter. Tubers from infected plants exhibit no visible symptoms of infection other than reduction in size.

Fig. 87. Stunting with reduced size and number of leaflets on plant infected with rugose mosaic (potato Y virus). (Courtesy of H. D. Thurston, Cornell University, Ithaca.)

Flexuous particles of the causal virus measure 11 × 730 nm and are inactivated by heating for 10 minutes at 55–60 C or storage *in vitro* for 48–72 hours. Infectivity in sap from diseased tobacco is lost by dilution greater than 10^{-3}. Transmission is readily accomplished in non-persistent fashion by several aphid species, although *Myzus persicae* may be the most efficient. The virus also is transmissible mechanically and infects a wide range of solanaceous plants, including tomato, pepper, several species of *Nicotiana* (jimson weed is immune), and some members of the *Chenopodiaceae* and *Leguminosae*. Several strains of the virus have been distinguished according to symptom severity in tobacco and other hosts.

Effective transmission by aphids in the field following short acquisition and infection feeding periods precludes successful control by application of insecticides and serves to emphasize the necessity of (susceptible) weed control, particularly solanaceous weeds, which can be reservoirs of virus. Dissemination of the disease other than localized transmission by aphids is accomplished through tubers from infected plants.

Tubers only from healthy plants should be used for seed in combination with routine application of insecticides for aphid control to reduce the probability of current-season infection and spread of the disease. Potatoes should not be planted near other (susceptible) solanaceous crops, especially those in which potato Y virus occurs, and a vigorous weed control program followed with particular attention to elimination of solanaceous weeds. Differences in varietal susceptibilities have been reported but have not been useful in control to date.

1. Bagnall, R. H. and R. H. E. Bradley. 1958. Resistance to virus Y in the potato. *Phytopathology 48:* 121–125.

2. Bald, J. G. 1945. The effect of rugose mosaic on the yield of potatoes. *Phytopathology 35:* 585–590.

3. Bawden, F. C. and B. Kassanis. 1946. Varietal differences in susceptibility to potato virus Y. *Ann. Appl. Biol. 33:* 46–50.

4. Bawden, F. C. and B. Kassanis. 1947. The behavior of some naturally occurring strains of potato virus Y. *Ann. Appl. Biol. 34:* 503–516.

5. Bawden, F. C. and B. Kassanis. 1951. Serologically related strains of potato virus Y that are not mutually antagonistic in plants. *Ann. Appl. Biol. 38:* 402–410.

6. Bawden, F. C. and F. M. L. Sheffield. 1944. The relationship of some viruses causing necrotic diseases of the potato. *Ann. Appl. Biol. 31:* 33–40.

7. Bradley, R. H. E. 1954. Studies of the mechanism of transmission of potato virus Y by the green peach aphid, *Myzus persicae* (Sulz.). *Can. J. Zool. 32:* 64–73.

8. Bradley, R. H. E., C. A. Moore and D. D. Pond. 1966. Spread of potato virus Y curtailed by oil. *Nature 209:* 1370–1371.

9. Delgado-Sanchez, S. and R. G. Grogan. 1966. Purification and properties of potato virus Y. *Phytopathology 56:* 1397–1404.

10. Delgado-Sanchez, S. and R. G. Grogan. 1970. Potato virus Y. *Commonwealth Mycol. Inst. Descriptions Plant Viruses* 37. 4 p.

11. Delhey, R. 1975. Zur natur der extremen virusresistenz bei der kartoffel. II. Das Y-virus. *Phytopathol. Z. 82:* 163–168.

12. Horvath, J. 1966. Studies on strains of potato virus Y. I. Strain C; II. Normal strain. *Acta Phytopathol. 1:* 125–138; 333–352.

13. Hutton, E. M. 1948. Some factors affecting localized and systemic necrotic reactions to virus Y in the potato. *Australian J. Sci. Res., Ser. B., 1:* 416–438.

14. Kahn, R. P. and R. L. Monroe. 1963. Detection of the tobacco veinal necrosis strain of potato virus Y in *Solanum cardenasii* and *S. andigenum* introduced into the United States. *Phytopathology 53:* 1356–1359.

15. Koch, K. L. 1933. The nature of potato rugose mosaic. *Phytopathology 23:* 319–342.

16. Muñoz, F. J., R. L. Plaisted and H. D. Thurston. 1975. Resistance to potato virus Y in *Solanum tuberosum* spp. *andigena. Am. Potato J. 52:* 107–115.

17. Nobrega, W. R. and K. Silberschmidt. 1944. Sobre uma provavel variante do virus "Y" da batatinha (*Solanum* virus 2, Orton) que tem a peculiaridade de provocar necroses em plantas de fumo. *Arq. Inst. Biol. 15:* 307–330.

18. Richardson, D. E. 1958. Some observations on the tobacco veinal necrosis strain of potato virus Y. *Plant Pathol. 7:* 133–135.

19. Salazar, L. F. 1969. Una raza del virus "Y" de la papa en el Perú. *Fitopatologia 4:* 26–33.

20. Silberschmidt, K. M. 1960. Types of potato virus Y necrotic to tobacco: history and recent observation. *Am. Potato J. 37:* 151–159.

21. Silberschmidt, K. M. 1961. The spontaneous occurrence of strains of potato viruses X and Y in South America. *Phytopathol. Z. 42:* 175–192.

22. Smith, K. M. 1931. On the composite nature of certain potato virus diseases of the mosaic group as revealed by the use of plant indicators, and selective methods of transmission. *Proc. Roy. Soc. London, Ser. B, 109:* 251–267.

23. Zimmerman-Gries, S. and F. Nitzany. 1964. Field spread and tuber transmission of potato virus Y in potato fields in Israel. *Phytopathol. Mediter. 3:* 14–18.

Spindle tuber

A description of spindle tuber of potato was published in 1922 from New Jersey (United States), but the infectious nature of the disease was reported a year later. Spindle tuber is known to occur in Canada, Europe (including Holland and Scotland), Russia and South Africa, in addition to the United States, where it is a major disease.

Plants affected by spindle tuber assume an exaggerated, upright habit, usually noticeable by midseason (Fig. 88). Leaflets may develop slight rugosity, and foliage may be somewhat darker green than in healthy plants, but no mottling of leaves is associated with spindle tuber. Foliar symptoms on plants infected in the field, i.e. current-season infection, are less obvious.

Tubers from spindle tuber plants are abnormally elongated and usually reduced in girth with excessive eyes (Fig. 88). Coloration of infected tubers from varieties with red skin is noticeably lightened.

The infectious agent of spindle tuber has been identified as free ribonucleic acid (RNA), i.e. "viroid," with properties (in sap from infected potato) that vary according to the method of preparation. Extraction from

Fig. 88. Reduced foliar growth with upright habit and petioles that join the stem at an acute angle characteristic of spindle tuber infection (healthy plant on left); elongated, "spindle-shaped" tubers are also characteristic (bottom row: healthy tubers). (Courtesy of E. K. Wade, University of Wisconsin, Madison.)

potato leaflets with 0.2 M NaCl yields a preparation with thermal inactivation at 75–80 C (10 minutes exposure) and dilution end-point between 10^{-2}–10^{-3}. Corresponding limits for phenol-treated extracts are 90–100 C and 10^{-3}–10^{-4}. Strains of the causal agent have been designated, including one identified as "unmottled curly dwarf strain." Members of several genera within the *Solanaceae*, including *Capsicum*, *Datura*, *Nicandra*, *Nicotiana*, *Petunia* and *Physalis*, are susceptible to infection but remain symptomless, while tomato is the only susceptible host to develop symptoms other than potato. Globe amaranth (*Amaranthaceae*) also is susceptible. The spindle tuber viroid, transmissible mechanically and in pollen and true seeds of potato as well as by several insects, including aphids, grasshoppers, flea beetles, leaf beetles and the Colorado potato beetle, has been found identical to the causal agent of citrus exocortis disease. Transmission of the infectious entity can be effected by the cutting knife also.

Satisfactory control can be achieved only through use of healthy tubers, although current-season infection and spread of the disease may be reduced by application of insecticides.

1. Diener, T. O. 1972. Potato spindle tuber viroid. *Virology 50:* 606–609.

2. Diener, T. O. and W. B. Raymer. 1969. Potato spindle tuber virus: a plant virus with properties of a free nucleic acid. II. Characterization and partial purification. *Virology 37:* 351–366.

3. Diener, T. O. and W. B. Raymer. 1971. Potato spindle tuber 'virus.' *Commonwealth Mycol. Inst. Descriptions Plant Viruses 66.* 4 p.

4. Fernow, K. H. 1967. Tomato as a test plant for detecting mild strains of potato spindle tuber virus. *Phytopathology 57:* 1347–1352.

5. Fernow, K. H., L. C. Peterson and R. L. Plaisted. 1970. Spindle tuber virus in seeds and pollen of infected potato plants. *Am. Potato J. 47:* 75–80.

6. Goss, R. W. 1926. Transmission of potato spindle-tuber by cutting knives and seed piece contact. *Phytopathology 16:* 299–303.

7. Goss, R. W. 1928. Transmission of potato spindle-tuber by grasshoppers (*Locustidae*). *Phytopathology 18:* 445–448.

8. Goss, R. W. 1931. Infection experiments with spindle tuber and unmottled curly dwarf of the potato. *Nebraska Agr. Expt. Sta. Res. Bull. 53.* 36 p.

9. O'Brien, M. J. and W. B. Raymer. 1964. Symptomless hosts of the potato spindle tuber virus. *Phytopathology 54:* 1045–1047.

10. Raymer, W. B. and T. O. Diener. 1969. Potato spindle tuber virus: a plant virus with properties of a free nucleic acid. I. Assay, extraction, and concentration. *Virology 37:* 343–350.

11. Raymer, W. B. and M. J. O'Brien. 1962. Transmission of potato spindle tuber virus to tomato. *Am. Potato J. 39:* 401–408.

12. Schultz, E. S. and D. Folsom. 1923. Transmission, variation and control of certain degeneration diseases of Irish potatoes. *J. Agr. Res. 25:* 43–117.

13. Semancik, J. S., D. S. Magnuson and L. G. Weathers. 1973. Potato spindle tuber disease produced by pathogenic RNA from citrus exocortis disease: evidence for the identity of the causal agents. *Virology 52:* 292–294.

14. Singh, R. P. 1971. A local lesion host for potato spindle tuber virus. *Phytopathology 61:* 1034–1035.

15. Singh, R. P. 1973. Experimental host range of potato spindle tuber "virus." *Am. Potato J. 50:* 111–123.

16. Singh, R. P. and R. H. Bagnall. 1968. Infectious nucleic acid from host tissues infected with the potato spindle tuber virus. *Phytopathology 58:* 696–699.

Spotted wilt (Marchitamiento, Necrose do topo)

Potato disease symptoms described from Brazil in 1938 (top necrosis) and from Argentina in 1941 (wilt) were attributed to infection by spotted wilt. The disease became severe in Australia in 1946, although it had been recognized there for several years.

Current-season infection manifested as irregular to circular necrotic brown spots on younger leaves, sometimes with concentric markings, extends to the growing point as necrotic streaking and wilt of the stem that results in death of the shoot apex. Necrosis may extend downward and kill the entire shoot, but often only one stem/plant will be killed. Plants that do survive infection later become dwarfed and develop a bunchy habit from growth of axillary buds and typically develop variable yellow-green ringspot mottle, foliar symptoms. Tubers infected late in their development may show practically no symptoms, while infection during early tuber formation causes distortion, pitting and cracking. Plants that arise from infected tubers develop a rosette appearance with thick, dark green leaves on stems with shortened internodes. Leaves also develop black, necrotic spotting, often with some necrotic rings. Plants that are not killed later develop yellow-green ringspot symptoms similar to symptoms from current-season infection. Yield from such plants is negligible.

1. Conroy, R. J., N. S. Shirlow, R. D. Wilson and E. J. Waring. 1949. Tuber transmission of the spotted wilt disease of potatoes. *Agr. Gaz. New S. Wales 60:* 101–103.

2. Costa, A. S. and J. Kiehl. 1938. Uma molestia de batatinha "necrose do tôpo" causada pelo virus de "vira-cabeça." *J. Agron (São Paulo) 1:* 193–202.

3. Hutton, E. M. 1947. Resistance in the potato to the spotted wilt virus. *J. Australian Inst. Agr. Sci. 13:* 190–192.

4. Ie, T. S. 1970. Tomato spotted wilt virus. *Commonwealth Mycol. Inst. Descriptions Plant Viruses 39.* 4 p.

5. Norris, D. O. 1951. Spotted wilt of potato. I. The field disease and studies of the causal virus; II. Tuber transmission and vector studies of the field disease. *J. Australian Inst. Agr. Res. 2:* 221–242; 243--260.

6. Norris, D. O. and J. G. Bald. 1943. Transmission of spotted wilt through potato tubers. *J. Australian Inst. Agr. Sci. 9:* 34–35.

7. Offermann, A. M. and E. R. Vitoria. 1941. Estudio sobre un virus productor del "marchitamiento apical" de la papa. *Rev. Argent. Agron. 8:* 105–113.

8. Silberschmidt, K. 1954. Potato viruses in the Americas. *Phytopathology 44:* 415–420.

9. Wilson, R. D. 1947. Spotted wilt disease of potatoes. *Agr. Gaz. New S. Wales 58:* 263–265.

Yellow vein

A graft-transmissible disease characterized by bright yellowing of leaf veins (Fig. 89) was observed in 1939 in glasshouse potatoes grown in England. The yellow vein disease has been found more recently in southern Colombia and Ecuador. Virus distribution within affected plants may be less than completely systemic, as indicated by irregular tuber transmission, although reduction in tuber yield of 50 percent or more has been determined.

Control can be effected presently only by selection of healthy seed.

1. Alba, V. R. 1950. Viropatogenos. *Conferencia Latino-Americana de Especialistas en Papa, Bogota:* 52–58.

2. Moreno, J. D. 1965. El virus del amarillamiento de las papas. *Cienc. Nat. 8:* 25–37.

3. Moreno, J. D. 1966. Incidencia del virus del amarillamiento de venas en papa en el Ecuador y su transmisión a través de los tubérculos. *Turrialba 16:* 15–24.

4. Silberschmidt, K. 1954. Potato viruses in the Americas. *Phytopathology 44:* 415–420.

5. Smith, K. M. 1972. *A textbook of plant virus diseases.* Academic Press, New York. 684 p.

Fig. 89. Leaf from potato plant infected with yellow vein virus disease.

S

SAGE: *Salvia officinalis* L.

Verticillium wilt

Non-flowering sage has been cultivated in the Vale of Evesham (England) since about the turn of the century. A "die-back" disorder was recognized for some 30 years before it was studied in 1964–66 and identified as *Verticillium* wilt. Losses attributable to wilt have not been assessed, but persistence of the pathogen in contaminated soils and serious disease potential from high inoculum concentration necessitate consideration of wilt as a disease of major importance.

Affected plants are first apparent about mid-season because of yellowed, rolled and drooping lower leaves that soon become brown and fall. Only a portion of the shoots on diseased plants may be affected, on which further leaf production and shoot elongation are retarded. Symptoms progress upwards on diseased shoots accompanied by brown discoloration of the vascular tissues and intensified external darkening of the stems.

The disease is caused by *Verticillium dahliae*, characterized by hyaline conidiophores with verticillate branching and 3–4 phialides 1–2.5 × 16–35 μm at each node. Hyaline, sub-cylindrical conidia are mostly aseptate and measure 1.4–3.2 × 2.5–8 μm, while microsclerotia 15–50 μm are variable in size and shape.

That the causal fungus may persist in the soil for more than 10 years in the absence of a sage crop has been demonstrated and serves to emphasize the desirability of avoiding the disease. Microsclerotia, produced in abundance in the bark of the lower stem and pith of affected plants, serve to perpetuate the fungus in soil in combination with its capability to survive by non-pathogenic infection of many host plants. Nearly all debris, whether moved by wind, rain or cultivation, can disseminate the fungus because of extensive systemic invasion of diseased plants. Cut stems during harvest also may be subject to infection from conidia.

Cuttings and planting stocks should be taken only from selected, healthy plants and cultivation reduced to a minimum, especially after the disease has appeared. Roguing infected plants as they appear will serve to slow the incidence of infection and keep the inoculum level low. Machinery should be thoroughly disinfected before use, particularly if it has been used previously in an infested area. Burning all debris, preferably *in situ*, will prevent incorporating additional inoculum into the soil.

1. Hawksworth, D. L. and P. W. Talboys. 1970. *Verticillium dahliae. Commonwealth Mycol. Inst. Descriptions Pathogenic Fungi Bacteria 256.* 2 p.

2. Horner, C. E. and H. L. Dooley. 1965. Propane flaming kills *Verticillium dahliae* in peppermint stubble. *Plant Dis. Reptr. 49:* 581–582.

3. Isaac, I. 1953. The spread of diseases caused by species of *Verticillium. Ann. Appl. Biol. 40:* 630–638.

4. Ryan, C. L. J. 1966. *Verticillium* wilt of sage (*Salvia officinalis* L.). *Ann. Appl. Biol. 58:* 71–79.

SWEET POTATO: *Ipomoea batata* (L.) Lam.

Soil rot

The disease of sweet potato known as soil rot was first reported from New Jersey (United States) in 1890. Subsequently, it has been found in other areas of this country (California, Delaware, Iowa) and generally throughout the southeastern states. It has been the most important disease of this crop in Iowa and Louisiana, although it apparently is not widely distributed in other parts of the world. Losses vary according to seasonal conditions and range from reduced quality only to almost complete destruction of plants and severe yield reduction.

Symptoms of soil rot are not easily confused with those of other sweet potato diseases and consist of plants with small, light green to yellow leaves that are noticeably smaller in size with little vine growth. Affected plants are easily pulled from the ground and many actually die before the growing season ends. The fibrous root system on diseased plants is destroyed, and dark, elongate lesions may be found on the stems below soil level as well as on larger roots. Lesions on the fleshy roots may enlarge to a cm or more in diameter with irregular margins and coalesce to girdle the root and cause decided malformation (Fig. 90). The surfaces of lesions on fleshy roots remain intact for a time but later break to expose conspicuous sunken pits and holes.

Soil rot is caused by *Streptomyces ipomoea* (Person & Martin) Waksman & Henrici, characterized by production of spiral-shaped conidia 0.9–1.3 × 1.3–1.8 μm in chains on aerial mycelium. Cardinal temperatures for growth *in vitro* are about 22, 32, and 42 C. Scant mycelial growth develops in culture media below pH 5.2, but the fungus grows well at pH 5.6–7.0.

The causal fungus persists in the soil, and infection probably results more often after plants have been set, although infection may result from

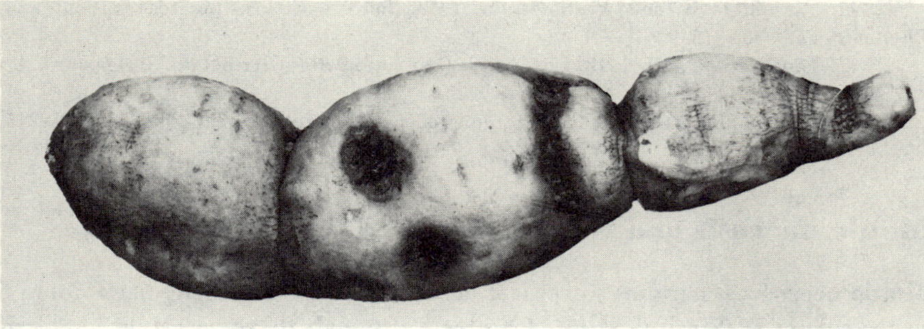

Fig. 90. Soil rot lesions on sweet potato. (Courtesy of W. J. Martin, Louisiana State University, Baton Rouge.)

infected roots used for propagating. Disease losses are more severe in seasons with low rainfall and in soils of reaction more alkaline than pH 5.0.

Broadcast application of sulfur 3–4 months before planting in quantity sufficient to reduce soil reaction to pH 5.0 (i.e. 460–735 kg/ha in some soils) has been found to satisfactorily reduce losses from soil rot. Such treatment has been effective for 4 years in controlled studies. Differences in susceptibility have been noted, and the variety Jasper was recently released as resistant in Louisiana. In-row, preplant application of soil fumigants (20 percent methyl isothiocyanate and 80 percent chlorinated C_3 hydrocarbons [Vorlex] at 38–47 ℓ/ha, or 2,3 dichloropropene and related chlorinated C_3 hydrocarbons with 15 percent chloropicrin [Telone C] at 112 ℓ/ha) served to increase yields from 1600–4600 kg/ha in Louisiana.

1. Adams, J. F. 1929. An actinomycete, the cause of soil rot or pox in sweet potatoes. *Phytopathology 19:* 179–190.

2. Cook, H. T. 1955. Sweetpotato diseases. *U. S. Dept. Agr. Farmers' Bull. 1059.* 26 p.

3. Elliott, J. A. 1916. The sweet potato "soil rot" or "pox," a slime mold disease. *Delaware Agr. Expt. Sta. Bull. 114.* 25 p.

4. Halsted, B. D. 1890. Some fungous diseases of the sweet potato. *N. J. Agr. Expt. Sta. Bull. 76.* 32 p.

5. Hooker, W. J. and L. E. Peterson. 1952. Sulfur soil treatment for control of sweet potato soil rot incited by *Streptomyces ipomoea. Phytopathology 42:* 583–590.

6. Martin, W. J. 1958. Reaction of sweet potato varieties and seedlings to soil rot. *Phytopathology 48:* 445–448.

7. Martin, W. J. 1969. Control of soil rot of sweetpotatoes by preplant, in-row application of fungicidal fumigants. *Phytopathology 59:* 1040.

8. Martin, W. J., T. P. Hernandez and T. P. Hernandez. 1975. Development and disease reaction of Jasper, a new soil rot-resistant sweetpotato variety from Louisiana. *Plant Dis. Reptr. 59:* 388–391.

9. Martin, W. J., L. G. Jones and T. P. Hernandez. 1967. Sweetpotato soil rot development in Olivier silt loam as affected by annual applications of lime or sulfur over a seven-year period. *Plant Dis. Reptr. 51:* 271–275.

10. Person, L. H. 1946. The soil rot of sweet potatoes and its control with sulfur. *Phytopathology 36:* 869–875.

11. Person, L. H. and W. J. Martin. 1940. Soil rot of sweet potatoes in Louisiana. *Phytopathology 30:* 913–926.

12. Taubenhaus, J. J. 1918. Pox, or pit (soil rot) of the sweet potato. *J. Agr. Res. 13:* 437–450.

13. Waksman, S. A. 1919. Cultural studies of species of *Actinomyces*. *Soil Sci. 8:* 71–215.

Mottle necrosis and rootlet rot

Mottle necrosis symptoms were first described in 1923, having been noted 6 years earlier in Virginia (United States), although there was unofficial evidence that the disease had occurred sporadically for 20 years or more. There is some indication that the disease described as "white rot" in 1890 actually was the same disease later named "mottle necrosis." Losses of 40 percent have been estimated for mottle necrosis, which has been reported throughout the United States where sweet potatoes are grown but usually is more severe in the northern production areas.

Two types of symptoms may be induced in sweet potato by infection with *Pythium* species. A rot of small, young rootlets that extends back 1–3 cm from the root tips can be severe in the seedbed. Other symptoms, referred to as mottle necrosis, develop on fleshy roots as small, brown, sunken superficial spots. Internal symptoms of mottle necrosis in fleshy roots may include a 1–3 mm band of necrotic tissue outside the vascular bundles, or random scattering of necrotic tissue throughout the root (as seen in cross-section) or a complete rot imparting consistency of soft cheese and a gray color to affected fleshy roots.

Several species of *Pythium* are infectious to sweet potato, but the diseases described here have been attributed largely to *P. ultimum*, *P. scleroteichum* Drechs. and *P. aphanidermatum.* Hyphae of the latter species measure 2.5–7 μm in diameter, with subspherical, intercalary or terminal oogonia 16–32 μm in diameter with smooth walls 0.5–1.2 μm thick. Clavate antheridia 4–7 × 9–16 μm numbering 1–5, sometimes from a branch of the oogonial hypha, wrap about each oogonium and yellowish oospores range from 11–26 μm in diameter with wall thickness of 0.8–1.4 μm.

Hyphae of *P. ultimum* are much branched and 1.7–6.5 μm in diameter with multinucleate, spherical conidia, mostly terminal (12–28 μm in diameter) but also intercalary (14–23 × 17–28 μm). Smooth oogonia 20–23 μm are terminal and spherical with curved antheridia that arise from the oogonial stalk. The yellowish, spherical oospores 15–18 μm always germinate by germ tubes (no zoospores).

The inflated, filamentous sporangia 4–20 × 50–1000 μm of *P. aphanidermatum* are borne on hyaline hyphae 2.8–7.3 μm in diameter. Reniform, biciliate zoospores measure 7.5 × 12 μm, and the spherical oogonia that are mostly terminal are 22–27 μm in diameter. Monoclinous antheridia, typically intercalary, are broadly clavate and measure 9–11 × 10–14 μm. Oospores 17–19 μm in diameter germinate by germ tubes.

Cardinal temperatures for growth *in vitro* for *Pythium* species that cause mottle necrosis of sweet potato range from 2–8, 29–32 and 39–42 C, but the most severe disease and decay symptoms occur from 12–15 C. Incubation of infected fleshy roots at 20 C or above results in the random discoloration typical of mottle necrosis, while incubation at lower temperatures results in a "cheesy" decay. Disease severity is greatest in seasons of high rainfall and more often on light, sandy soils.

Rotation of bedding sites and planting fields on a 3-year or more cycle has been recommended, and cultivars quite susceptible to damage from *Pythium* diseases should be avoided.

Rootlet rot, similar to that described above, can be caused by *Rhizoctonia solani*. Infection results in a soft, black rot of affected rootlets on which the cortex is loosened and easily separated from the stele and which often extends from the root tip to the main stem of the plant.

1. Cook, H. T. 1955. Sweetpotato diseases. *U. S. Dept. Agr. Farmers' Bull. 1059.* 26 p.

2. Drechsler, C. 1934. *Pythium scleroteichum* n. sp. causing mottle necrosis of sweetpotatoes. *J. Agr. Res. 49:* 881–890.

3. Harter, L. L. 1927. *Pythium* rootlet rot of sweet potatoes. *J. Agr. Res. 29:* 53–55.

4. Harter, L. L., J. I. Lauritzen and J. L. Weimer. 1923. Mottle-necrosis of sweet potatoes. *Phytopathology 13:* 145–146.

5. Harter, L. L. and J. L. Weimer. 1929. A monographic study of sweet potato diseases and their control. *U. S. Dept. Agr. Tech. Bull. 99.* 117 p.

6. Harter, L. L. and W. A. Whitney. 1927. Mottle necrosis of sweet potatoes. *J. Agr. Res. 34:* 893–914.

7. Waterhouse, G. M. and J. M. Waterston. 1964. *Pythium aphanidermatum. Commonwealth Mycol. Inst. Descriptions Pathogenic Fungi Bacteria 36.* 2 p.

White rust

Light colored spots on leaves precede the appearance of clusters of raised, pustular lesions that rupture to expose white, granular, powdery masses of spores on the lower surfaces of affected leaves and sometimes on stems. The diseased tissues may become brown and necrotic and fall, leaving holes in the leaves, but the disease is usually not responsible for significant defoliation. The disease is caused by *Albugo ipomoeae-panduratae* (Schwein.) Swing., which produces chains of hyaline sporangia 10–21 × 12–22 μm that germinate directly or by zoospore formation. Occurrence of white rust is widespread throughout the world and has been reported to be more severe in cool, moist weather, but incidence of disease usually does not warrant attempts at control. Pathotypes of the fungus and resistance in sweet potato selections have been reported.

1. Ciferri, R. 1928. Osservazioni sulla specializzazione dell' "*Albugo ipomoeae-panduratae*" (Schw.) Sw. *Nuovo Gior. Bot. Ital. 35:* 112–134. Also *Rev. Appl. Mycol. 8:* 11. (1929).

2. Cook, H. T. 1955. Sweetpotato diseases. *U. S. Dept. Agr. Farmers' Bull. 1059.* 26 p.

3. Martin, W. J. 1967. Sweet potato diseases and their control. *Proc. Intern. Symp. Trop. Root Crops (St. Augustine, Trinidad) 2:* 1–10.

4. Mukerji, K. G. and C. Critchett. 1975. *Albugo impomoeae-panduratae. Commonwealth Mycol. Inst. Descriptions Pathogenic Fungi Bacteria 459.* 2 p.

5. Weber, G. F. and E. West. 1930. Diseases of sweet potatoes in Florida. *Florida Agr. Expt. Sta. Bull. 212.* 40 p.

Black rot

Black rot disease of sweet potato was first described from New Jersey (1890) but had been observed there several years earlier. It has been reported since that time from several other locations in the United States and from many other countries and probably occurs wherever sweet potato is grown. Losses can be heavy, particularly in storage, but the disease does occur in the field, too.

Infection of plants, which become noticeably stunted with yellow foliage, is confined to the underground portions, first as small, black stem lesions which may enlarge and later cause the stems to become completely rotted. Production of adventitious roots above the affected portions of stems permits the plant to live, but blackening of the stem may continue upward to the soil level to impart symptoms described as "black shank." Infections on fleshy roots appear as slightly sunken, circular lesions of various size which may enlarge to involve much of the surface of the root (Fig. 91). The affected areas remain rather shallow, however, and infections usually do not penetrate beyond the vascular ring. Color of lesions ranges from a dark, metallic green when moist to a gray-black when dry.

The disease is caused by *Ceratocystis fimbriata*, characterized by slender, hyaline, lateral conidiophores that range to 160 μm in length, through the open ends of which are produced a succession of cylindrical, hyaline conidia 4–5.5 × 11–25 μm. Mostly obovate, brown, thick-walled chlamydospores 6–13 × 9–18 μm are produced in terminal chains. Globose, brown to black perithecia 140–220 μm in diameter have long, slightly tapered necks to 900 μm, through which elliptical, hyaline, "hat-shaped" (a gelatinous sheath forms the hat "brim") ascospores 2.5–5.5 × 4.5–8 μm are released.

The fungus that causes black rot may persist in the soil for a short time but apparently does not survive for more than 2 years. Isolates of the fungus from various host plants have been found to differ in pathogenic capabilities.

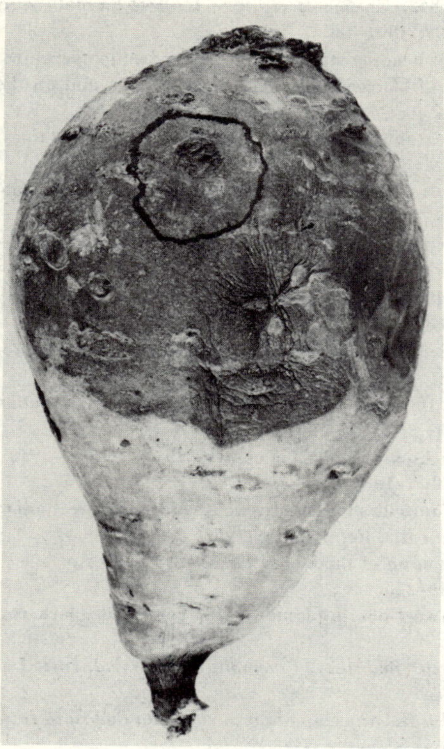

Fig. 91. Black rot infection of sweet potato. (Courtesy of W. J. Martin, Louisiana State University, Baton Rouge.)

Infection of plants may result from diseased seed-pieces and occurs over a range of soil temperatures from 10–30 C (optimum about 25 C). Disease severity was found to increase according to soil moisture content, up to approximately 60 percent field capacity, and can continue to develop and cause serious losses of stored roots.

Selection of healthy roots for propagating purposes is strongly recommended, although the fungus can be killed in infected roots by exposure to hot air treatments (50 C for 6 hours, 43 C for 24 hours or 35 C for 5 days) not lethal to roots. Dipping roots in borax (2.5 percent) before bedding has been recommended but more recent experimentation has revealed that dips of 2-(4-thiazolyl)-benzimidazole (0.1 kg/15 ℓ water) or ferbam (0.1 kg/20 ℓ water) were much more effective and their activity enhanced when the treating solutions were heated to 49 C (5 minute treatments) or 54 C (1 minute treatments). Useful resistance in some cultivars has been reported and crop rotation on a 2-year, or more, schedule should prove beneficial.

1. Cheo, P. C. 1953. Varietal differences in susceptibility of sweet potato to black rot fungus. *Phytopathology 43:* 78–81.

2. Cook, H. T. and L. L. Harter. 1942. Chemicals effective for sweet potato seed treatment. *Virginia Truck Expt. Sta. Bull. 109:* 1801–1807.

3. Daines, R. H. 1959. The influence of bedding stock, source, plant bed temperature, and fungicide treatment on the development of black rot of sweetpotato sprouts and on the resulting crop during storage. *Phytopathology 49:* 249–251.

4. Daines, R. H. 1962. Sprout dip treatments for control of black rot of sweetpotatoes. *Plant Dis. Reptr. 46:* 253–255.

5. Daines, R. H. 1971. The control of black rot of sweet potatoes by the use of fungicide dips at various temperatures. *Phytopathology 61:* 1145–1146.

6. Daines, R. H., I. A. Leone and E. Brennan. 1962. Control of black rot by prebedding heat treatment of sweetpotato roots. *Phytopathology 52:* 1138–1140.

7. Kushman, L. J. and J. S. Cooley. 1949. Effect of heat on black rot and keeping quality of sweet potatoes. *J. Agr. Res. 78:* 183–190.

8. Martin, W. J. 1954. Varietal reaction of *Ceratostomella fimbriata* in sweet potato. *Phytopathology 44:* 383–384.

9. Martin, W. J. 1965. Elimination of the black rot fungus from diseased sweet potato roots by a dry heat treatment. *Phytopathology 55:* 1066–1067.

10. Martin, W. J. 1967. Sweet potato diseases and their control. *Proc. Intern. Symp. Trop. Root Crops (St. Augustine, Trinidad) 2:* 1–10.

11. Martin, W. J. 1971. Evaluation of fungicides for effectiveness against the sweetpotato black rot fungus, *Ceratocystis fimbriata. Plant Dis. Reptr. 55:* 523–526.

12. Martin, W. J. 1972. Further evaluation of thiabendazole as a sweetpotato "seed" treatment fungicide. *Plant Dis. Reptr. 56:* 219–223.

13. Miles, L. E. 1934. Treatment of sweet potato plants for the control of black rot. *Phytopathology 24:* 1227–1236.

14. Morgan-Jones, G. 1967. *Ceratocystis fimbriata. Commonwealth Mycol. Inst. Descriptions Pathogenic Fungi Bacteria 141.* 2 p.

15. Olson, E. O. and W. J. Martin. 1949. Relationship of *Ceratostomella fimbriata* from *Hevea* rubber trees and sweet potato. *Phytopathology 39:* 17.

16. Person, L. H., E. O. Olson and W. J. Martin. 1948. Effectiveness of fungicides in controlling black rot of sweet potatoes. *Phytopathology 38:* 474–479.

17. Pontis, R. E. 1951. A canker disease of the coffee tree in Colombia and Venezuela. *Phytopathology 41:* 178–184.

Scab

Brown, circular lesions on leaves and elongate lesions on petioles and stems were reported in 1943 to occur in Brazil and Guam. The causal fungus, characterized by globose asci 10–12 × 15–16 μm, each with 4–6 hyaline, septate ascospores 3–4 × 7–8 μm, was named *Elsinoe batatas* (Saw.) Viegas & Jenkins.

1. Jenkins, A. E. and A. P. Viegas. 1943. Stem and foliage scab of sweet potato (*Ipomoea batatas*). *J. Wash. Acad. Sci. 33:* 244–249.

Rust

Coleosporium ipomoeae (Schw.) Burr., which produces uredia and telia on sweet potato and pycnia and aecia on several species of *Pinus*, is more commonly found in tropical than temperate climates. Catenulate, elliptical urediospores, that appear yellow to orange *en masse*, measure 13–21 × 18–27 μm, while darker teliospores are 19–23 × 60–80 μm and develop thick, gelatinous walls. Evidence of physiologic races of the fungus has been noted.

1. Harter, L. L. and J. L. Weimer. 1929. A monographic study of sweet potato diseases and their control. *U. S. Dept. Agr. Tech. Bull. 99.* 117 p.

2. Hedgecock. G. C. and N. R. Hunt. 1917. The peridermium belonging to *Coleosporium ipomoeae. Phytopathology* 7: 67.

3. Laundon, G. F. and A. F. Rainbow. 1971. *Coleosporium ipomoeae. Commonwealth Mycol. Inst. Descriptions Pathogenic Fungi Bacteria 282.* 2 p.

Sclerotial blight and Circular spot

Usually plants in small areas of sweet potato seedbeds are affected by *Sclerotium rolfsii*, and infected plants become yellow and stunted before they

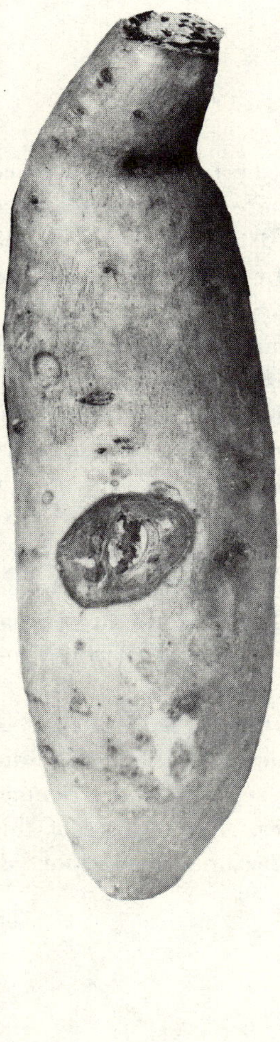

Fig. 92. Circular spot lesions on cv. Centennial sweet potato. (Courtesy of W. J. Martin, Louisiana State University, Baton Rouge.)

are killed. Soon after infection, a white clump of mycelium that encircles the stem can be seen at the soil line, where a rapid rotting of the stem develops. Embedded in the mycelium clumped around the stems of affected plants are nearly spherical, light brown sclerotia. Infection may extend to fleshy roots to cause "circular spot" rotting symptoms (Fig. 92). Isolated, nearly round leaf lesions to 5 mm or more in diameter also may develop on leaves and often a single sclerotium-like body can be found in the center of each lesion.

Rotation of seedbeds and use of clean roots for propagating purposes has been recommended, and there is some evidence that application of 2, 6-dichloro-4-nitroaniline is an effective control.

1. Barry, J. R., and W. J. Martin. 1967. Effects of plant-bed applications of 2, 6-dichloro-4-nitroaniline on control of sclerotial blight and on plant production in sweet-potatoes. *Plant Dis. Reptr. 51:* 191–194.

2. Higgins, B. B. 1927. Physiology and parasitism of *Sclerotium rolfsii* Sacc. *Phytopathology 17:* 417–448.

3. Martin, W. J. 1953. Circular spot, a disease of sweet potato roots. *Phytopathology 43:* 432–433.

4. Mordue, J. E. M. 1974. *Corticium rolfsii. Commonwealth Mycol. Inst. Descriptions Pathogenic Fungi Bacteria 410.* 2 p.

Stem rot and Wilt

Prevalence of wilt disease in the United States has been recognized since 1890, and it has occurred in nearly every state where sweet potatoes are grown. Annual losses from wilt have been estimated to range from 10–50 percent in several states, but nearly all of the plants in some fields have been killed. The disease has been found also in Japan.

Symptoms on affected plants, whether first seen in the seedbed or after transplanting into the field, include dull green coloration of the youngest leaves, followed by interveinal chlorosis and some distortion of the laminae. Older leaves soon fall, with subsequent stunting and wilting of the entire plant before it collapses and dies. Purplish vascular discoloration in the whitened stem just below the soil line is particularly noticeable and may extend some considerable distance upward through the vine and downward into the fleshy roots. In the late stages of disease development, the cortex of the stem may split, exposing blackened vascular tissue where the causal fungus often sporulates. Such splitting of the cortex can extend into the vine, as well as in the opposite direction into the fleshy roots.

Wilt is caused by *Fusarium oxysporum* f. *batatas* (Wr.) Snyder & Hansen, that produces abundant aseptate, hyaline, ovoid microconidia 2.5–4 × 6–15 μm but few hyaline, septate (2–3 septa), fusiform macroconidia 3.5–5.5 × 25–33 μm and chlamydospores. A pathotype of this fungus, which causes only a surface rot of sweet potatoes, has been reported.

"Foot rot" of sweet potato, caused by *F. solani* f. *batatas* McClure, also has been described.

The causal fungus of stem rot disease can survive for several years in the soil in the absence of sweet potato plants. There has been reported infection of sweet potato plants by a closely related form of the fungus that causes a similar disease in tobacco. Infection occurs largely through wounds that expose vascular tissues and can take place in the seedbed or in the production field. Soil moisture appears not to seriously influence infection or disease development, which can occur throughout the range of soil temperatures under which sweet potato will grow but is optimum at about 30 C. Dissemination of the fungus can be accomplished through surface drainage water or transport of contaminated soil, but most often occurs by sale or exchange of infected plants or seed roots.

Strict attention should be given to selection of healthy roots for propagation of plants and rotation of crops to permit planting on soil not previously planted to sweet potatoes, tobacco, tomatoes or cabbage. Differences in susceptibility of cultivars, but nothing approaching immunity to stem rot infection, have been reported, and practical field control can be achieved by planting cultivars with intermediate resistances. Severity of disease development was noticeably reduced by dipping cuttings in 2-(4-thiazolyl)-benzimidazole (6.5 kg/100 ℓ water) up to 96 hours after inoculation in controlled experiments.

1. Harter, L. L. and J. L. Weimer. 1919. The surface rot of sweet potatoes. *Phytopathology 9:* 465–470.

2. Harter, L. L. and W. A. Whitney. 1927. Relation of soil temperature and soil moisture to the infection of sweet potatoes by the stem-rot organisms. *J. Agr. Res. 34:* 435–441.

3. Harter, L. L. and W. A. Whitney. 1927. The comparative susceptibility of sweet-potato varieties to stem rot. *J. Agr. Res. 34:* 915–919.

4. Hendrix, F. F., Jr. and L. W. Nielsen. 1958. Invasion and infection of crops other than the Forma Suscept by *Fusarium oxysporum* f. *batatas* and other Formae. *Phytopathology 48:* 224–228.

5. Holliday, P. 1970. *Fusarium oxysporum* f. sp. *batatas. Commonwealth Mycol. Inst. Descriptions Pathogenic Fungi Bacteria 212.* 1 p.

6. Hughes, M. B., C. E. Steinbauer, M. T. Deonier and H. B. Cordner. 1963. Preliminary studies on the inheritance of wilt resistance in sweetpotato. *Proc. Am. Soc. Hort. Sci. 83:* 623–628.

7. Hwang, L., Y. S. Chen and Y. Y. Hwang. 1956. A preliminary study of sweet potato wilt and its control. *Acta Phytopathol. Sinica 2:* 97–113. Also *Rev. Appl. Mycol. 37:* 179. (1958).

8. Lauritzen, J. I. 1926. A strain of Yellow Jersey sweet potato resistant to surface rot (*Fusarium oxysporum* W. & C.). *J. Agr. Res. 33:* 1091–1094.

9. Martin, W. J. 1972. Effect of thiabendazole on development of stem rot of sweet potato. *Phytopathology 62:* 776.

10. McClure, T. T. 1949. Mode of infection of the sweet-potato wilt *Fusarium. Phytopathology 39:* 876–886.

11. McClure, T. T. 1950. Anatomical aspects of the *Fusarium* wilt of sweet potatoes. *Phytopathology 40:* 769–775.

12. McClure, T. T. 1951. *Fusarium* foot rot of sweet-potato sprouts. *Phytopathology 41:* 72–77.

13. Smith, T. E. and K. J. Shaw. 1943. Pathogenicity studies with fusaria isolated from tobacco, sweet potato and cotton. *Phytopathology 33:* 469–483.

14. Smith, S. N. and W. C. Snyder. 1971. Relationship of inoculum density and soil types to severity of *Fusarium* wilt of sweet potato. *Phytopathology 61:* 1049–1051.

15. Struble, F. B., L. S. Morrison and H. B. Cordner. 1966. Inheritance of resistance to stem-rot and to root-knot nematode in sweet potato. *Phytopathology 56:* 1217–1219.

Scurf

Scurf of sweet potatoes was described in 1890 from New Jersey (United States) but has been reported since that time from several other locations, including Florida and Hawaii, as well as Japan and Australia. It probably occurs wherever sweet potatoes are grown. Actual damage to fleshy roots is relatively minor, but salability of diseased roots is reduced considerably.

Infection occurs on all underground plant parts with no symptoms produced on aerial portions of plants, complicating assessment of damage prior to harvest. Small roots and rootlets become discolored, and fleshy roots develop dark colored areas that may coalesce in severe infections to cover most of the root surface (Fig. 93). Penetration remains quite superficial without rotting, but affected roots may shrink more than healthy roots because of excessive loss of moisture from lesioned areas.

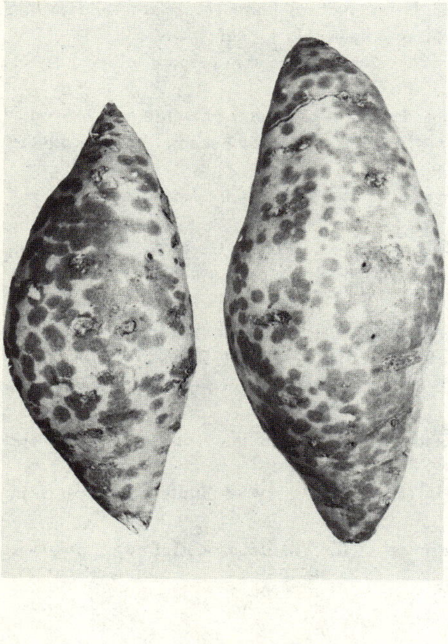

Fig. 93. Scurf symptoms on cv. Centennial sweet potato. (Courtesy of W. J. Martin, Louisiana State University, Baton Rouge.)

Monilochaetes infuscans Ell. & Halst. ex Harter produces hyaline mycelium that darkens with age and from which simple, septate conidiophores 40–175 μm arise. Subhyaline conidia 4–7 × 12–20 μm are borne successively from the apices of conidiophores.

The fungus that causes scurf persists in storage on infected roots and in the field on infected debris for at least a year. Incidence of the disease is greater on heavy soils and in seasons with high rainfall. Inoculum from diseased seed roots assures infection of rooted sprouts, which may occur after transplanting by direct penetration of the epidermis. Disease incidence in sprouts is more severe when seedbed temperature is low (24 C).

Sweet potatoes should not be planted more frequently than in a 3-year crop rotation plan. Seedbeds also should be relocated on soil not previously planted to sweet potatoes. Maneb, benomyl, ferbam and captan have been used to reduce infection on both seed roots and sprouts with some success, but best control has been achieved with dip treatments of 35–40 C.

1. Baldwin, R. E. 1972. Control of sweet potato scurf. *Phytopathology 62:* 667.

2. Cook, H. T. 1948. Control of sweet potato scurf by vine cuttings. *Phytopathology 38:* 568.

3. Cook, H. T. 1955. Sweetpotato diseases. *U. S. Dept. Agr. Farmers' Bull. 1059.* 26 p.

4. Daines, R. H. 1970. Effect of plant bed, prebedding air, and fungicide dip temperature in controlling scurf on sweetpotatoes. *Phytopathology 60:* 1474–1476.

5. Harter, L. L. 1916. Sweet potato scurf. *J. Agr. Res. 5:* 787–791.

6. Kushman, L. J. and E. M. Hildebrand. 1948. Hot water sprout treatment, a promising control for scurf and black rot of sweetpotatoes. *Plant Dis. Reptr. 52:* 475–477.

7. Martin, W. J. 1967. Sweet potato diseases and their control. *Proc. Intern. Symp. Trop. Root Crops (St. Augustine, Trinidad) 2:* 1–10.

8. Nielsen, L. W. and D. E. Yen. 1966. Resistance in sweetpotato to the scurf and black rot pathogens. *New Zeal. J. Agr. Res. 9:* 1032–1041.

9. Nugent, T. J. 1950. The relative importance of the various control measures for sweet potato scurf. *Phytopathology 40:* 873.

10. Weber, G. F. and E. West. 1930. Diseases of sweet potatoes in Florida. *Univ. Florida Agr. Expt. Sta. Bull. 212.* 40 p.

Foot rot

Originally described from Virginia in 1913, foot rot became serious there and rather widespread throughout other sweet potato growing areas of the United States until control measures were devised.

Infection commonly first appears as dark spots on the stem near or just below the soil line, enlarging slowly, finally to girdle the stem and extend a few inches above the soil line. Affected plants, usually still in the seedbed when infected, become stunted and sometimes are killed but more often persist until mid-season or later. Leaves turn yellow on diseased plants that wilt but may survive for varying periods from adventitious root formation and even produce sizeable vines. Fleshy root production is severely restricted, but the infection often extends to roots formed before the plant is killed.

Pycnidia (about 300 μm in diameter), produced by *Plenodomus destruens* Harter, are dark colored and irregular in size and shape. Aseptate, oblong pycnidiospores are hyaline and measure 3–4 × 7–10 μm, while hyaline, aseptate stylospores produced in the same pycnidia are 5–15 μm in length.

Infection typically occurs in the seedbed from diseased roots used for propagating purposes or in the field from debris remaining from infected plants. Dissemination of the fungus in the field is largely by water-borne spores.

Rotation of seedbed and field planting sites, combined with use of clean roots for propagating purposes, usually precludes serious losses from foot rot. The use of vine cuttings in the seedbed assures freedom of transplants from inoculum on seed roots.

1. Harter, L. L. 1913. The foot-rot of the sweet potato. *J. Agr. Res. 1:* 251–274.
2. Harter, L. L. 1913. Foot-rot, a new disease of the sweet potato. *Phytopathology 3:* 243–245.

(Texas) Root rot

Fleshy roots may be affected by a firm, brown dry rot readily identified by external, brown strands of mycelium on roots of affected plants. Vines, too, can be invaded for some distance from infection sites below ground. The disease, caused by *Phymatotrichum omnivorum*, occurs in areas of various sizes in fields of the southern states (United States) and in Mexico, where some losses have been quite high. The fungus can survive in the soil between crops, although it does not withstand freezing temperatures, and on infected roots in storage. Only crop rotation, particularly with sweet clover or grasses, including cereals, has been recommended for control, in addition to use of healthy seed roots.

1. Cook, H. T. 1955. Sweetpotato diseases. *U. S. Dept. Agr. Farmers' Bull. 1059.* 26 p.

Java black rot

Sometimes Java black rot affects plants in the seedbed, particularly on plants subjected to environmental extremes, but it is more often a disease that appears in stored roots. The causal fungus is *Diplodia tubericola* (Ell. & Ev.) Taub., which produces black, globose pycnidia 250–350 μm in diameter with hyaline, aseptate spores 11–14 × 18–20 μm that later turn dark and become uniseptate.

1. Cook, H. T. 1955. Sweetpotato diseases. *U. S. Dept. Agr. Farmers' Bull. 1059.* 26 p.

2. Martin, W. J. 1967. Sweet potato diseases and their control. *Proc. Intern. Symp. Trop. Root Crops* (St. Augustine, Trinidad) 2: 1–10.

Dry rot

Caused by *Diaporthe batatatis* Harter & Field, dry rot is primarily a storage problem in fleshy roots, but stems and vines in the field and young plants in the seedbed can be infected. Both pycnidia (60 × 110–130 μm) and (later) perithecia (120–370 μm in diameter) are produced in the same stromata. Aseptate, hyaline, pycnidiospores measure 3–5 × 6–8 μm, while stylospores, often found in the same pycnidium, are filiform, aseptate, hyaline and 16–30 μm in length. Clavate to cylindrical asci are 7–12 × 23–38 μm, and each enclose 8 nearly elliptical, hyaline, uniseptate ascospores 4–6 × 8–12 μm constricted at the septa.

1. Daines, R. H., E. Breenan and D. A. Leone. 1960. Effect of plant bed temperature and sweet potato dip treatments on incidence of sweet potato sprout decay caused by *Diaporthe batatatis*. *Phytopathology 50:* 186–187.

2. Harter, L. L. and E. C. Field. 1913. A dry rot of sweet potatoes caused by *Diaporthe batatatis*. *U. S. Dept. Agr. Bur. Plant Ind. Bull. 281.*

Stemphylium foliage blight and root rot [*Pleospora herbarum* (Fr.) Rab. =*Stemphylium botryosum* Wallr.]

Brown patches on leaves and production of numerous perithecia on dead petioles and vines are characteristic. A soft, brown, moist rot develops in stored roots infected only through injuries.

1. Booth, C. and K. A. Pirozynski. 1967. *Pleospora herbarum. Commonwealth Mycol. Inst. Descriptions Pathogenic Fungi Bacteria 150.* 2 p.

2. DuToit, J. J. 1951. *Pleospora herbarum* (Pers.) Rabh. (conidial stage *Stemphylium botryosum* Wallr.) on sweet potatoes. *Dept. Agr. S. Afr. Sci. Bull. 301.* 47 p. Also *Rev. Appl. Mycol. 30:* 486 (1951).

Alternaria leaf spot (*Alternaria* sp.)

Alternaria foot rot (*Alternaria solani*)

Pathogenicity of the leaf spot fungus has not been demonstrated beyond question.

1. Ellis, M. B. and I. A. S. Gibson. 1975. *Alternaria solani. Commonwealth Mycol. Inst. Descriptions Pathogenic Fungi Bacteria 475.* 2 p.

2. Taubenhaus, J. J. 1925. A new foot rot of the sweet potato. *Phytopathology 15:* 238–240.

3. Weber, G. F. and E. West. 1930. Diseases of sweet potatoes in Florida. *Univ. Florida. Agr. Expt. Sta. Bull. 212.* 40 p.

Cercospora leaf spot (*Cercospora bataticola* Ciferri & Bruner)

Mostly round leaf spots with dark borders and lighter centers that measure 2–8 mm in diameter are typical of *Cercospora* leaf spot. Conidiophores are lightly colored, septate but not branched and 3–5 × 40–140 μm, and the olivaceous, obclavate conidia measure 3–6 × 20–160 μm. This disease may be confined to the West Indies.

1. Chupp, C. 1953. *A monograph of the fungus genus* Cercospora. Ithaca, N. Y. 667 p.

2. Ciferri, R. and S. C. Bruner. 1931. *Cercospora bataticola* n. sp., parasite of the sweet potato in America. *Phytopathology 21:* 93–96.

Phyllosticta leaf blight [*Phyllosticta batatas* (Thuem.) Cke.]

Angular to circular leaf spots, at first yellowish and later white to brown with necrotic centers and pycnidia 100–125 μm in diameter and ovate, aseptate pycnidiospores 2.5–5.8 × 3.4–9.3 μm, characterize *Phyllosticta* leaf blight. The disease is more serious in southern United States and humid areas, i.e. Caribbean islands.

1. Cook, H. T. 1955. Sweetpotato diseases. *U. S. Dept. Agr. Farmers' Bull. 1059.* 26 p.

2. Martin, W. J. 1967. Sweet potato diseases and their control. *Proc. Intern. Symp. Trop. Root Crops (St. Augustine, Trinidad) 2:* 1–10.

Septoria leaf spot (*Septoria bataticola* Taub.)

White leaf spots to 8 mm in diameter with brown borders were observed in Delaware (United States). Hyaline, septate (3–7 septa), filiform, curved to straight pycnidiospores to 0.5 μm in width and 15–80 μm in length were produced in dark pycnidia 70–130 μm in diameter.

1. Taubenhaus, J. J. 1914. Recent studies of some new or little known diseases of the sweet potato. *Phytopathology 4:* 305–320.

Mineral deficiency symptoms

Foliar disfiguration, including twisting of petioles and distortion of the young tissues at the tops of plants, is typical of *boron* deficiency. Yellowing and premature leaf drop also occur, in addition to the skin of poorly shaped fleshy roots being rough and leathery. Internal, darkened necrotic areas develop, often in the cambial region. Control is readily accomplished by soil application of borax before planting.

Plants deficient in *iron* develop severe chlorosis with entire laminae turning almost white, and only the networks of veins remain green. Vine growth is reduced.

Deficiency of *manganese* results in symptoms similar, but less severe, to those described for plants given insufficient iron.

Interveinal chlorosis on leaves with only somewhat restricted vine growth is caused by lack of adequate *magnesium.* Yield is not noticeably affected.

Symptoms of *calcium* deficiency appear late and include small, chlorotic leaf areas that become necrotic as the leaves become leathery. Reduction in leaf size is a late symptom followed by cessation of growth, necrosis of the growing point and subsequent axillary growth.

Phosphorus deficiency causes reduced vine growth with darkened laminae and purplish coloration of main veins on the under surfaces of leaves.

Interveinal chlorosis, first apparent in the tips of older leaves and which extends progressively around the margins of laminae, is characteristic of *potassium* deficiency. The chlorosis of laminae becomes more general and severe around leaf margins, until only the main secondary veins at the base of the leaf remain green and interveinal necrosis develops. Fleshy root production is seriously reduced.

Vine growth with shortened internodes is severely restricted by lack of sufficient *nitrogen*, and older leaves first exhibit general chlorosis that later extends to young leaves. Petioles and stems develop dark purple pigmentation that also may be evident on fleshy roots but yield is not seriously affected.

1. Cibes, H. and G. Samuels. 1957. Mineral-deficiency symptoms displayed by sweet-potato plants grown under controlled conditions. *Univ. Puerto Rico Agr. Expt. Sta. Tech. Paper 20.* 22 p.

2. Nielsen, L. W. 1965. Blister, a new disease of sweet-potato. *Plant Dis. Reptr. 49:* 97–101.

3. Nusbaum, C. J. 1946. Internal brown spot, a boron deficiency disease of sweet potato. *Phytopathology 36:* 164–167.

4. Spence, J. A. and N. Ahmad. 1967. Plant nutrient deficiencies and related tissue composition of the sweet potato. *Agron. J. 59:* 59–62.

Brazilian sweet potato mosaic

A mosaic disease of sweet potato was first observed in Campinas in 1953, where it reappeared in serious proportions in 1963. Particles of the causal virus measure 15 × 850 nm, and transmission in non-persistent fashion is accomplished by *Myzus persicae* and *Aphis gossypii.* Chlorotic, localized lesions develop on *Chenopodium amaranticolor*, *C. murale* L. and *C. quinoa*, while *Amaranthus* sp. also is susceptible to the virus, which is considered a member of the potato Y virus group.

1. Costa, A. S., E. W. Kitajima and E. S. Normanha. 1973. Mosaico da batata-doce, causado por virus do grupo Y. *Fitopatologia 8:* 7.

2. Kitajima, E. W. and A. S. Costa. 1973. Morfologia e aspectos intracelulares do virus do mosaico da batata doce. *Fitopatologia 8:* 9.

Feathery mottle complex

Mosaic symptoms in sweet potato were reported from Arkansas (United States) in 1920, but disease incidence did not justify serious attention. More than 20 years later (1944), losses in African countries were considerably greater.

A number of aphid-transmitted sweet potato diseases were described and named subsequent to the original report of the "feathery mottle virus" disease in 1945 from Maryland. A "mosaic disease" of sweet potato was described from Puerto Rico in 1955, "internal cork" disease was reported from South Carolina in 1946, "mosaic virus A" from Africa in 1957 and "leafspot virus" and "ringspot virus" 3 years later from the United States and Israel, respectively. "Russet crack" disease was not reported until 1964, although it had been seen in New Jersey 3 years earlier. All of these diseases were distinguished by symptom descriptions, as the common names suggest, which often were characteristic only in one or a few specific cultivars of sweet potato. Further complication was added to this situation by naturally occurring multiple virus infections in some sweet potato cultivars. Losses attributable to virus diseases have not been accurately assessed but have been sufficient in some locations to discourage farmers from planting sweet potatoes.

Symptoms originally ascribed to the *feathery mottle* disease, noted in one plant of a cultivar imported from Russia in 1934, included definite yellowing along the veins on young leaves or small, indiscrete yellowish spots 1–2 mm in diameter, often accompanied by vein clearing. Affected leaves later developed an irregular blotchy mottle with "feathery" patches of yellowing near veins but little distortion of leaves somewhat reduced in size with slight rugosity. Diseased plants developed shortened internodes, but no necrosis appeared on leaves, stems or on roots that were smaller in size than

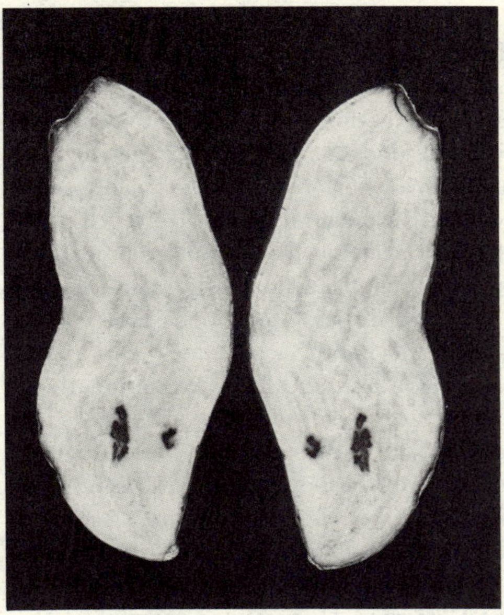

Fig. 94. Internal cork (feathery mottle) lesions in cv. Goldrush sweet potato. (Courtesy of W. J. Martin, Louisiana State University, Baton Rouge.)

from healthy plants. "Leaf pucker" and "vein-clearing" symptoms later described from Israel may have been different manifestations of mottling and leaf rugosity noted earlier.

Internal cork symptoms were described as necrotic areas irregular in size and shape but clearly delimited in thin slices of fleshy roots (Fig. 94). After exposure to air for several hours, the dark colored necrotic tissues faded to brown. Although spots occurred singly or in groups, no external symptoms were evident, except that cavities sometimes developed on the surface of roots over places where necrotic spots formed internally near the periphery of the root. Foliage symptoms associated with internal cork disease consisted entirely of late-season appearance of irregularly circular, purple rings 0.5–3 cm in diameter with green centers. *Ringspot virus* disease symptoms consisted of similar but smaller (4–5 mm in diameter) leaf symptoms on plants that were reduced in size and became chlorotic.

The disease named *russet crack* was characterized by small, cortical cracks in dark lesions that extended in bands of various width partly or completely around the girth of fleshy roots. One or more such bands developed on a single fleshy root, or the whole surface of the root was affected, but the inner tissues remained unaffected. Roots sometimes were constricted to some extent under bands, while mildly affected roots developed only small, dark spots with longitudinal cracks. Feeder roots also exhibited shallow cracks and discoloration that often girdled the root. Discolored lesions appeared on underground stems and, more infrequently, stems above ground level. Dark spots, streaks or ringspots with irregular margins and light colored centers occurred on foliage.

Each of the disease syndromes described probably is a manifestation of a single virus disease, herein designated "feathery mottle." Differences in symptom expression are considered to result from infection of specific host cultivars by different virus isolates or strains. Not all of the causal agents have been characterized adequately for critical comparisons. However, most of the diseases have been transmitted by grafting, by aphids, particularly *Myzus persicae*, in a stylet-borne manner, and mechanically from infected foliage with some irregularity attributable, in part, to a virus-inactivating material present in sweet potato leaves. Virus particles extracted from russet crack plants are elongated and measure 876 nm, and 844 nm from feathery mottle plants. Infectivity has been limited to species of *Ipomoea*, with some individual species and cultivars, e.g. *I. nil* (L.) Roth (Scarlet O'Hara) and *I. setosa* Ker., particularly useful as indicator plants, but infection of *I. batata* from (mechanical) inoculation has been somewhat erratic.

Attempts to eliminate infection from fleshy roots by subjecting them to various heat treatments have not been completely successful. Only roots from healthy plants of cultivars least affected by localized form(s) of the disease should be used for propagating purposes.

1. Adsuar, J. 1955. A mosaic disease of sweetpotato, *Ipomoea batatas*, in Puerto Rico. *J. Agr. Univ. Puerto Rico 39:* 49–50.

2. Alconero, R. 1971. Sweetpotato virus infections in Puerto Rico. *Plant Dis. Reptr. 55:* 902–906.

3. Campbell, R. N., D. H. Hall and N. M. Mielinis. 1974. Etiology of sweet potato russet crack disease. *Phytopathology 64:* 210–218.

4. Daines, R. H. and W. J. Martin. 1964. Russet crack, a new virus disease of sweetpotatoes. *Plant Dis. Reptr. 48:* 149–151.

5. Doolittle, S. P. and L. L. Harter. 1945. A graft-transmissible virus of sweet potato. *Phytopathology 35:* 695–704.

6. Hansford, C. G. 1944. A probable virus disease of sweet potato. *East African Agr. J. 10:* 126–127.

7. Hildebrand, E. M. 1956. Sweet potato internal cork virosis indexed on Scarlett O'Hara morning glory. *Science 123:* 506–507.

8. Hildebrand, E. M. 1959. Sweet potato ringspot virus associated with internal cork virus. *Phytopathology 49:* 524.

9. Hildebrand, E. M. 1960. The feathery mottle virus complex of sweet potato. *Phytopathology 50:* 751–757.

10. Hildebrand, E. M. 1964. Heat treatment for eliminating internal cork viruses from sweetpotato plants. *Plant Dis. Reptr. 48:* 356–358.

11. Hildebrand, E. M. 1967. Russet crack—a menace to the sweet potato industry. I. Heat therapy and symptomatology. *Phytopathology 57:* 183–187.

12. Hildebrand, E. M. 1968. Russet crack—a menace to the sweet potato industry. II. Natural transmission. *Plant Dis. Reptr. 52:* 967–971.

13. Hildebrand, E. M. 1969. Russet crack—a menace to the sweet potato industry. III. Role of fungi in the root-lesion phase of russet crack. *Plant Dis. Reptr. 53:* 103–107.

14. Hildebrand, E. M. and F. F. Smith. 1958. Aphid transmission of a virus associated with sweet potato internal cork and masked in feathery mottle-infected sweet potatoes. *Plant Dis. Reptr. 42:* 1148–1153.

15. Kantack, E. J., W. J. Martin and L. D. Newsom. 1958. Transmission of internal cork of sweet potato by the cotton aphid, *Aphis gossypii. Science 127:* 1448.

16. Kantack, E. J., W. J. Martin and L. D. Newsom. 1960. Relation of insects to internal cork of sweet potato in Louisiana. *Phytopathology 50:* 447–449.

17. Loebenstein, G. and I. Harpaz. 1960. Virus diseases of sweet potatoes in Israel. *Phytopathology 50:* 100–104.

18. Martin, W. J. 1962. Susceptibility of certain *Convolvulaceae* to internal cork, tobacco ringspot, and cucumber mosaic viruses. *Phytopathology 52:* 607–611.

19. Martin, W. J. 1970. The reproduction of russet crack in Jersey Orange sweetpotatoes by grafting on plants affected with either sweetpotato virus leaf spot or internal cork. *Phytopathology 60:* 1302.

20. Martin, W. J. and E. J. Kantack. 1960. Control of internal cork of sweet potato by isolation. *Phytopathology 50:* 150–152.

21. McClean, A. P. D. and P. J. Klesser. 1947. Mottle-leaf disease of the sweet potato. *Farming S. Afr. 22:* 897–900.

22. Nusbaum, C. J. 1946. Internal cork, a new disease of sweet potato of unidentified cause. *Phytopathology 36:* 18–23.

23. Rankin, H. W. and J. H. Girardeau. 1958. Transmission by *Myzus persicae* (Sulz.) of the internal cork virus of sweet potatoes. *Plant Dis. Reptr. 42:* 581–582.

24. Salama, F. M., C. M. Lyman and U. G. Whitehouse. 1966. Isolation of the internal cork virus from sweet potato leaves having the purple ring structure by use of differential filtration. *Phytopathol. Z. 57:* 89–95.

25. Sheffield, F. M. L. 1957. Virus diseases of sweet potato in East Africa. I. Identification of the viruses and their insect vectors. *Phytopathology 47:* 582–590.

26. Sheffield, F. M. L. 1958. Virus diseases of sweet potato in East Africa. II. Transmission to alternative hosts. *Phytopathology 48:* 1–6.

27. Webb, R. E. 1954. Some hosts of feathery mottle virus of sweet potato. *Plant Dis. Reptr. 38:* 448–452.

28. Webb, R. E. and R. H. Larson. 1954. Mechanical and aphid transmission of the feathery mottle virus of sweet potato. *Phytopathology 44:* 290–291.

29. Williams, A. S. 1954. Development of foliage symptoms associated with internal cork of sweet potato. *Phytopathology 44:* 334.

Little leaf, Witches' broom, Ishuku-byo, Weed variant

A disease presumed to be virus-induced (transmitted by leafhoppers and grafting) and characterized by excessive proliferation of shoots from leaf axils, dwarfing and failure to produce fleshy roots was reported from the Ryukyu Islands in 1951. "Little leaf" symptoms also were reported from Papua New Guinea in 1967 and "weed variant" symptoms were reported from the Solomon Islands in 1975. Similar symptoms were noted in Maryland (United States) in a cultivar (Plant Introduction 308200) introduced from Tonga (via New Zealand) and ascribed to a mycoplasma-like organism (100–1000 nm in diameter), despite the presence of virus particles similar to those associated with russet crack disease. Proliferation symptoms did not develop for 2 years following treatment of cuttings or plants with symptoms with the calcium salt of oxytetracycline, either as soaks (100–500 μg/ml for 48–72 hour treatments) or in irrigation water (1000 μg/ml for 6 weeks).

1. Dabek, A. J. and D. B. Gollifer. 1975. Mycoplasma-like organisms in *Ipomoea batatas* in the Solomon Islands. *PANS 21:* 384–387.

2. Kahn, R. P., R. H. Lawson, R. L. Monroe and S. Hearon. 1972. Sweet potato little-leaf (witches' broom) associated with a mycoplasmalike organism. *Phytopathology 62:* 903–909.

3. Lawson, R. H., R. P. Kahn, S. Hearon and F. F. Smith. 1970. The association of mycoplasmalike bodies with sweetpotato little-leaf (witches' broom) disease. *Phytopathology 60:* 1016.

4. Summers, E. M. 1951. "Ishuku-byo" (dwarf) of sweetpotato in Ryukyu Islands. *Plant Dis. Reptr. 35:* 266–267.

5. Summers, E. M. 1951. Witches broom (ishuku-byo) of sweet potatoes in the Ryukyu Islands. *Plant Dis. Reptr. 35:* 494.

6. Van Velsen, R. J. 1967. "Little leaf" a virus disease of *Ipomoea batatas* in Papua and New Guinea. *Papua New Guinea Agr. J. 18:* 126–128.

Mild mottle

Mild, systemic, mottle symptoms in sweet potato in Kenya, Tanzania and Uganda have been attributed to a virus transmissible mechanically, by dodder and by whitefly. Particles of the virus are filamentous, straight, 800–950 nm in length and are not infectious (in sap from infected tobacco) after exposure to 60 C (for 10 minutes), storage *in vitro* for 7 days or dilution greater than 2×10^{-4}. The virus, which is similar in many respects to virus B described by Sheffield, is infectious on a wide range of plants, including members of the Chenopodiaceae, Leguminosae and Solanaceae, as well as the Convolvulaceae.

1. Hollings, B., O. M. Stone and K. R. Bock. 1976. Purification and properties of sweet potato mild mottle, a white-fly borne virus from sweet potato (*Ipomoea batatas*) in East Africa. *Ann. Appl. Biol. 82:* 511–528.

2. Sheffield, F. M. L. 1957. Virus diseases of sweet potato in East Africa. I. Identification of the viruses and their insect vectors. *Phytopathology 47:* 582–590.

3. Sheffield, F. M. L. 1958. Virus diseases of sweet potato in East Africa. II. Transmission to alternative hosts. *Phytopathology 48:* 1–6.

Mosaic

Symptoms in sweet potato described as "mosaic" were reported from Arkansas (United States) in 1920 and later noted from Kansas, having been observed there every year from 1927 to 1959. The disease severely affected plants, but incidence was low and of no economic concern to growers.

Affected plants are noticeably stunted with greatly reduced vine growth and misshapen leaves that show mottle symptoms. Fleshy roots produced on infected plants are smaller and more uniformly shaped than on healthy plants.

The causal virus is a strain of tobacco mosaic infectious by mechanical inoculation to a number of solanaceous plants, including tomato, tobacco, pepper, petunia and species of *Physalis*, as well as globe amaranth and *Vinca minor* L. In sap from infected tobacco, the virus was inactivated by exposure to temperatures between 85–90 C, dilution greater than 10^{-6} or storage *in vitro* longer than 3 years. The rod-like particles, shadowed with chromium, measured 41 × 300 nm.

Incidence has not warranted control measures, but roots from infected plants should not be used for propagating material.

1. Elmer, O. H. 1957. Sweet potato mosaic. *Phytopathology 47:* 9.

2. Elmer, O. H. 1960. Etiology and characteristics of sweet potato mosaic. *Phytopathology 50:* 744–749.

3. Rosen, H. R. 1920. The mosaic disease of sweet potatoes. *Arkansas Univ. Fayetteville Agr. Expt. Sta. Bull. 167.* 10 p.

4. Rosen, H. R. 1926. The mosaic disease of sweet potatoes with special reference to its transmissibility. *Arkansas Univ. Fayetteville Agr. Expt. Sta. Bull. 213.* 16 p.

5. Sill, W. H., Jr., S. B. Lal and M. E. del Rosario. 1960. Additional evidence that sweetpotato mosaic virus is a strain of tobacco mosaic virus. *Phytopathology 50:* 709–711.

Vein mosaic

A decline disease of sweet potato described from Argentina in 1973 is characterized by general chlorosis and stunting of affected plants in the field. Leaves exhibit a definite vein clearing and often a mottle pattern between main veins.

The causal virus consists of flexuous rods 13 × 761 nm, inactivated by exposure to 62–65 C, dilution between 10^{-4} and 10^{-5} or storage longer than 10–20 days *in vitro* at 22 C. Transmission is effected by *Myzus persicae* in a nonpersistent manner after 5 minute acquisition and infection feedings. The virus also can be transmitted by mechanical inoculation with highest rate of infection (70–80 percent) by preparation of inoculum to contain 0.03 M potassium phosphate buffer (pH 7.6), activated charcoal (1 percent) and sodium diethyl-dithiocarbamate equal to half the weight of fresh leaf tissue. Infectivity is restricted to members of the *Convolvulaceae*.

1. Nome, S. F. 1973. Sweet potato vein mosaic virus in Argentina. *Phytopathol. Z. 77:* 44–54.

2. Nome, S. F., T. A. Shalla and L. J. Petersen. 1974. Comparison of virus particles and intracellular inclusions associated with vein mosaic, feathery mottle, and russet crack diseases of sweet potato. *Phytopathol. Z. 79:* 169–178.

Yellow dwarf

Sweet potato diseases of undetermined etiology were reported from Africa in 1939 and 1944 and from Georgia (United States) in 1954, although the disease in Georgia had been known to occur there since 1945. Symptomatology in both locations may have been complicated because of multiple infections, but yellowing of veins in young leaves and shortening of internodes on affected plants prompted later application of the common name "yellow dwarf." This name is used here to distinguish the disease of sweet potato transmitted by the whitefly *Bemisia tabaci* from another disease of sweet potato, i.e., mosaic, caused by a completely different virus and first reported under this common designation some 20 years before yellow dwarf was recognized. Attempts to characterize the causal agent of yellow dwarf disease may have been complicated by use of contaminated inocula, but there is some evidence that globe amaranth, *Datura ferox* L., tomato and tobacco, in addition to species of *Ipomoea*, are susceptible to infection. The economic importance of yellow dwarf alone as a disease of sweet potato remains to be established.

1. Borders, H. I. and T. J. Ratcliff. 1954. A mosaic of sweetpotato in plant beds and fields in Georgia. *Plant Dis. Reptr. 38:* 6–9.

2. Girardeau, J. H. 1958. The sweetpotato whitefly, *Bemisia inconspicua* (Q), as a vector of a sweetpotato mosaic in south Georgia. *Plant Dis. Reptr. 42:* 819.

3. Girardeau, J. H., Jr. and T. J. Ratcliffe. 1960. The vector-virus relationship of the sweetpotato whitefly and a mosaic of sweetpotatoes in south Georgia. *Plant Dis. Reptr. 44:* 48–50.

4. Sheffield, F. M. L. 1957. Virus diseases of sweet potato in East Africa. I. Identification of the viruses and their insect vectors. *Phytopathology 47:* 582–590.

5. Sheffield, F. M. L. 1958. Virus diseases of sweet potato in East Africa. II. Transmission to alternative hosts. *Phytopathology 48:* 1–6.

T

TEA: *Camellia sinensis* (L.) O. Kuntze

Scab diseases

Symptoms of "mottle" scab disease were found in Sri Lanka in 1917 and described 5 years later. The same disease was found in India in 1917, in

Brazil in 1939, soon thereafter in Tanganyika, and 4 years later in Uganda. Also in 1939, a "white scab" disease was reported from Japan and additional specimens of this disease were collected a year later in Guatemala. Two earlier collections of tea scab, presumed to be white scab, had been made in Florida (1893) and Cuba (1922).

Only full grown leaves develop symptoms of mottle scab, which begin as yellow-green spots that become dark brown and finally black. All the different colors may be seen in developing lesions, and this coloration imparts a mottled appearance, from which the common name for the disease is derived. Lesions are irregular or angular in shape, usually numerous and vary considerably in size but may coalesce to involve the whole leaf. As the lesions develop, the affected tissues become raised and eventually bulge upward, become brittle and crack but are usually confined to the upper leaf surfaces.

Symptoms of white scab develop on both surfaces of young leaves and on twigs as small, dark purple to brown spots that are first raised and later become depressed. As the centers of the lesions become sunken, coloration changes to light brown and later gray-white as the spots enlarge to 1–3 mm and assume a circular shape. Margins of spots remain dark colored and clearly demarcated, although lesions on lower leaf surfaces often coalesce to form irregular blotches that feel rough to the touch. Light colored lesions also form on stems and are more numerous on stem surfaces exposed to direct sunlight.

Elsinoë theae Bitancourt & Jenkins is the causal agent of mottle scab and produces intraepidermal ascomata 15–18 × 200–300 μm, in which are produced asci 12–20 × 14–22 μm, each with 8 hyaline, 4-celled ascospores 3–7 × 10–14 μm. Numerous, hyaline, spherical (0.5–1 μm diameter) microconidia are produced.

Erumpent ascomata of *Elsinoë leucospila* Bitancourt & Jenkins (*Sphaceloma theae* Kurosawa), which causes white scab, measure 50–80 × 50–100 μm with globose asci 20–50 μm in width, each with 8 hyaline, 2–3 septate ascospores 5–7 × 13–16 μm. Acervuli with erumpent sporodochia characterize the imperfect (*Sphaceloma*) stage.

Control measures have not been devised.

1. Bitancourt, A. A. and A. E. Jenkins. 1939. "*Elsinoë theae* n. sp.," agente da verrugose do chá. *Arq. Inst. Biol. 10:* 193–198.

2. Chaudhuri, H., V. S. Kapur and L. L. Bhatia. 1937. Diseases of the tea bush in the Kangra Valley. *Indian J. Agr. Sci. 7:* 565–573.

3. Hirokawa, S. 1965. On the causal fungus of the leaf spot disease of tea plant. *Japanese Tea Res. Sta. Study Tea 30:* 31–38.

4. Jenkins, A. E. and A. A. Bitancourt. 1946. Duas verrugoses do chá, causadas por "*Elsinoë*," e sua distribuição. *Arq. Inst. Biol. 17:* 67–72.

5. Kurosawa, E. 1939. *Sphaceloma theae* n. sp. *Ann. Phytopathol. Soc. Japan 9:* 131–132.

Blister blight

The culture of tea in countries other than Japan, where the tea plant was taken from China about AD 800, began in the Brahmaputra Valley of northeast India about 1840. Less than 20 years later (1858), blister blight disease was recognized but given little attention, and the culture of tea, encouraged by the disaster of coffee rust in 1877, spread to Sri Lanka, Iran, Georgian Russia and, much later, Turkey. Tea growing in Java also was successful for the first time, just before the turn of the century, when the first spread and serious outbreaks of blister blight occurred in northeastern India in 1895 and 1900. Despite the appearance of the disease in Formosa (1912), Italy (1925) and Indochina (1930), no serious consideration was given blister blight, until it quickly spread throughout southern India and across Sri Lanka in about 2 months time in late 1946. Sumatra with 10,000 ha of tea suffered the same fate 3 years later. Intense research programs soon assured the continued culture of tea in Sri Lanka (and elsewhere), but even recent estimates of 30 percent crop losses verify the potential serious consequences of blister blight, which remains the most important leaf disease of tea.

Evidence of infection is appearance of generally round, translucent spots lighter in color than surrounding healthy tissues and sometimes slightly pinkish. These spots gradually enlarge to 6–12 mm and become more conspicuous as concave depressions on the upper leaf surfaces and corresponding convex "blisters" on the lower leaf surfaces, although these positions of concavity and convexity may occasionally be reversed. Coloration of the lower, convex tissues turns from gray-brown to white, and the surfaces of the blisters rupture to expose velvety, sporulating tissues. After spores have been released, lesioned tissues are often invaded by other pathogens and subsequently become blackened, drop out and later result in defoliation, sometimes more serious than that caused by the original blister blight infection.

Infection is largely confined to young, expanding leaves less than 30 days old but also may involve young, green stems (Fig. 95). Diseased tissues, sometimes involving buds, are covered by white mycelium and are killed, resulting in death also of shoot parts above the infection sites.

Causal agent of blister blight is *Exobasidium vexans* Massee, an obligate parasite that forms compact mycelial strands into hymenia with clavate basidia interspersed among sterile hyphae. Usually 2 thin-walled, long, oval to slightly curved, septate, hyaline basidiospores 3.5–5.6 × 14–17 μm are borne on each basidium. Basidiospores, which may be thick-walled under some conditions, are covered by an extremely adhesive layer of mucilage.

Dissemination of the fungus is achieved via the basidiospores, which are produced in great abundance (a million or more spores may be produced in a single blister), are quite light and readily air-borne and strongly adhere to any surface because of the mucilaginous covering. However, basidiospores

Fig. 95. Blister blight lesions on twig and leaves of tea. (Courtesy of Tea Research Institute, Talawakelle.)

do not withstand unfavorable conditions well and are killed by exposure to 35 C or 30 minutes direct sunlight and require 50 percent relative humidity to remain viable. Germination of basidiospores has been reported to occur at 80 percent relative humidity, but prior exposure to light may alter the speed of the germination process. Free water, even as a thin film, promotes germination, which may begin within 2 hours after the basidiospore is released but also may be delayed for as long as 4–5 days or until conditions are favorable. Germination occurs more quickly at 25 C than at 15 C. Formation of germ tubes and appressoria immediately precedes infection, which usually follows within 24 hours with subsequent development of intercellular hyphae and intracellular, lobed haustoria. Incubation may persist for 3–10 days before translucent spots appear, and 6–9 days may elapse before the rupture of blisters and sporulation begins. Thus, the entire life cycle of the fungus may vary in length from 11–28 days.

Each blister results from a single infection and, although growth of the fungus is restricted inside the leaf, blisters may be so numerous as to appear to coalesce, and the whole undersurface of the leaf may be affected. Infections on young stems do not result in noticeable blisters, but ramification of the mycelium may reach the pith.

Prevalence of basidiospores in the air has been found to follow a diurnal cycle, with the greatest concentration of spores occurring between midnight and 0400 hours. However, heavy rain may alter this periodicity by causing an increase in spore concentration soon after rainfall begins, but prolonged periods of rain or drought may cause a drop in quantity of airborne spores.

Elevation is quite influential in determining severity of blister blight. Lower temperatures and more intense solar insolation at high elevations retard blight development, but more interruptions in periods of full sunshine and longer duration of rainy periods both serve to enhance disease severity.

Much attention has been given to control of blister blight and environmental factors contributing to disease development examined extensively. Recent experimentation has verified that better field control is achieved with copper fungicides, because of better tenacity under high rainfall conditions, than organic fungicides, many of which are more toxic to the fungus under laboratory conditions. Either dusts or sprays, depending on availability of water, can be used successfully, but application should be made soon after plucking to reduce copper content in the next harvest to an acceptable level. Copper oxide (50 percent wettable powder) at a rate of 461 g/ha, applied often as a low-volume spray at 7–10 day intervals, provides satisfactory control, as does copper oxychloride (4–6 percent) at 4.6 kg/ha on a 5–8 day schedule. Disease forecasting based on hours of sunshine offers promise of more economical disease control but should be tried first on an experimental basis.

1. De Silva, R. L. 1965. Investigations with new fungicides for the control of blister blight (*Exobasidium vexans* Massee) on tea. *Tea Quart. 37:* 64–71.

2. De Silva, R. L. 1965. The use of nickel-chloride and "Perezin" for the control of blister blight (*Exobasidium vexans* Massee) on tea. *Tea. Quart. 36:* 191.

3. De Silva, R. L., S. Murugiah and T. V. Saravanapavan. 1972. Losses of tea crops caused by *Exobasidium vexans* Massee on tea. I. Losses on unshaded seedling tea. *Tea. Quart. 43:* 140–146.

4. De Weille, G. A. 1960. Blister blight (*Exobasidium vexans*) in tea and its relationship with environmental conditions. *Neth. J. Agr. Sci. 8:* 183–210.

5. Eden, T. 1949. Recent advances in tea research. II. Pathological problems. *World Crops 1:* 132–135.

6. Gadd, C. H. and C. A. Loos. 1948. The basidiospores of *Exobasidium vexans. Trans. Brit. Mycol. Soc. 31:* 229–233.

7. Gadd, C. H. and C. A. Loos. 1949. The fungus *Exobasidium vexans. Tea Quart. 30:* 54–61.

8. Gadd, C. H. and C. A. Loos. 1950. Further observations on the spore growth of *Exobasidium vexans. Trans. Brit. Mycol. Soc. 33:* 19–21.

9. Kerr, A. and R. L. De Silva. 1969. Epidemiology of tea blister blight (*Exobasidium vexans*). *Tea Quart. 40:* 9–18.

10. Kerr, A. and W. R. F. Rodrigo. 1967. Epidemiology of tea blister blight (*Exobasidium vexans*). III. Spore deposition and disease prediction; IV. Disease forecasting. *Trans. Brit. Mycol. Soc. 50:* 49–55; 609–614.

11. Kerr, A. and N. Shanmuganathan. 1966. Epidemiology of tea blister blight (*Exobasidium vexans*). I. Sporulation. *Trans. Brit. Mycol. Soc. 49:* 139–145.

12. Loos, C. A. 1951. The causative fungus. *Tea Quart. 22:* 63–72.

13. Mulder, D. and R. L. De Silva. 1960. A forecasting system for blister-blight control, based on sunshine records. *Tea Quart. 31:* 56–67.

14. Portsmouth, G. B. 1961. Blister blight disease of tea. *Outlook Agr. 3:* 81–88.

15. Portsmouth, G. B. and C. A. Loos. 1949. Blister blight—a review. *Tea Quart. 20:* 77–84.

16. Shanmuganathan, N. and P. V. Arulpragasam. 1966. Epidemiology of tea blister blight (*Exobasidium vexans*). II. The diurnal and seasonal periodicity of spores in the air over a tea estate. *Trans. Brit. Mycol. Soc. 49:* 219–226.

17. Shanmuganathan, N. and T. V. Sarávanapavan. 1974. Comparative effectiveness of copper and organic fungicides against tea leaf blister blight (*Exobasidium vexans*). *Plant Dis. Reptr. 58:* 928–932.

18. Shaw, D. E. 1965. Condition resembling blister blight of tea on tea seedlings in quarantine in New Guinea. *FAO Plant Protect. Bull. 13:* 56–64.

19. Subba Rao, M. K. 1946. Blister blight of tea in South India. *Planters' Chron. 41:* 462–463.

20. Thompson, A. 1949. "Blister blight" of tea. *Malayan Agr. J. 32:* 25–27.

21. Tubbs, F. R. 1947. A leaf disease of tea new to Ceylon. *Tea Quart. 19:* 43–45.

22. Venkata Ram, C. S. and V. Jayaraman. 1961. The relation of interval and rate of application of nickel chloride and other inorganic and organic fungicides to the infection of tea plants by *Exobasidium vexans* Massee. *Tea Trade Ind. 10:* 199–204.

23. Visser, T., N. Shanmuganathan and J. V. Sabanayagam. 1961. The influence of sunshine and rain on tea blister blight, *Exobasidium vexans* Massee, in Ceylon. *Ann. Appl. Biol. 49:* 306–315.

Collar and branch canker

A dieback of young shoots in Sri Lanka was originally described in 1927 and the causal nature established 8 years later. Erratic occurrence of the disease was largely responsible for its being ignored, until more serious incidence erupted after 1957. Canker remains a minor disease in Sri Lanka and India, although affected plants often are killed.

Diseased plants are noticeable from a distance because of the yellow and brown leaves on dead branches or bushes, which are a marked contrast to the surrounding green foliage. Lesions at the bases of dead branches or bushes result from slow progress and death of the bark near the site of infection, and the margins of such cankers are characteristically raised from callus formation (Fig. 96). Affected tissues are often darkened in color, sunken and irregular in shape.

The causal fungus is *Phomopsis theae* Petch, that produces dark, globose pycnidia and 2 types of conidia, one ovoid, hyaline and aseptate (alpha), often extruded in yellowish cirri, and the second (beta) filiform and curved. (The pathogen has been identified by some investigators as *Leptothyrium theae* Petch.)

Occurrence of the disease is largely restricted to young plantings 2–8 years old but is quite sporadic. It occurs more frequently at elevations of

Fig. 96. Dieback and collar canker of tea. (Courtesy of Tea Research Institute, Talawakelle.)

1200 m or above, and there are some indications of differences in clonal susceptibilities. Dissemination of spores is primarily by water droplets, and infection can occur only through wounds, but some of the most severe cases of canker have occurred after exceptionally dry weather and on poor soils. Further, canker development has been found to be favored by low soil moisture.

Clones locally known to be especially susceptible to canker should not be used for new plantings. Care should be exercised to avoid unnecessary injury to plants during cultivation, and pruning injuries should be protected by application of fungicide.

1. Kasai, K., S. Hirokawa, S. Takaya and H. Kibushi. 1965. Factors affecting the occurrence of die-back and canker diseases of tea. *Japanese Tea Res. Sta. Study Tea 30:* 39–58.

2. Punithalingam, E. and I. A. S. Gibson. 1972. *Phomopsis theae. Commonwealth Mycol. Inst. Descriptions Pathogenic Fungi Bacteria 330.* 2 p.

3. Shanmuganathan, N. 1965. Collar and branch canker in young tea caused by *Phomopsis theae* Petch. *Tea Quart. 36:* 14–21.

4. Shanmuganathan, N. and W. R. F. Rodrigo. 1966. Studies on collar and branch canker of young tea (*Phomopsis theae* Petch). I. Recent observations on disease incidence. *Tea Quart. 37:* 221–228.

5. Shanmuganathan, N. and W. R. F. Rodrigo. 1967. Studies on collar and branch canker of young tea (*Phomopsis theae* Petch). II. Influence of soil moisture on the disease. *Tea Quart. 38:* 320–330.

Red root

It has been suggested that red root of tea has occurred in Sri Lanka since tea was first planted there, although when the disease was first noted is uncertain. The disease also is prevalent in India. Because there are a number of root diseases of tea and because of the early nature of these diseases, losses directly attributable to red root have not been determined, but it does occur more widely than the other root diseases and is the most difficult to control.

Infection and early disease development typically go undetected until there is no possibility to save the plant. Visible, above-ground symptoms appear only after the disease is far advanced in the root systems of plants that are then moribund. Infection in many cases has already spread to adjacent plants, which in turn show no symptoms until the disease is so advanced that they are dying.

The causal fungus is *Poria hypobrunnea* Petch (= *Poria hypolateritia*) characterized by resupinate, annual carpophores 3–6 mm in thickness with pores round to angular and 8–11 mm (Fig. 97). Coloration of the pore surface changes from yellow-white through pink to brown and finally gray-purple. Clavate basidia measure 4.5–5 × 9–10.5 μm and ovoid basidios-

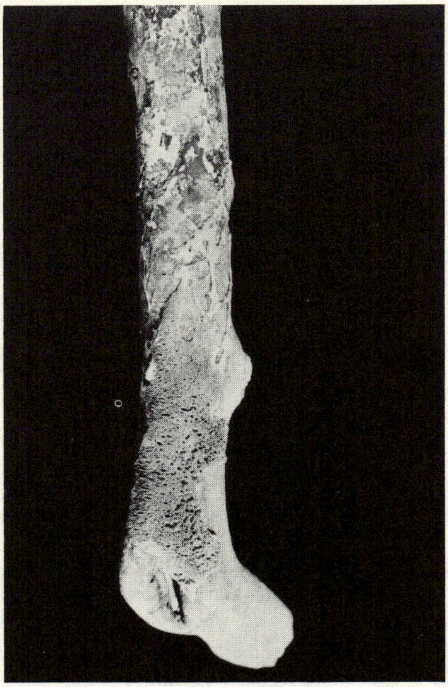

Fig. 97. Red root infection of tea. (Courtesy of Tea Research Institute, Talawakelle.)

pores 3.5–5 × 4–5 μm. Cystidia are hyaline, thin-walled and 5–10 × 15–45 μm.

Infection results from exposure to roots or root debris from infected (jungle or other tea) plants or mycelium from soil or debris that remains after clearing for planting or uprooting infected tea plants.

Control can be effected by soil fumigation with methyl bromide (0.2 kg/10 m^2) with the cover left in place for 48 hours. Care should be taken not to reinfect treated soil with soil, roots or debris from infected areas.

1. Agnihothrudu, V. 1963. Some important root diseases, red rust and *Poria*. *Tocklai Expt. Sta. Newsletter 10:* 9–12.

2. Harler, C. R. 1973. Pests and diseases of tea. *World Crops 25:* 245–251.

3. Mulder, D. and W. Redlich. 1962. Red root disease in tea. *Commonwealth Phytopathol. News 8:* 43–44.

4. Pegler, D. N. and I. A. S. Gibson. 1972. *Poria hypobrunnea. Commonwealth Mycol. Inst. Descriptions Pathogenic Fungi Bacteria 322.* 2 p.

5. Shanmuganathan, N. 1964. Recent developments in the control of *Poria* root disease. *Tea Quart. 35:* 22–31.

6. Shanmuganathan, N. 1969. Studies on the parasitism and control of tea root disease fungi in Ceylon. *Tea Quart. 40:* 19–25.

7. Shanmuganathan, N. and S. R. A. Fernando. 1967. Some observations on *Poria* control by soil fumigation with methyl bromide. *Tea Quart. 38:* 311–319.

8. Shanmuganathan, N. and W. W. Redlich. 1965. Control of *Poria* root disease with methyl bromide. *Tea Quart. 36:* 144–150.

Phloem necrosis

The first published account of phloem necrosis of tea in Sri Lanka appeared in 1939, although the disease was thought to have been present for several years because this unproductive condition of plants even then was rather general throughout tea estates at the higher elevations. It was slowly increasing in prevalence and causing sizeable losses before being brought to attention publicly at a conference of planters. Estimates of plant infections have ranged to 70 percent in some fields (average incidence of 25 percent in all estates above 1800 m elevation), of which 15 percent may be entirely unproductive. The disease is known to occur only in Sri Lanka and only at the higher elevations there.

A number of symptoms may be found on plants affected by phloem necrosis disease, but the complete syndrome is complex, and many individual symptoms are not completely reliable for diagnosis. Diseased plants assume a gradual decline, become stunted and unproductive and often have some curled leaves, shortened internodes and stems that exhibit zigzag growth (Fig. 98). Leaf size is noticeably reduced, and leaves often are arranged in a single plane on the stem, rather than being distributed more or less evenly all around the stem, as on healthy bushes. The production of flowers, fruits and seeds is early and abundant on diseased plants. The only diagnostic symptom of the disease is necrosis of the phloem that can be found

Fig. 98. Zig-zag growth typical of phloem necrosis disease of tea. (Courtesy of Tea Research Institute, Talawakelle.)

in roots, stems, leaf petioles and midribs. Severity of the necrosis varies from minor, indistinct flecking to an irregular, blotchy spotting, seen best by peeling off bark and observing the inner surface of the phloem on the bark that was removed. Necrosis of the phloem in petioles and leaf midribs can be seen in tangential cuts but can be distinguished reliably from non-pathogenic "false necrosis" only in petiole sections. Except in severe cases, necrotic tissue indicative of phloem necrosis disease infection is restricted to the outer, peripheral phloem close to the pericycle. In severe infections, the pericycle may be affected, and necrosis may also spread in the phloem toward the cambium.

Phloem necrosis disease has been transmitted only by grafting, which has prompted the assumption that the disease is virus-induced. There has been noted some variation in severity of symptoms expressed by different clonal materials, but even symptomless clones have been found to be infected and capable of transmitting the disease through grafts.

Occurrence of phloem necrosis is restricted almost entirely to elevations above 1200 m, and symptoms are generally more severe above 1650 m altitude. Symptoms often are restricted to one part of the plant only, and diseased plants taken to sea level or placed under constant temperature of about 28 C have produced new growth without symptoms. There has been no evidence of soil transmission of the disease, and replants have remained healthy after being placed immediately at sites where diseased plants have been rogued. Diseased plants in the field occur in small, localized patches of various sizes that together may form a much larger area in which most bushes may be infected but always with some healthy bushes interspersed. It has been suggested that spread of the disease occurs in a discontinuous, random manner at some distance from an original center of infection, as well as from plant to plant in the row.

Control measures have not been adequately developed, and the possibility of satisfactory disease tolerance in clonal selections has not been thoroughly investigated. Elimination of infection by exposure of diseased plants, or propagating materials, to 28 C may offer possible usefulness under special conditions. Routine roguing of affected plants and replanting with healthy stocks provide practical control at present. Propagating materials should be taken only from stocks with no symptoms of phloem necrosis disease.

1. Bond, T. E. T. 1944/1947. The "phloem necrosis" virus disease of tea in Ceylon. I. Introductory account, symptoms, and transmission by grafting; II. Field observations and effect on yield; III. Further characterization of necrosis in the leaf. *Ann Appl. Biol. 31, 34:* 40–47, 300–310; 517–525.

2. Gadd, C. H. 1939. A virus disease of tea. *Tea Quart. 12:* 110–130.

3. Mulder, D. and R. L. De Silva. 1961. The relation between air temperature and the occurrence of symptoms of phloem necrosis virus disease. *Commonwealth Mycol. Conf. Rep. 6:* 112–115.

Oil spot

The condition of tea referred to as "oil spot" was observed in Sri Lanka in 1955 and reported 4 years later after it had been found in only a few estates, with the highest incidence of affected plants no greater than 5 percent.

Numerous, small, translucent spots develop on the undersurfaces of young, tender leaves and later turn dark gray and then brown with translucent halos seen in transmitted light (Fig. 99). The spots are irregular in size and shape, may coalesce and cause extensive defoliation. New leaves that follow on affected shoots are reduced in size and fall, too, with the progressive defoliation eventually resulting in dieback of shoots and killing the diseased part of the plant. Discoloration of the wood in the main branches below diseased shoots is typical.

An unidentified fungus has been isolated from the discolored wood below affected shoots. Symptoms induced in rooted cuttings grown in culture filtrates of the fungus have prompted the supposition that a toxin produced by the fungus causes oil spot disease. Similar leaf symptoms also were caused in rooted cuttings grown in water extracts from discolored wood.

Fig. 99. Oil spot disease lesions on tea leaves. (Courtesy of Tea Research Institute, Talawakelle.)

Occurrence of the disease remains restricted to a few estates, all situated at elevations above 1500 m, with some evidence that severe pruning predisposes plants to infection. Incidence of the disease has not necessitated serious attempts at control but preliminary experiments have shown foliar applications of fungicides to be ineffective.

1. Mulder, D. and N. Shanmuganathan. 1959. Oilspot of tea leaves—a new disease? *Tea Quart. 30:* 44–45.

2. Mulder, D., N. Shanmuganathan and P. V. Arulpragasam. 1963. Oil spot disease in tea. *Nature 199:* 92.

3. Shanmuganathan, N. and D. Mulder. 1963. Further observations on the oil spot disease of tea. *Tea Quart. 34:* 184–187.

Mineral deficiency symptoms

Leaves near the ends of twigs become typically distorted by uneven growth of the 2 halves, resulting in "sickle-shaped" leaves on plants deficient in *zinc*. Margins of affected leaves are sometimes wavy or rugose, and an indistinct chlorosis can be seen, particularly in the interveinal areas, on leaves that are often hard, dull, and darker in color than on plants supplied sufficient zinc. Internodes near the apical ends of twigs are noticeably shortened, and severely affected twigs become dormant with apical leaves distorted. Foliar application(s) of zinc sulfate (4.6–9.2 kg/ha depending on severity) have given earlier and less expensive correction of zinc deficiency than soil treatment. It has been recommended that more than one application be made at the lower dosage rate rather than one application at the higher dosage. Foliar sprays may be included with fungicide or insecticide treatments.

Mature leaves are the first to develop symptoms of *magnesium* deficiency, that consist of progressive interveinal chlorosis and result in a "herring-bone" pattern of distinct, dark green main veins on chlorotic laminae. Often there develops a well-defined, green, inverted "V" at the base of chlorotic leaves with or without the herring-bone pattern. There is no obvious effect on growth or flushing of plants deficient in magnesium, which can be readily corrected by foliar application(s) of magnesium sulfate.

Bushes deprived of sufficient *potassium* have only a few leaves at the ends of branches and thin, twiggy shoots. There is hardly any flush growth, and many leaves exhibit marginal scorching. Plantings deficient in potassium for a number of years are spotty because some bushes usually have died.

Symptoms of *boron* deficiency have been described only from pot culture experiments and include development of brown spots first in the petioles and later in the laminae of young leaves. Spots enlarge along leaf midribs and petioles with obvious suberization, while affected leaves become severely distorted and droop to near vertical but remain turgid. Internodes are reduced to less than one cm before growing points on twigs are killed. Leaves that have matured before deficiency was acute develop suberized cracks on the upper surface of the midribs.

Deficiency of *sulfur* has been considered the cause of "yellows" symptoms in tea, which include gradual decrease in size of leaves that all exhibit various degrees of mottling but with veins darker green than other parts of leaves. In late stages of the disease, new leaves are quite small and chlorotic with necrotic margins. Shortened internodes on stems that are noticeably thinner than normal and develop dieback, often with some axillary growth at the bases, characterize chronic deficiency, which can be corrected by soil application of organic or inorganic materials containing available sulfur.

1. Chenery, E. M. 1958. Boron deficiency in tea. *Nature 181:* 426.

2. Chenery, E. M. and J. Shoenmaekers. 1959. Magnesium deficiency in East African tea. *East Afr. Agr. J. 25:* 25–27.

3. De Haan, I. 1941. Deficiency symptoms in tea caused by a shortage of the most important mineral nutrients, with the exception of potassium. *Arch. Tea Cultiv. 15:* 1–24.

4. De Haan, I. and A. F. Schoorel. 1940. Potassium deficiency in tea growing. *Arch. Tea Cultiv. 14:* 43–81.

5. Forbes, A. P. S. 1942. Some observations on the "yellows" sulphur deficiency disease of tea. *Nyasaland Agr. Quart. J. 2:* 20–26.

6. Portsmouth, G. B. 1953. Potash deficiency in tea. *Tea Quart. 24:* 79–81.

7. Southern, P. J. 1969. Nutritional studies of tea in the Territory of Papua and New Guinea. *Dept. Agr. Stock Fisheries Port Moresby Res. Bull. 6.* 50 p.

8. Storey, H. H. and R. Leach. 1933. A sulphur-deficiency of the tea bush. *Ann. Appl. Biol. 20:* 23–56.

9. Tolhurst, J. A. H. 1954. Magnesium and manganese deficiencies in the nutrition of the tea bush. *Tea Quart. 25:* 84–86.

10. Tolhurst, J. A. H. 1962. Zinc deficiency of tea in Ceylon. *Tea Quart. 33:* 134–147.

11. Venkataram, C. S. 1968. Zinc deficiency in tea in southern India. *Planters' Chron. 63:* 111-112.

OTHER DISEASES

Anthracnose (*Gloeosporium theae-sinensis* Miyake)

1. Abe, T. and M. Kono. 1959. Studies on the anthracnose of tea bush. IV. On the sporulation, germination and infection of conidia of the anthracnose fungi. *Kyoto Prefectural Univ. Fac. Agr. Sci. Rep. 11:* 38–43.

Armillaria root rot (*Armillariella mellea*)

1. Goodchild, H. A. 1958. *Armillaria mellea* (Vahl.) Fr. *Tea Res. Inst. East Afr. Pam. 16:* 19–30.

2. Leach, R. 1939. Biological control and ecology of *Armillaria mellea* (Vahl.) Fr. *Trans. Brit. Mycol. Soc. 23:* 320–329.

3. Pegler, D. N. and I. A. S. Gibson. 1972. *Armillariella mellea. Commonwealth Mycol. Inst. Descriptions Pathogenic Fungi Bacteria 321.* 2 p.

"Bitten-off" root disease

Deterioration of seedling roots is caused by culturing in alkaline soil.

1. Gadd, C. H. 1940. "Bitten-off" disease of tea seedlings. *Tea Quart. 13:* 54–58.

Black rot [*Corticium invisum* (Petch) and *C. theae* (Bernard)]

1. Eden, T. 1958. *Tea.* Longmans, Green & Company, Ltd., London. 205 p.
2. Harler, C. R. 1973. Pests and diseases of tea. *World Crops 25:* 245–251.

Branch canker (*Botryobasidium salmonicolor*)

1. Thompson, A. 1941. Branch canker of tea. *Malayan Agr. J. 29:* 152–154.

Brown leaf spot (*Calonectria colhounii* Peerally; *Cylindrocladium colhounii* Peerally)

1. Peerally, M. A. 1973. *Calonectria colhounii* sp. nov., a common parasite of tea in Mauritius. *Trans. Brit. Mycol. Soc. 61:* 89–93.

Cercospora leaf spot [*Cercospora theae* (Cav.) B. de Haan]

1. Plakidas, A. G. 1959. *Cercospora theae* on *Camellia sasanqua* and *Thea sinensis. Plant Dis. Reptr. 43:* 668–669.

Cercosporella disease [*Calonectria theae* Loos=*Cercosporella theae* Petch.; *Candelospora theae* (Petch) Wakefield ex Gadd]

1. Eden, T. 1958. *Tea.* Longmans, Green & Company, Ltd., London. 205 p.
2. Loos, C. A. 1950. *Calonectria theae* n. sp.—the perfect stage of *Cercosporella theae* Petch. *Trans. Brit. Mycol. Soc. 33:* 13–18.

Cylindrocladium root rot (*Cylindrocladium camelliae* Venkataramani and Venkata Ram)

1. Venkataramani, K. S. and C. S. Venkata Ram. 1961. A new species of *Cylindrocladium* parasite on tea roots. *Current Sci. 30:* 186.

Decline (*Calonectria kyotensis* Terashita; *Cylindrocladium floridanum* Sobers & Seymour)

1. Peerally, A. 1972. A new disease of tea. *Rev. Agr. Sucrière l'Ile Maurice 51:* 115–117.
2. Peerally, A. 1972. A decline of tea bushes associated with a root rot. *Rev. Agr. Sucrière l'Ile Maurice 51:* 147–152.

Defoliation

1. De Jong, P. 1950. "Defoliation": a disorder of tea in S. India. *Planters' Chron. 45:* 155–158; 174–180.

"Edagare" disease

1. Uegaki, K. and M. Yasutomi. 1959. Studies on "Edagare" disease of "Benihomore" and some other varieties of tea plant. *Proc. Crop Sci. Soc. Japan 28:* 222–223.

Horse hair blight (*Marasmius equicrinis* Müller)

1. Abe, T. and M. Kono. 1957. On the horse hair blight fungus on the tea bush and its cultural characters. *Saikyo Univ. Fac. Agr. Sci. Rep. (1957):* 41–44.

Leaf tumors

Malformation of true leaves on tea seedlings (Fig. 100) has been reported from India, Sri Lanka, Africa, Australia, Papua New Guinea and Japan, but the cause has not been determined.

1. Shaw, D. E. 1965. Condition resembling blister blight of tea on tea seedlings in quarantine in New Guinea. *FAO Plant Protect. Bull. 13:* 3–11.
2. Shaw, D. E. and W. M. Burnett. 1968. Investigation into the cause of leaf tumours of tea seedlings. *Papua New Guinea Agr. J. 19:* 167–192.

Phyllosticta leaf spot (*Phyllosticta theaefolia*)

1. Hirokawa, S. 1965. On the causal fungus of the leaf spot disease of tea plant. *Japanese Tea Res. Sta. Study Tea 30:* 31–38.

Pruning-cut infection (*Chaetospermum camelliae* Agnihothrudu)

1. Agnihothrudu, V. 1961. Notes on fungi from north-east India. X. *Chaetospermum camelliae* sp. nov. on tea [*Camellia sinensis* (L.) O. Kuntze]. *Mycopathol. Mycol. Appl. 16:* 113–116.

Fig. 100. Non-infectious leaf tumor disorder of tea. (Courtesy of D. E. Shaw, Department of Agriculture, Stock and Fisheries, Konedobu.)

Purple root rot (*Helicobasidium compactum*)

1. Satyanarayana, G. and P. C. Sharma. 1965. Purple root rot of tea. *Tocklai Expt. Sta. Newsletter 12:* 43.

Red rust (*Cephaleuros parasiticus* Karst)

1. Sharma, P. C. 1965. Red rust of tea. *Tocklai Expt. Sta. Newsletter 12:* 72–74.

Root rot [*Ustulina zonata* (Lév.) Sacc.]

1. Agnihothrudu, V. 1961. Rhizosphere microflora of tea [*Camellia sinensis* (L.) O. Kuntze] in relation to the root rot caused by *Ustulina zonata* (Lév.) Sacc. *Soil Sci. 91:* 133–137.

Stem canker (*Botryodiplodia theobromae*)

1. De Silva, R. L. 1965. Stem cankers in clonal tea in the low country. *Planters' Chron. 60:* 60–61, 67.

Thorny stem blight [*Tunstallia aculeata* (Petch) Agnihothrudu var. *Kesabii*]

1. Agnihothrudu, V. 1961. Notes on fungi from north-east India. VII. *Tunstallia* gen. nov. causing the "thorny stem blight" of tea [*Camellia sinensis* (L.) O. Kuntze]. *Phytopathol. Z. 40:* 227–282.

White root rot (*Rosellinia necatrix* Prill.) (Fig. 101)

1. Sivanesan, A. and P. Holliday. 1972. *Rosellinia necatrix. Commonwealth Mycol. Inst. Descriptions Pathogenic Fungi Bacteria 352.* 2 p.

2. Takaya, S. 1965. Inoculation experiments of white root rot fungus [*Rosellinia necatrix* (Hart) Berl.] to the tea plant and some considerations on the ecology of the fungus. *Japanese Tea Res. Sta. Study Tea 31:* 38–53.

TEJPAT: *Cinnamomum tamala* Fr. Nees.

Rust (*Aecidium cinnamomi* Rac.)

1. Goswami, R. N. and S. Bhattacharjee. 1973. Rust, a new disease of tejpat. *Current Sci. 42:* 257.

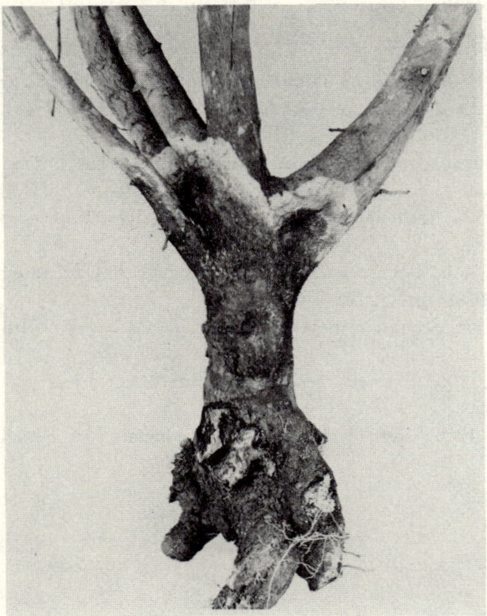

Fig. 101. White root rot (*Rosellinia* sp.) of tea. (Courtesy of Tea Research Institute, Talawakelle.)

TURMERIC: *Curcuma longa* L.

Taphrina leaf spot

A disease commonly found on the leaves of turmeric in many areas of India was described in 1911 as "leaf spot." Despite severe infections, growers did not consider the disease of major import.

Poorly defined spots, first yellowish in color, develop on both leaf surfaces but are more frequent on the upper leaf surfaces. Coloration of lesions intensifies to golden yellow and later brown with golden halos as spots become 1–2 mm in diameter and frequently coalesce. As the spots mature, the cuticle is ruptured by fungal growth to expose mature spores.

The causal fungus is *Taphrina maculans* Butler with clavate asci 6.5–10 × 20–30 μm, each with 8 ovoid to oblong ascospores 1–2.5 × 6.5 μm. (*Zingiber casumunar* Roxb. also was found infected by *T. maculans* in India.) Numerous large and characteristic haustoria invade epidermal cells of the host, and it is asci produced from clumped ascogenous cells in the epidermis that enlarge to rupture the leaf surface, permitting ascospore release. Discharged ascospores can cause secondary infections directly, which typically result in more profuse spotting than from the primary inoculum.

Differences in susceptibility of cultivars to leaf spot have been reported and cvs. China and Jaweli and a variant clone of *C. longa*, in addition to the species *C. amada* Roxb., were found to be immune to infection.

1. Ahmed, L. and N. B. Kulkarni. 1968. Studies on *Taphrina maculans* Butler, inciting leaf spot of turmeric (*Curcuma longa* L.). I. Isolation of the pathogen. *Mycopathol. Mycol. Appl. 34:* 40–46.

2. Butler, E. J. 1911. The leaf spot of turmeric (*Taphrina maculans* sp. nov.). *Ann. Mycol. 9:* 36–39.

3. Pavgi, M. S. and R. Upadhyay. 1964. Artificial culture and pathogenicity of *Taphrina maculans* Butler. *Sci. Cult. 39:* 558–559.

4. Reddy, G. S., V. Dakshinamurti and S. S. Sarma. 1963. Note on varietal resistance against leaf spot diseases in turmeric. *Andhra Agr. J. 10:* 146–148.

5. Upadhyay, R. and M. S. Pavgi. 1966. Secondary host infection by *Taphrina maculans* Butler. *Phytopathol. Z. 56:* 151–154.

6. Upadhyay, R. and M. S. Pavgi. 1967. Varietal resistance in turmeric to leaf spot disease. *Indian Phytopathol. 20:* 29–31.

7. Upadhyay, R. and M. S. Pavgi. 1974. Control of leaf spot of turmeric. *Riv. Patol. Veg. 10:* 153–161.

Colletotrichum leaf spot

The first report of *Colletotrichum* leaf spot of turmeric was from Coimbatore (India) in 1917, but the disease was not described until more than 40 years later (1954). Annual incidence of the disease, apparently confined to peninsular districts in India, has varied in severity.

Mostly oval to oblong spots, variable in size to 3.8 × 5.1 cm, develop on leaves and may coalesce to involve most of a leaf, which then is killed. The centers of spots are nearly white in color with concentric rings of small, black acervuli, while the margins of lesions are brown with indefinite, yellow halos around the lesions.

Symptoms are caused by *Colletotrichum capsici*, characterized by stromata 60–120 μm in diameter and brown, tapering setae 50–145 μm in length. Conidia are hyaline, aseptate, curved, measure 3–4 × 17–31 μm and appear pink *en masse*. The fungus is pathogenic on a number of other plants, including pepper, gram, eggplant and cabbage (*Brassica oleracea* var. *capitata* L.).

Dissemination of the conidia is largely by air after they are dislodged from acervuli by splashing water. Disease development is most severe on plants 4 months old or older during periods of continuous high humidity.

Differences in varietal susceptibility have been noted, but satisfactory control can be obtained with foliar application of zineb spray (0.1 kg/60 ℓ water).

1. Kandaswamy, T. K. 1958. Control of the leaf spot disease of turmeric caused by *Colletotrichum capsici* (Syd.) Butl. and Bisby. *Madras Agr. J. 45:* 55–60.

2. Mordue, J. E. M. 1971. *Colletotrichum capsici. Commonwealth Mycol. Inst. Descriptions Pathogenic Fungi Bacteria 317.* 2 p.

3. Ramakrishnan. T. S. 1954. Leaf spot disease of turmeric (*Curcuma longa* L.) caused by *Colletotrichum capsici* (Syd.) Butl. & Bisby. *Indian Phytopathol. 7:* 111–117.

4. Reddy, G. S., V. Dakshinamurti and S. S. Sarma. 1963. Note on varietal resistance against leaf spot diseases in turmeric. *Andhra Agr. J. 10:* 146–148.

Phyllostica leaf spot (*Phyllosticta zingiberi*) (See Ginger—*Phyllosticta* leaf spot)

Small, round to irregular, white spots that later turned brown and coalesced were described from plants grown on the Poona Agricultural College farm (India) in 1956–57. Dark pycnidia 23–94 × 25–90 μm, with one-celled, hyaline pycnidiospores 1.5–4 × 3–5.8 μm, developed in necrotic leaf tissues.

1. Summanwar, A. S. and V. P. Bhide. 1962. Leaf spot of turmeric (*Curcuma longa* L.) caused by *Phyllosticta zingiberi* Ramakrishnan. *J. Indian Bot. Soc. 41:* 313–315.

2. Ramakrishnan. T. S. 1942. A leaf spot disease of *Zingiber officinale* caused by *Phyllosticta zingiberi* n. sp. *Proc. Indian Acad. Sci. 15 B:* 167–171.

Pythium root rot (*Pythium aphanidermatum*; *P. graminicola*)

1. Middleton, J. T. 1943. The taxonomy, host range and geographic distribution of the genus *Pythium. Mem. Torrey Bot. Club 20:* 1–171.

2. Ramakrishnan, T. S. and C. K. Sowmini. 1954. Rhizome and root rot of turmeric caused by *Pythium graminicolum* Sub. *Indian Phytopathol. 7:* 152–159.

3. Waterhouse, G. M. and J. M. Waterston. 1964. *Pythium aphanidermatum. Commonwealth Mycol. Inst. Descriptions Pathogenic Fungi Bacteria 36.* 2 p.

4. Waterhouse, G. M. and J. M. Waterston. 1964. *Pythium graminicola. Commonwealth Mycol. Inst. Descriptions Pathogenic Fungi Bacteria 38.* 2 p.

Sphaceloma leaf spot (*Sphaceloma curcumae* Thirumalachar)

Small, brownish, mostly round lesions 2–4 mm on the upper leaf surfaces of *Curcuma* sp. were observed in Mysore State (India) in 1945. Subcuticular acervuli measured 31–50 μm in diameter, but conidia were not observed.

1. Thirumalachar, M. J. 1947. Some new *Sphaceloma* diseases of economic plants in Mysore. *Trans. Brit. Mycol. Soc. 31:* 1–6.

U

ULLUCU: *Ullucus tuberosus* Caldas

Rust (*Aecidium cantensis*) (See Potato—Rusts)

1. Bazán de Segura, C. 1975. *Enfermedades de cultivos frutícolas y hortícolas.* Lima, Peru. 276 p.

V

VANILLA: *Vanilla planifolia* Andrews

Root rot

The first published account of root rot of vanilla was from Puerto Rico in 1927, although the disease had been recognized there for almost 10 years. It had been observed also in Java and the Seychelles Islands by 1925 and now occurs throughout the world wherever the vanilla of commerce is grown. Root rot is the most important disease of vanilla and was responsible for disappearance of the vanilla industry in Puerto Rico. Losses vary depending on age of the plants at infection, vigor of plants before infection, the severity of infection and the length of time affected vines survive.

Browning and killing of vine roots, most of which are partly aerial, is the first visible symptom, although the underground portions of affected roots often have been destroyed by that time. As root damage progresses, plant growth is terminated and dieback of the growing tip occurs, followed by loss of green coloration in stems and leaves. Successive formation of roots, which are soon killed upon reaching the soil, may permit the plant to survive for a time, even years in some cases, but diseased plants typically perish eventually.

Fusarium oxysporum Schlecht. f. sp. *vanillae* (Tucker) Gordon is responsible for root rot and has been found to cause dieback of terminal buds but does not infect stems or leaves nor invade the vascular system of affected plants. *F. solani* also has been found to cause root rot and wilt of vanilla plants. It has been suggested recently that the primary causal agent of root rot in vanilla is *Rhizoctonia solani*, which predisposes roots to infection by *F. oxysporum vanillae*, an important, but secondary, pathogen.

Persistence of *F. oxysporum vanillae* in soil prompted the search for resistance and discovery of *V. phaeantha* Reich., which is effectively resistant. Hybridizations with *V. planifolia* to improve quality of fruits from the resistant plant have been accomplished but progress in breeding for root rot resistance and improved fruit production and quality has been less than completely satisfactory.

1. Alconero, R. 1968. Infection and development of *Fusarium oxysporum* f. sp. *vanillae* in vanilla roots. *Phytopathology 58:* 1281–1283.

2. Alconero, R. 1969. Mycorrhizal synthesis and pathology of *Rhizoctonia solani* in vanilla orchid roots. *Phytopathology 59:* 426–430.

3. Alconero, R. and A. G. Santiago. 1969. Mycorrhizal infections of mature portions of vanilla roots by *Rhizoctonia solani* as a predisposing factor to infection by *Fusarium oxysporum* f. sp. *vanillae. Phytopathology 59:* 1521–1524.

4. Alconero, R. and A. G. Santiago. 1969. Fusaria pathogenic to vanilla. *Plant Dis. Reptr. 53:* 854–857.

5. Delassus, M. 1963. La lutte contre la fusariose du vanillier par methodes génétiques. *Agron. Trop. 18:* 245–246.

6. Irvine, J., T. Theis, F. A. Jimenez and R. H. Freyre. 1964. Techniques for evaluating reactions of vanilla to *Fusarium* root rot. *Phytopathology 54:* 827–831.

7. Theis, T. and F. A. Jimenez. 1957. A vanilla hybrid resistant to *Fusarium* root rot. *Phytopathology 47:* 579–581.

8. Tucker, C. M. 1927. Vanilla root rot. *J. Agr. Res. 35:* 1121–1136.

Mildew (*Phytophthora jatrophae* Jens.)

"Powdery mildew" (*Cephaleuros virescens*)

Anthracnose (*Calospora vanillae* Massee)

Apoplexie (Blast disease)

(Madagascar) Root rot

1. Stehle, H. 1952. Le vanillier et sa culture. III. Maladies, insectes et affections diverses du vanillier. *Fruits 7:* 253–260.

Y

YAMS: LESSER YAM, *Dioscorea esculenta* (Lour.) Burk; WEST AFRICAN YAM, *D. rotundata* Poir.; ASIAN GREATER YAM (RATALU) *D. alata* L.

Rust [*Goplana dioscoreae* (Berk. & Br.) Cummins; *Uredo dioscoreae-alatae* Racib.]

Described from leaves of *D. alata* collected at Manila in 1912, rust is characterized by internal basidia (teliospores that become 4-celled by formation of septae) submersed in a gelatinous matrix. A sterigma is produced from each cell of the basidium and extends to the surface of the matrix, where a single basidiospore is liberated.

1. Cummins, G. B. 1935. Notes on some species of the *Uredinales*. *Mycologia 27:* 605–614.

Smut (*Urocystis dioscoreae* H. Sydow & P. Sydow)

Pustular sori consisting of black, dusty spore masses composed of spore balls, each containing a central, brown, globose spore 10.5–14.5 × 14–19

μm, surrounded by a discontinuous layer of ellipsoid, light brown, sterile peripheral cells 4.5–8.5 × 7.5–10.5 μm, were observed on leaves and petioles of *Dioscorea deltoidea* Wall.

1. Mirza, J. H. 1968. *Urocystis dioscoreae* from West Pakistan. *Mycologia 60:* 439–440.

Cercospora leaf spots

Cercospora cylindrata Chupp & Lindner

Sporulation on lower leaf surfaces of *Dioscorea* sp. was noted as loose clusters of conidiophores 4.5–6 × 90–290 μm with brown, cylindrical, conidia 4–6.5 × 25–75 μm with 3–5 septae.

Cercospora pachyderma Sydow

1. Chupp, C. and D. H. Lindner. 1937. Notes on Chinese *Cercosporae*. *Mycologia 29:* 26–33.

Cercospora ubi Rac.

Reported on leaves of *Dioscorea alata* from Taihoku, Formosa, in 1929.

1. Yamamoto, W. 1934. *Cercospora*-arten aus Taiwan (Formosa). II. *J. Soc. Trop. Agr. (Taiwan) 6:* 599–608.

Anthracnose

Small, brown spots on leaves and stems of *Dioscorea alata* and *D. batatas* Decne. have been reported from Formosa (1929) and India. Spots on stems may coalesce and cause external blackening, but coalescence of leaf spots induces withering and shedding of leaves, which constitutes a characteristic symptom of the disease. Severely affected plants become completely dried, and no rhizomes are formed.

The causal fungus is *Colletotrichum gloeosporioides*, which produces hyaline mycelium 2.9–6.5 μm in diameter. Conidia 4–6 × 11–19 μm are oblong to cylindrical, borne singly on simple, hyaline conidiophores, and hyaline when viewed separately but appear pinkish *en masse*. Growth in culture is excellent at 25 C with abundant production of acervuli (without setae), which are quite variable but occur irregularly in nature. Only species of *Dioscorea* were susceptible to infection under controlled conditions in India.

Infection and development of symptoms can be completed in 96 hours. High humidity favors infection, and the disease spreads rapidly in periods of continuous rain.

Bi-weekly applications of zineb (0.2 percent) on plants beyond 2 months of age have proven beneficial in India.

1. Bhatnagar, M. K. 1968. Combating yam disease in Rajasthan. *Indian Hort. 12:* 35.

2. Goto, K. 1929. On the black-spot disease of *Dioscorea alata. J. Soc. Trop. Agr. (Taiwan) 1:* 301–313.

3. Mordue, J. E. M. 1971. *Glomerella cingulata. Commonwealth Mycol. Inst. Descriptions Pathogenic Fungi Bacteria 315.* 2 p.

4. Prasad, N. and R. D. Singh. 1960. Anthracnose disease of *Dioscorea alata* L. *Current Sci. 29:* 66–67.

Sobrefoliación (*Passalora* sp.)

Abnormal proliferation of leaves and shoots of young plants in Mexico was found to be caused by a fungus with dark, septate, branched conidiophores 4–7 × 99–161 μm and dark, uniseptate obclavate conidia 4–7 × 11–19 μm, identified as a species of *Passalora*.

1. De Bauer, L. I. and S. Romero. 1975. A new fungus disease of yam (*Dioscorea* sp.) in Mexico. *Plant Dis. Reptr. 59:* 1011–1013.

2. Hughes, S. J. 1953. Some foliicolous hyphomycetes. *Can. J. Bot. 31:* 560–576.

Green-banding disease

Incidence of the green-banding disease, characterized by broad, dark green lengthwise bands along main veins of slightly puckered and rugose leaves with elongated, yellow-green interveinal bands increased from 10 to 95 percent in a single year (1963–64) in Puerto Rico. No data were collected, but yields from infected plants were noticeably reduced.

The disease was transmitted mechanically, by the cotton aphid (*Aphis gossypii*) and by tuber grafts but not by contaminated cutting knives. Plants of *Crotalaria striata* DC and *Nicotiana glutinosa*, as well as *Dioscorea composita* Hemsl. and *D. floribunda* Mart. & Gal., developed systemic symptoms, and tobacco was infected but remained symptomless following experimental inoculation. Neither pepper, cucumber, tomato, bean nor cowpea was infected by mechanical inoculation, but *D. spiculiflora* Hemsl. and *D. friedrichsthalii* Kunth were found infected naturally.

Partial control of the disease has been achieved through breeding. Severity of infection varied among the 4 species of *Dioscorea* observed (*D. floribunda* and *D. friedrichsthalii*, quite susceptible; *D. composita*, intermediate in response; and *D. spiculiflora*, only mildly affected). However, the responses of hybrids did not always reflect these same differences, i.e. plants of *D. spiculiflora* × *D. floribunda* were susceptible, but plants of *D. spiculiflora* × *D. friedrichsthalii* were more tolerant of infection. Control has been reported by tri-weekly foliar application of demeton (21 ml concentrate/4 ℓ water).

1. Alconero, R. 1969. Control of *Dioscorea* green-banding virus with demeton foliar sprays. *Plant Dis. Reptr. 53:* 450–452.

2. Martin, F. W. and N. E. Delfel. 1967. The tubers, sapogenins, and virus resistance of *Dioscorea* species hybrids. *Am. J. Bot. 54:* 1158–1162.

3. Ruppel, E. G., H. Delpin and F. W. Martin. 1966. Preliminary studies on a virus disease of a sapogenin-producing *Dioscorea* species in Puerto Rico. *Univ. Puerto Rico J. Agr. 50:* 151–157.

Internal brown spot

A condition described as "internal brown spot" was discovered in yams shipped from Barbados to England in 1965, but there is some evidence that the disease had been recognized more than 30 years earlier. Occurrence of the disease remains confined to Barbados and limited to the cultivar "White Lisbon" (syn. Crop Lisbon, Lisbon) of *Dioscorea alata*.

Symptoms in freshly harvested yam tubers consist of small (0.5–2 mm), hard, dark nodules embedded in the non-vascular tissues. They occur in greater frequency near the bud ends of tubers and often are progressively smaller toward the stem ends. The nodules, which are easily removed from tuber tissues, enlarge during storage. Plants grown from tubers with internal brown spot symptoms are stunted with mottled and distorted leaves.

The disease is presumed to be caused by a virus because of the presence in affected plants of bacilliform particles 29 × 130 nm, in addition to the flexuous particles 13 × 750 nm found also in apparently healthy plants. Transmission of the disease has not been achieved to date.

Cultivars of *D. alata* other than White Lisbon are not known to be affected by internal brown spot. Planting materials should be taken only from healthy plants.

1. Coursey, D. G. 1967. Internal brown spot—a condition in yams in Barbados. *J. Agr. Soc. Trinidad Tobago 67:* 473–475, 477, 479, 481–482.

2. Harrison, B. D. and I. M. Roberts. 1973. Association of virus-like particles with internal brown spot of yam (*Dioscorea alata*). *Trop. Agr. (Trinidad) 50:* 335–340.

Mosaic

A disease of yams (*Dioscorea rotundata*) first reported in Puerto Rico in 1952 was described as "mosaic." Similar symptoms had been noted previously from Puerto Rico and Africa (Sierra Leon) about 1936. All of the plants observed exhibited symptoms of leaf mottling with some distortion of the laminae, but not all leaves were so affected. Diseased plants were somewhat smaller than healthy plants.

The disease was considered to be caused by a virus and was transmitted (infrequently) by rubbing to yam, tobacco, pepper, cucumber and occasionally by tuber-graft in yam, as well as by pricking with a needle through an infected leaf wrapped around the stem or a petiole of a healthy yam plant. Mechanical transmission was effected more easily with sap from infected tobacco. The virus remained infectious after exposure for 10 minutes to 50 C, but not 60 C, and dilution to 10^{-3} but not 10^{-4}. Infectivity in sap from infected plants was lost in less than 24 hours when stored at room temperature but was maintained for more than 24 days in air-dried leaves in the laboratory. No insect vector of the disease was identified but the disease was transmitted through planting material taken from diseased plants.

Control procedures have not been developed, but planting materials should not be taken from infected plants.

1. Adsuar, J. 1955. A mosaic disease of the yam, *Dioscorea rotundata*, in Puerto Rico. *Univ. Puerto Rico J. Agr. 39:* 111–113.

OTHER DISEASES

Leaf spots

Cercospora contraria Sydow

Cercospora annulata Cke.

Cercospora dioscorea Ellis & Martin

Colletotrichum capsici

Sclerotium rolfsii

Mycosphaerella dioscoreicola [*M. dioscoreae* (Tass.) Tomilin ?]

Pestalotiopsis cruenta (Sydow) Steyaert

Phleospora sp.

Phyllosticta dioscoreicola (*P. dioscoreae* Cke. ?)

Phyllosticta dioscoreae Cke.

Phyllosticta graffiana Sacc.

Leaf blight

Rhizoctonia solani

Witches' broom

Phylleutypa dioscoreae (Wakef.) Petr.

Crown gall

Agrobacterium tumefaciens (Smith & Townsend) Conn.

Tuber rots

Black, dry rots

Rosellinia bunodes

Sphaerostilbe repens **Berk. & Cke.**

Soft rots

Botryodiplodia theobromae

Fusarium solani

Fusarium oxysporum **Schlecht.**

Lasiodiplodia **sp.**

Penicillium sclerotigenum **Yamamoto**

Rhizopus nodosus **Namysl.**

Root rots

Armillaria mellea

Rhizoctonia solani

1. Baudin, P. 1956. Maladies parasitaires des Ignames en Cote d'Ivoire. *Rev. Mycol. (Paris, Suppl. Colon. 2) 21:* 87–111. Also see *Rev. Appl. Mycol. 36:* 569 (1957).
2. Choudhury, S. 1946. A *Rhizoctonia* leaf blight of *Dioscorea*. *Current Sci. 15:* 81–82.
3. Waitt, A. W. 1963. Yams, *Dioscorea* species. *Field Crops Abstracts 16:* 145–157.

4. Yamamoto, W. 1934. Species of *Cercospora* from Taiwan (Formosa). II. *J. Soc. Trop. Agr. (Taiwan) 6:* 599–608.

YAM BEAN (SINCAMAS): *Pachyrrhizus erosus* (L.) Urban

Cercospora leaf spot (*Cercospora canescens*)

Mostly round to irregular, brown leaf lesions with gray centers were observed on leaves of yam bean in India in 1959. Multiseptate, hyaline conidia 1.5–3 × 24–270 μm were borne on lightly colored, multiseptate conidiophores 3.5–5 × 45–160 μm.

1. Mohanty, N. N. and B. C. Behera. 1961. Leaf spot of *Pachyrrhizus angulatus* Rich. *Sci. Cult. 27:* 254.

2. Mulder, J. L. and P. Holliday. 1969. *Cercospora canescens. Commonwealth Mycol. Inst. Descriptions Pathogenic Fungi Bacteria 462.* 2 p.

Mosaic

A systemic mottle disease described from the Philippines in 1932 was presumed to be caused by a virus that was mechanically transmissible and seed-borne.

1. Fajardo, T. G. and J. Marañon. 1932. The mosaic disease of sincamas *Pachyrrhizus erosus* (Linnaeus) Urban. *Philippine J. Sci. 48:* 129–142.

INDEX